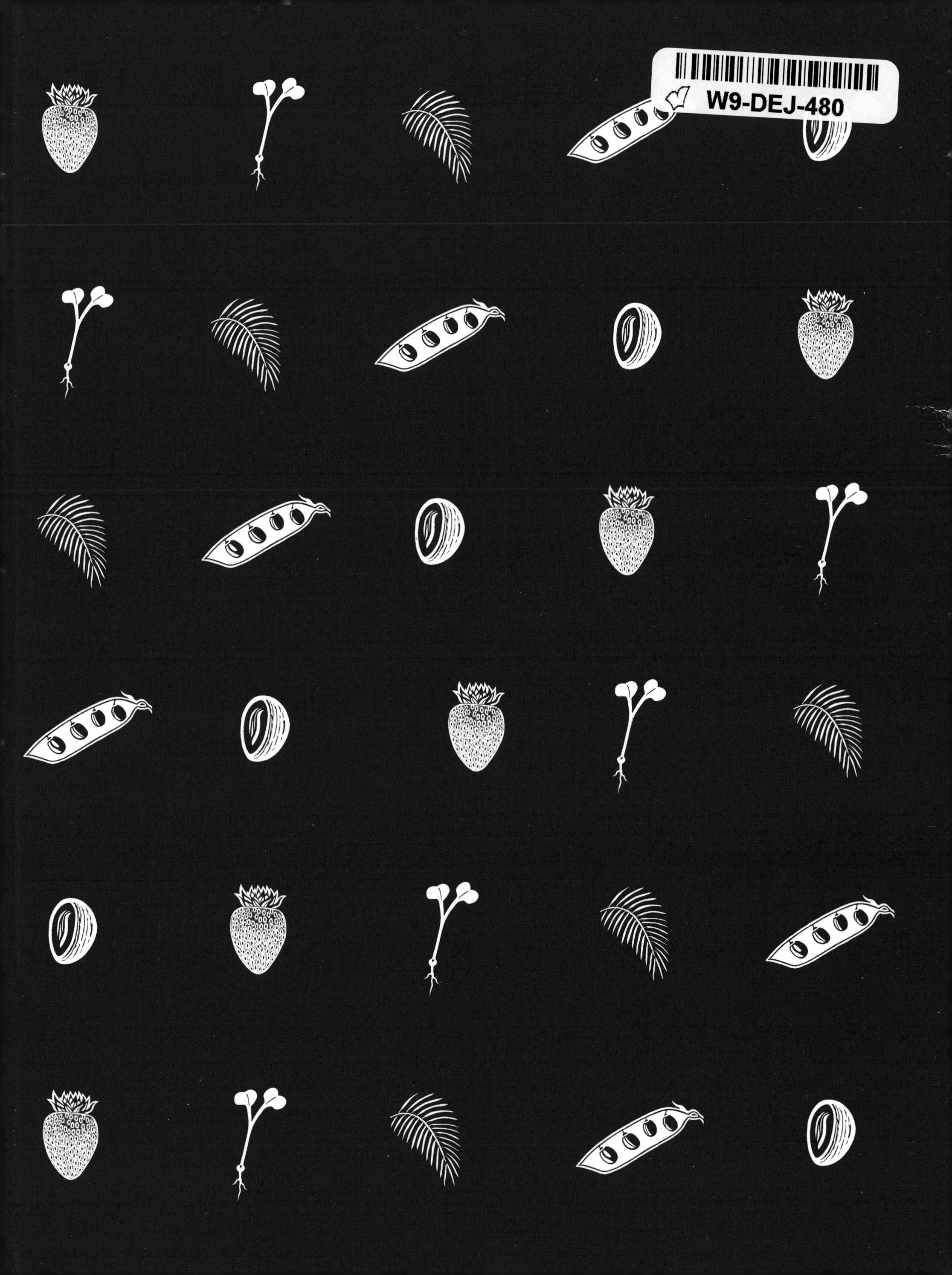
W9-DEJ-480

THE NEW OXFORD BOOK OF FOOD PLANTS

J. G. Vaughan and C. Geissler

THE NEW OXFORD BOOK OF FOOD PLANTS

Illustrated by B. E. Nicholson

With additional illustrations by Elisabeth Dowle and Elizabeth Rice

Oxford New York Tokyo
OXFORD UNIVERSITY PRESS
1997

Oxford University Press, Great Clarendon Street, Oxford OX2 6DP

Oxford New York
Athens Auckland Bangkok Bogota Bombay Buenos Aires
Calcutta Cape Town Dar es Salaam Delhi Florence Hong Kong
Istanbul Karachi Kuala Lumpur Madras Madrid Melbourne Mexico City
Nairobi Paris Singapore Taipei Tokyo Toronto Warsaw

and associated companies in
Berlin Ibadan

Oxford is a trade mark of Oxford University Press

Published in the United States
by Oxford University Press Inc., New York

A catalogue record for this book is available from the British Library

Library of Congress Cataloging-in-Publication Data
Vaughan, J. G. (John Griffith)
The new Oxford book of food plants / J. G. Vaughan and C. A. Geissler;
illustrated by B. E. Nicholson; with additional illustrations by
Elisabeth Dowle and Elizabeth Rice.
1. Food crops. I. Geissler, Catherine Alison. II. Title.
SB175.V38 1997 633—dc21 97-6803
ISBN 0 19 854825 7 (hbk)

Typeset by Footnote Graphics, Warminster, Wilts
Printed in China

Foreword

David Bellamy

'The world on which we all depend was "the way it was" thanks to its mantle of green plants. Sadly, all this is changing due to the greeds and needs of 5.8 billion human beings.'

'After tripling between 1950 and 1990, world production of grains, which today staples all human diets, including feedstock beef, hogs, poultry, eggs, and farmed fish, has not shown a net increase since 1990, largely because these grains cannot effectively use more fertilizer.'

'We are what we eat.'

Those three statements of fact show the timely importance of this book.

Unfortunately, today we all take our food for granted; the shelves of our hypermarts groaning with products of the plant kingdom, ripped from the sustainable cultures of the world, have divorced us from the back-ache of plough and hoe and so separated us from direct contact with the cultivation in good heart of the Earth's most endangered resource, soil.

This is a book of our times that must help change those attitudes by making us realize our debt to the plant kingdom and to the indigenous people who have, over some 10 000 years, become cultured as they selected, bred, and cultivated the plants that feed us all.

Read all about the plants that feed us ; this will not only educate, enthral, and entertain, but in so doing will tempt your palates, stimulate your digestive juices, and, I hope, your resolve to put your weight behind the cause of conservation.

Each and every one of the plants described in this book started life as a 'weed', a wild plant growing as part of the natural vegetation of this planet. The true creativity of humankind reached into this genetic treasure house to producee the myriad tastes and textures of the cultivars that sustainably supported our forebears. Unfortunately, many of those original cultivars were lost as commercial gain overcame common sense. More worryingly, the diversity of the natural populations of those original 'weed' species have been, and are being, lost at an ever increasing pace. So much so that many now face extinction, and with them goes any hope of a sustainable future.

John Vaughan was one of the members of staff at the then Chelsea College of Science and Technology who inspired me to a life of fascination as a botanist. He taught me plant anatomy, a subject usually guaranteed to switch most students off. Not so John Vaughan's vision, which related to the structure of the plants that feed the world:—leaves, the powerhouse of life in water and on earth; the structure of stems and roots, pathways for transport of water, sugar, and hormones; sections of seeds and fruits, triple-wrapped packages of energy-rich carbohydrates and essential proteins, their sculptured cuticles readily recognized logos of the cultivars that feed us. Their turgor and patination as good as any sell by date. *The new Oxford book of food plants* is spiced with all that and much more, and so should enrich university and school libraries, the kitchen worktop, and the coffee and dining tables of our homes.

An interesting and important development in this book compared to its predecessor is the account by Catherine Geissler of the contemporary situation concerning human nutrition with particular emphasis on food plants, adding to the fascination of this story of the food we all eat and enjoy.

'Culture is not life in its entirety, but just the moment of security, strength and clarity'; so teaches the wisdom of Buddha. Cultivars, the diversity of food plants that have sustained humankind since we moved from the tooth and claw of the hunter–gatherer to the sooth and law of agriculture, is the basis of all human life, for we are what we eat.

Preface

In 1969 *The Oxford book of food plants* appeared with illustrations by B. E. Nicholson and text by S. G. Harrison, G. B. Masefield, and M. Wallis. The purpose of the book was to describe the origin, distribution, structure, cultivation, utilization, and nutritive value of the world's common food plants, as well as some lesser-known species. This book achieved great success and has been found useful by professional scientists as well as the general reader.

Since the original publication there have been many developments in this subject area. For reasons of health there is now considerable emphasis on the inclusion of food plants in the human diet because of their particular contribution of fibre, unsaturated fats, certain vitamins, and some other constituents, in contrast to animal products. With modern and rapid transportation, food plants from all regions of the globe can now be delivered in fresh condition to the consumer. The growth of ethnic communities in certain countries has increased the variety of food plants for sale and given rise to an interest by other segments of the population. Plant breeding, including genetic engineering, has vastly increased the quality of plant material available for consumption.

For these reasons it is considered timely to produce a new version of *The Oxford book of food plants*. In the main, the original format and grouping of plants have been retained. However, several new species are described and all the original entries have been revised and updated. A new Introduction has been written, covering plant origins and dispersal and giving an overview of all the plant groups in the order in which they occur in the book. A very significant new chapter on nutrition and health has been added at the end, setting out the basic facts in an accessible and responsible way. This is followed by a series of tables summarizing the composition and caloric values of the major food plants (per 100 g), with additional tables on vitamins, fats, and food consumption patterns in selected countries.

The original illustrations have been reproduced in full and unaltered. For plants not included in the earlier version, two artists have contributed new paintings with the greatest care and skill. They are Elizabeth Rice (buckwheat, quinoa, lotus 'root', wing bean) and Elisabeth Dowle (tamarind, lupin, guar, grass pea, red delicious and gala apples, comice and nashi pears, tiger nuts, kiwifruit, prickly pear, mooli radish, coriander leaf, Chinese cabbage, various squashes).

The available literature concerning food plants is, of course, enormous. In the first version of the book no references were given but, in this version, some titles for recommended reading are provided to allow interested readers to delve deeper into the subject. An index of plant names (both scientific and vernacular) and general subjects is included.

The authors of the original book had hoped that it would prove a useful introduction to a topic of great importance to the human race, and their hopes were certainly justified. In view of the recent increased interest in the nutritional role of food plants, this revised version of *The Oxford book of food plants* should be of interest to the lay public, cooks, and gardeners, as well as biologists, nutritionists, dietitians, food scientists, and students at universities, colleges, and schools.

J. G. V. & C. A. G.

London
September 1996

Acknowledgements

We acknowledge with many thanks the assistance given by the following: Dr B. F. Bland (John K. King & Sons Ltd, Colchester, UK); Ryvita Co. (UK); A. D. Evers (Campden & Chorleywood Food Research Association, UK); Dr N. Galwey (Cambridge University); Professor Sir Ghillean Prance (Director), J. S. Keesing, D. Field, Dr D. N. Pegler, Frances Cook, C. Foster, Dr N. Taylor, C. Shine, Dr G. Lewis (Royal Botanic Gardens, Kew); Linda Seaton (Lockwood Press); Professor L. J. G. Van der Maesen, Dr J. J. Bos, Dr P. C. M. Jansen, Dr R. Van den Berg (Wageningen University, The Netherlands); Natural Resources Institute, Chatham, UK; Dr P. Richards (Tate & Lyle); Dr M. Jenner (Speciality Sweeteners, UK); Dr A. J. Vlitos (International Sugar Association); Dr J. Smartt (Southampton University); G. Milford (Rothamsted Experimental Station, UK); G. P. Gent (Processors and Growers Research Organization, Peterborough, UK); M. Day (National Institute of Agricultural Botany, Cambridge, UK); Professor J. S. G. Reid (Stirling University, UK); Dr Joan Morgan; Jean Hodges (Horticulture Research International, East Malling, UK); Saphir Produce (UK); Dr Ann Butler (Institute of Archaeology, London University); J. England (Royal Horticultural Society, Wisley, UK); Marine Biological Association, Plymouth, UK; Professor J. G. Hawkes (Birmingham University); Dr P. Mills (Horticultural Research International, Wellesbourne, UK); Dr Pamela W. Ewan (Medical Research Council, Cambridge, UK); Professor A. Bender, Dr Pat Judd, Dr A. R. Leeds, Dr P. Emery, Dr Helen Wiseman, Dr W. Turnbull, Charlotte Townsend (Department of Nutrition and Dietetics, King's College, London University); M. Halna (France).

We would also like to thank the staff of Oxford University Press for editorial and technical guidance and Liz Moor (King's College) for typing the manuscript.

Financial assistance from Salamon & Seaber (London) is gratefully acknowledged.

Special thanks must be given to Jane Coper who allowed the excellent illustrations produced by her mother (B. E. Nicholson) to be re-published in this book.

Acknowledgements from the 1969 edition

The food plants illustrated in this book have been drawn from live specimens in the main. This was made possible by the generous and skilled collaboration of many university departments, nurserymen, societies, institutions, and individuals. Among the many to whom we owe especial gratitude are the Director and Staff of The Royal Botanic Garden, Kew; the Keeper and Staff of the University Botanic Garden, Oxford; the Welsh Plant Breeding Station, Aberystwyth; Staff of the Fairchild Tropical Garden, Miami, Florida; the Director of the Division of Tropical Research, Tela Railroad Company, La Lima, Honduras; the Director of the Botanical Garden, University of California (Berkeley); Hong Kong University Botany Department; the Professor of Botany, University of Ghana, Legon; the Southern Circle of the Botanical Survey of India; and A. Thornton Jackson who procured many specimens from Malaysia.

Nurserymen in Great Britain were generous in providing plants, fruits, and seeds, particular Scott's Nurseries (Merriott, Somerset); Sutton & Sons (Reading); Alexander Brown (Perth); Blanchard's Nurseries (Ludwell); and the late Margaret Brownlow of the Herb Farm (Seal, Kent).

Among those who grew plants especially to provide specimens for the illustrator are the Department of Agricultural Science, University of Oxford, in whose tropical glasshouses many plants were grown from seeds obtained mainly through the enthusiastic endeavours of Oxford University Press Branches in Nigeria, India,

California, Pakistan, East Asia, and Australia; members of the Shaftesbury Gardens Society; members of Guy's Marsh Borstal Institution (Dorset); Jane Gate; and A. L. Pears.

Sources of advice, information, and practical help which ranged from lending books or colour photographs to seasonal shopping in Soho include: The Royal Horticultural Society; the Director of the Botanical Gardens and National Herbarium, Melbourne; C. T. C. Tatham of the Wine Society; Thompson and Morgan Ltd.; Ambrose Dunston; Lucie Rie; and Sue Thompson.

The plan of the book was originally drawn up by Michael Wallis, who also wrote the text of the fruit. S. G. Harrison, Keeper of the Botany Department, National Museum of Wales (Cardiff), wrote the text of food plants of temperate regions, and Geoffrey Masefield, Lecturer in Agriculture at Oxford University, wrote both the text on tropical items and the general articles on the history, distribution, and nutritional value of food plants. Barbara Nicholson, the sole illustrator, also designed the jacket. We wish to thank the platemakers for lavishing appropriate care and skill on the 95 colour-plates.

Contents

xii Glossary

xiv Introduction

2 *Grain crops*

16 *Sugar crops and sago*

22 *Oil crops*

30 *Nuts*

36 *Exotic water plants*

38 *Legumes*

52 *Fruits*

118 *Beverage crops*

122 *Vegetable fruits*

138 *Spices and flavourings*

148 *Herbs*

160 *Salad crops*

164 *Leaf vegetables*

172 *Stem, inflorescence, and bulb vegetables*

180 *Root vegetables*

190 *Tropical root crops*

194 *Seaweeds*

196 *Mushrooms*

198 *Wild food plants*

200 Nutrition and health

214 Nutrition tables

227 Recommended reading

229 Index of plant names

237 Subject index

Glossary

Achene. A small, dry, 1-seeded, indehiscent fruit (Fig. 2).
Alternate. Arranged spirally or alternately, not in whorls or opposite pairs (Fig. 3).
Anther. Pollen-bearing part of a stamen (Fig. 2).
Appressed. Pressed against another organ but not united with it.
Aril. An extra covering to the seed in some plants.
Awn. A bristle-like projection from the tip or back of the glumes in some grasses (Fig. 5).
Axillary. In the axil, the angle between the stem and the leaf-stalk (Fig. 3).
Bipinnate. Double pinnate; with the primary divisions again divided (Fig. 4).
Bract. A modified leaf beneath a flower, or part of an inflorescence (Fig. 3).
Bracteole. A small or secondary bract.
Calyx. The outer whorl of floral parts (Sepals) which may be free or united (Fig. 2).
Campanulate. Bell-shaped.
Carpel. Unit of an ovary or fruit.
Caryopsis. Grain or grass-fruit, in which the seed-coat is united with the ovary wall.
Corm. The base of a stem swollen with reserve materials into a bulbous shape.
Corolla. The second or inner whorl of a flower, consisting of free or united petals (Fig. 2).
Corymb. A broad, flattish inflorescence in which the outer flowers open first (Fig. 1).
Crenate. With blunt teeth (Fig. 3).
Crenulate. With small, blunt teeth.
Culm. The flowering stem of a grass.
Cupule. A little cup.
Cyme. A repeatedly branching inflorescence, with the oldest flower at the end of each branch (Fig. 1).
Decumbent. Lying on the ground but with the ends curving upwards.
Decurrent. Extending downwards below the point of attachment.
Dichotomous. Equally forked (Fig. 1).
Disc Floret. A flower in the centre of the flower-head (of the Daisy family) (Fig. 2).
Dorsal. On the back or outer face.
Drupe. A fleshy, indehiscent fruit with a stone usually containing 1 seed (e.g. a plum) (Fig. 2).
Endosperm. Part of a seed containing most of the reserves.
Fascicle. A compact cluster.
Floret. An individual flower in a dense inflorescence, as in the Daisy and Grass families.
Glabrous. Not hairy.
Glume. Basal bracts in grass spikelets (Fig. 5).
Indehiscent. Not opening along any definite lines to shed its seeds.
Inflorescence. The flowering region or mode of flowering of a plant.
Involucre. A number of free or united bracts, surrounding or just below one or more flowers or fruits (Fig. 3).
Keel. Lower petal or fused petals ridged like the keel of a boat, as in the pea family (Fig. 3).
Lanceolate. Lance-shaped; narrow and tapering towards the tip (Fig. 4).
Node. A joint on the stem where a leaf is (or was) attached (Fig. 3).
Oblanceolate. Inverted lanceolate, the broadest part above the middle (Fig. 4).
Ovary. The female part of a flower, enclosing the ovules (Fig. 2).
Palmate. With three or more lobes or leaflets radiating like fingers from the palm of a hand (Fig. 4).
Panicle. A brached inflorescence (Fig. 1).
Papilionaceous. Type of flower characteristic of the pea family (Fig. 3).
Pappus. Modified calyx of the Compositae commonly either membranous or in the form of a 'parachute' of hairs (e.g. Dandelion) (Fig. 2).
Perianth. The sepals and petals of a flower (Fig. 2).
Petiole. Leaf-stalk (Fig. 3).
Pinnate. A compound leaf with more than 3 leaflets arranged in 2 rows on a single common stalk (Fig. 4).
Pinnatifid. Pinnately lobed, but not completely divided into leaflets (Fig. 4).
Pubescent. Covered with short, soft hairs.
Raceme. An unbranched inflorescence with the individual flowers stalked (Fig. 1).
Rachis. Main axis of an inflorescence.
Receptacle. The end of the flower-stalk on which the parts of the flower are borne (Fig. 2).
Rhizome. A more or less swollen stem, wholly or partially underground.
Scabrous. Rough to the touch.
Septum. A partition.
Serrulate. With minute, forward-pointing teeth (Fig. 3).
Sessile. Not stalked.
Silicula. Like a siliqua, but short and broad in proportion to its length.
Siliqua. A fruit characteristic of the Wallflower family, elongated and pod-like, but with a central partition and opening from below by 2 valves (Fig. 2).
Spadix. A spike bearing flowers (sometimes sunken), enclosed in a spathe (Fig. 1).
Spathe. Large, sheathing bract (Fig. 1).
Spike. An unbranched, elongated flower-head, bearing stalkless flowers (Fig. 1).
Spikelet. Unit of a grass flower-head, usually with 2 glumes and 1 or more florets (Fig. 5).
Spore. Minute reproductive body of a non-flowering plant.
Stamen. Male (pollen-bearing) part of a flower (Fig. 2).
Standard. The broad, upper petal of a flower of the pea family (Fig. 3).
Stigma. The part of the female organ of the flower which receives the pollen (Fig. 2).
Stipule. A scaly or leaf-like outgrowth at the base of the petiole (Fig. 3).
Style. The connecting portion between stigma and ovary (Fig. 2).
Syncarp. A multiple fruit made up of small fruits united together (Fig. 2).
Ternate. Divided or arranged in threes (Fig. 4).
Tuber. Swollen, underground part of a stem or root.
Umbel. An inflorescence with branches radiating like the ribs of an umbrella (Fig. 1).
Valve. Segment of a dehiscent fruit.
Wings. Lateral petals characteristic of flowers of the pea family (Fig. 3).

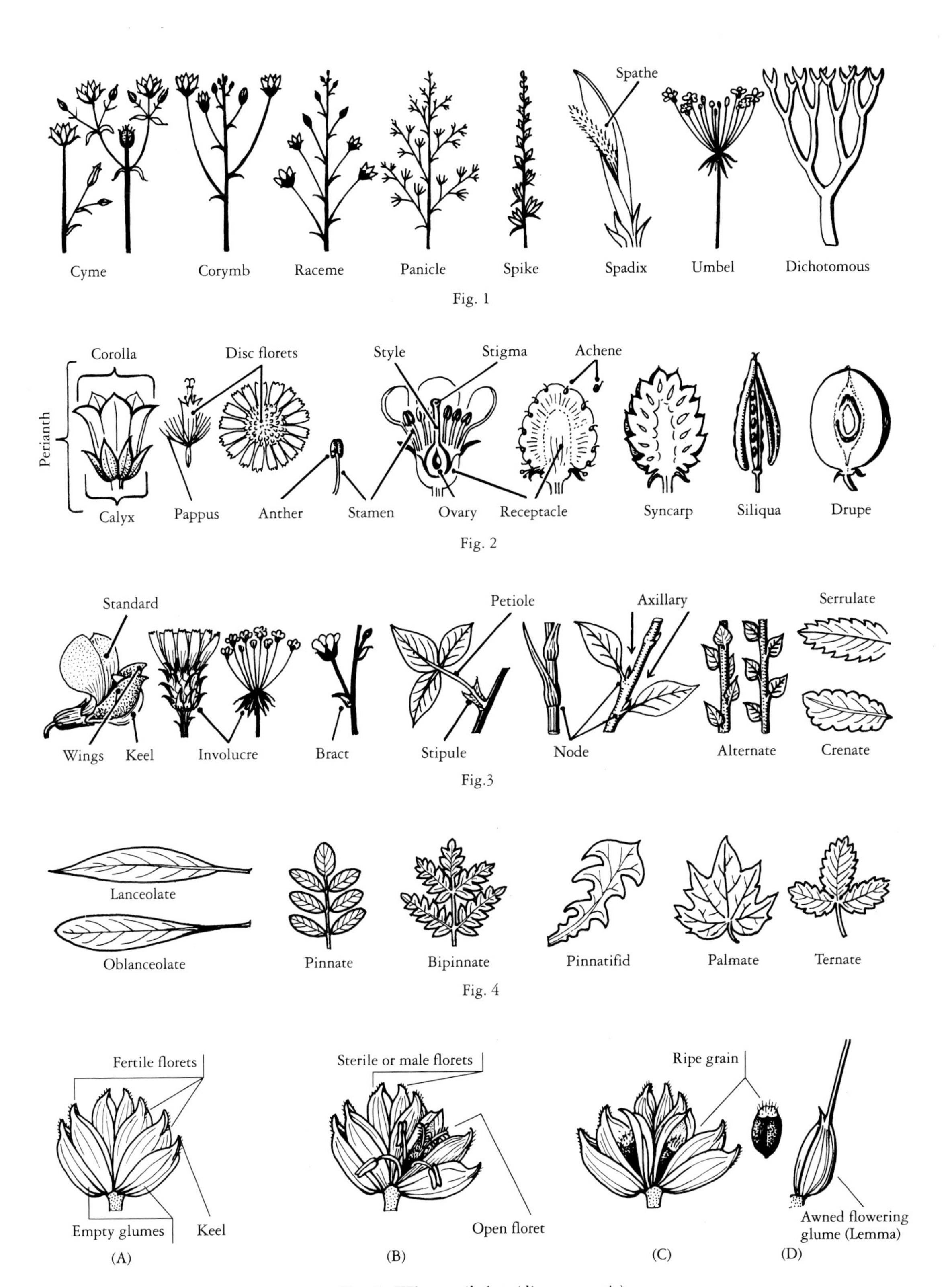

Fig. 5. Wheat spikelets (diagrammatic).
(A) Complete spikelet. (B) Complete spikelet, with one fertile floret exposed, showing ovary and stamens.
(C) Mature spikelet with ripe grains. (D) single-awned flower.

Introduction

The purpose of this book is to provide accurate and attractive illustrations, and textual descriptions, of the plants that serve the human race for food. In this opening chapter, we survey all the major plant groups in the same order as they appear in the book, with a few words at the start on plant classification and naming, and on the domestication and spread of edible plants.

The text descriptions, which are opposite the illustrations to which they refer, aim to provide for each plant information on its origin and geographical distribution; the parts used for food and their treatment and nutritional value; and other features of special interest.

The plants have been arranged in groups, sometimes but not always botanically related, according to the kind of food they provide. Thus, cereal crops come first and are followed by sugar crops, oil crops, nuts, and legumes. Later groups include, amongst others, fruits, spices, salad plants, leaf vegetables and root crops. Wild plants which provide food have been included as well as cultivated ones, and the aim has been to omit no plant which plays any significant part in nourishing the human race. The impossible goal of including every plant ever used for food has not, however, been attempted in this book. The omissions are of slight significance, but include as particular groups: the many scores of fungi which are eaten in different parts of the world; probably some dozens of minor tropical fruits which are occasionally gathered from wild plants or more rarely planted in very small numbers; and a large number of trees and herbaceous plants whose leaves are occasionally cooked and eaten in various regions of the world.

Plant classification

Plants (and animals) are classified into groups using certain biological criteria. The basic unit of classification is the 'species' (sp.), the plural being also 'species' (spp.). Similar species are grouped to form a 'genus', the plural being 'genera'. Similar genera form a 'family'. Plants are given Latin names to allow international understanding, common names often being different in different languages. Thus the scientific name for potato is *Solanum tuberosum* where *Solanum* is the genus, *tuberosum* the species. Potato, together with many other species, belongs to the family Solanaceae. In a perfect situation, a species should have one scientific name but, sometimes, an alternative name or synonym (syn.) is available. Some synonyms are given in this book. According to the particular classification, there are also some alternative names for families, e.g. Cruciferae or Brassicaceae; Gramineae or Poaceae. Some species have originated as hybrids. Thus peppermint is regarded as a hybrid of watermint and spearmint and is designated as either *Mentha* × *piperita* or *Mentha aquatica* × *Mentha spicata*. Food (and other economic) plants have been and are subjected to intense selection and plant breeding. A form within a species produced by such processes and maintained in cultivation is referred to as a 'cultivar' (cv.), e.g. *Solanum tuberosum* cv. King Edward. The term 'cultivar' is, in agriculture and horticulture, often erroneously referred to as 'variety'.

The domestication of food plants

In earliest times food plants (fruits, nuts, leaves, roots) were collected from the wild by the hunter–gatherers, as demonstrated by many archaeological remains, e.g. apples and hazel-nuts. This activity is still practised today but in a number of ways. Certain peoples, e.g. some Australian aborigines and North American Indians, collect food plants from the wild, but these will probably include staples—wild yams in the case of the Australian aborigines. In other countries there is harvesting of wild mushrooms, nuts, fruits, and other types of food plants. However, these often constitute additions to a diet of cultivated food plants and recently there has been renewed interest in this activity, frequently with plants used in previous times. Some plants growing under wild or semi-wild conditions, e.g. Brazil-nut, wild rice (*Zizania*), shea-nut (*Butyrospermum*), provide food materials of commercial importance. Of course, in times of famine and scarcity, wild plants, not normally employed as food, are utilized.

Plants were domesticated and brought into cultivation (the beginnings of agriculture and horticulture) for very good reasons. They could be protected from pests, weeds could be removed, and harvesting could be carried out in a controlled area. No doubt at this early stage human selection of crops with favourable traits (e.g.non-shattering of seed heads, uniform ripening of seeds, large fruits and seeds, easy storage, and uniform germination) started to take place.

Sometimes the wild ancestors of cultivated food plants are easily identified but this is not always the case, neither is it always possible to establish the time of domestication. If domestication is recent (e.g. kiwifruit, wild rice), then much relevant information is available, but many food plants are of ancient origin; consequently a number of techniques are employed to try to elucidate their ancestry. Archaeological investigations are widely utilized. Cooking converts plant material into charcoal, which resists decomposition and often retains detailed structural characteristics that may allow identification. Identification of imprints of grains and other plant parts on pottery may also be possible. Food plant material may be preserved by desiccation in very dry areas. Such material has notably been found in the pyramids and tombs of ancient Egypt. Preservation can also take place under waterlogged conditions. For instance, corpses of 'bog people' have been retrieved in Denmark, Holland, Germany, and the UK, and their stomach contents examined for food materials. Although digestion has taken place, there may well be some resistant plant remains found in preserved human faeces, or 'Copralites', which are presumably a direct indication of the food consumed. Chemical tests can also identify food plant material or food substances. A most important advance in this field has been the development of radiocarbon (^{14}C) dating which can now be applied to very small samples of food plant remains and give a dating to the millennium, or possibly the century.

Sometimes it is not obvious which wild plants evolved into present-day cultivated food plants. It is possible that the ancestral species are no longer living. However, a popular approach has been to compare the botanical characteristics (structural, genetical, and biochemical) of a cultivated plant with possible wild living relatives and thus form an opinion concerning the wild progenitor.

Some cultivated food plants (like water-cress, blackberry, hazel, carrot, and parsnip) are exactly like the wild species, or very similar, except that their edible parts are particularly well-developed. Other living and important food plants, e.g. brown or Indian mustard (*Brassica juncea*) and bread wheat (*Triticum aestivum*), have no immediate wild progenitors. Their evolution is regarded as having taken place by a process of hybridization between species which have not always been properly identified. Information on the origin of the various food plants will be given later in the book.

As stated earlier, human selection of crops with favourable traits no doubt took place at the beginnings of agriculture and horticulture. This developed, thousands of years later, into the science of 'plant breeding'. The purpose of plant breeding is to improve crops (food and others) and may have various objectives. Improvements of an agronomic nature may be desired, such as the production of dwarf or semi-dwarf forms (to facilitate harvesting), increased yield, or resistance to insects and other pests. Breeding, in the case of food plants, has sometimes been aimed at the improvement of food constituents, such as the essential amino acid lysine in maize (corn) and sugar (sucrose) in sugar beet.

Until quite recently plant breeding only involved the hybridization of forms within a crop and germination of the resulting seeds to produce plants, some of which might have improved characters. This process obviously takes some time but, as far as food plants are concerned, has achieved some very important results, e.g. the production of 'hybrid' maize (corn) in the Corn Belt of North America and high-yielding wheat cultivars in India—the so-called Green Revolution. Home, as well as commercial growers can now obtain F_1 seed of food plants showing 'hybrid vigour'. Care must be taken with the so-called 'miracle' cultivars because if one is grown to the virtual exclusion of others, and is then attacked by some type of disastrous disease, there could be a shortage of varied genetical material to rectify the situation. For these reasons, collections or banks of different cultivar seeds or plants are established throughout the world to retain varied breeding material. In more recent times other techniques have been included in plant breeding, such as the use of the chemical colchicine to increase genetic variation, and tissue culture methods, including hybridization at the cell level. The most recent addition to crop (including food) improvement has been the use of genetic engineering, where either the genes of a plant are altered or new genes are introduced. With this technique, insect resistance has been induced in cotton (an oilseed plant) and the rate of softening in tomatoes has been considerably reduced, thus leading to a longer shelf-life and better texture. There is no doubt that with genetic engineering many more changes can be made to food plants, although the consumer will have to be convinced that the technique is acceptable. Notwithstanding recent developments, much plant breeding is still carried out with the established methods of hybridization.

Dispersal of food plants around the world

The dispersal of food plants around the world is a fascinating subject with many facets. In the main part of this book, and under the headings for the individual crops, some detailed information will be given so that the object of this section is to survey the main phases of food plant dispersal.

Certain cereals and pulses (legumes) were domesticated in very ancient times. In about 8000 BC in the Fertile Crescent of the Near and Middle East (present-day Syria, Iran, Iraq, Turkey, Jordan, Israel), wheats, barley, lentil, pea, bitter vetch, chick-pea, and possibly faba bean, were brought into cultivation by the Neolithic people. These crops spread from the point of origin. Archaeological evidence indicates that the wheats, and some of the legumes, had reached Greece by 6000 BC and evidence of their presence within that millennium has been found in the Danube basin, the Nile valley, and the Indian subcontinent (Pakistan). Dispersal continued through Europe, the crops reaching Britain and Scandinavia in 4000–2000 BC. There was quite a hiatus in this dispersal until the sixteenth and following centuries when, following the exploration and colonization of various countries, wheat species were taken to North and South America, South Africa, Australia, and New Zealand (see p. 2). Another example of the dispersal of an ancient cereal concerns maize (corn), of which the earliest known material (5500 BC) comes from Mexico. It spread through the Americas, and Columbus, possibly after his first voyage of 1492, brought the seed back to Europe. The plant was dispersed rapidly through parts of Europe and Africa. Maize is now cultivated in almost every continent.

The expansion of empires has contributed largely to the spread of food crops. In ancient times the Romans (*circa* 600 BC to AD 500), through their conquests, introduced a number of food plants (e.g. carrots, globe artichokes, garlic, onions, lettuce, almonds, chestnuts, and walnuts) throughout Europe, sometimes including Britain. The Moors (eighth to eleventh centuries) brought rice, Seville orange, lemon, aubergine, and spinach, to Spain.

The first voyage of Columbus (1492), and the exposure of the New World to Europeans, had an enormous effect on the interchange of food plants between the Old and New Worlds. It has been stated that Columbus brought back food plants such as maize and sweet potato (*Ipomoea batatas*), not seen before in Europe, although it seems difficult to obtain exact evidence. On his return to the Caribbean, Columbus took back crops such as wheat, barley, sugar cane, and grape vine.

The development of the Spanish and Portuguese empires in Central and South America had a profound influence on the exchange of food plants. There are many examples to choose from. The potato (*Solanum tuberosum*) came to Europe in the sixteenth century, introduced by the Spanish from South America. It was not introduced into North America until a later date. Other important crops brought by the Spanish (possibly started by Columbus) to Europe in the sixteenth century were tomato (*Lycopersicon esculentum*) and sweet potato. A mystery surrounds the latter crop because, although it arose in tropical America, it appeared in Polynesia in pre-Columbian times, and no satisfactory explanation for this migration has emerged. Cassava, with its origin in tropical America, was taken by the Portuguese to Africa in the sixteenth century and, because of its tolerance to adverse conditions, is now widespread throughout the tropics. The groundnut, a native of South America, was carried by the Portuguese from Brazil to West

Africa in the sixteenth century; the Spaniards took the crop in the opposite direction to the Philippines and it then spread to India, Malaysia, and the Far East. During their conquest of the Incas, the Spaniards introduced European crops such as wheat and barley, and somewhat forcibly inhibited the cultivation of the ancient Inca food plants. This ancient cultivation is now being resuscitated.

The settlers from Europe transported a vast array of food plants (cereals, fruit trees, and vegetables) to North America. Some crops (e.g. cowpea, okra, and a few yam species) are reputed to have been brought to the New World through the slave trade.

This dispersal of food plants around the world has led to some interesting developments. Hot *Capsicum* peppers, also said to have been brought back by Columbus after his first voyage, were being cultivated in India by the middle of the sixteenth century. They virtually replaced the native black pepper (*Piper nigrum*) in curries in India. Macadamia or Queensland nut, a native of Australia, has its greatest commercial development in Hawaii. More palm oil is produced from the *Elaeis* palm grown in Malaysia than in its native West Africa. Soya bean, an ancient and important food plant in the Far East, was introduced into North America late in the nineteenth century as a fodder plant. Since then it has become of world importance as an oilseed and source of protein. A twentieth-century phenomenon has been the development of kiwifruit, originally from China, by plant breeders in New Zealand, into a product of significant importance. Most food plants have now become dispersed from their centres of origin.

Grain crops

Grain crops, or cereals, are by far the most important sources of plant food for the human race. On a world basis, they provide two-thirds of the energy and half the protein of the diet. However, the proportion of food energy supplied depends on the location, ranging from 25 per cent in the USA to as much as 85 per cent in regions where only small amounts of other foods (e.g. meat) are available, or where incomes are extremely low. Bread (leavened and unleavened), pasta, noodles, porridges, breakfast foods, starch, protein, oils, beer, spirits, and animal feed are produced from the grain; the plants may be utilized as forage, silage, hay, or straw. With increasing affluence there is a trend to move away from the utilization of coarse grains to finer products.

All cereals, except the so-called pseudo-cereals such as buckwheat and quinoa, belong to the grass family (Gramineae or Poaceae). The grass inflorescence, or head, is made up of units known as spikelets which contain the flowers or florets within scales or glumes. The 'grain' is strictly speaking the fruit (caryopsis) of the plant and develops from the fertilized ovary. Although botanically incorrect, the term 'seed' is often applied in commerce to the grain. The true seed is within the outer layers of the grain. In some cereal species, threshing completely separates the grain from the rest of the plant (chaff and straw); in others the grain, after thrashing, remains within a small number of glumes, often known as chaff, hulls, or husk.

Cereals are important in international food trade, being the main component of exports and imports. The cereal grain is an important dietary source of carbohydrate, fibre (insoluble and soluble), protein, certain vitamins (the B complex and E), and minerals. Starch is by far the major carbohydrate, constituting on average about 75 per cent of the weight of the grain. The protein content varies between 10 and 15 per cent, with brown rice and certain millets being at the lower end of the scale, while oats and wheats are at the upper end. Although cereal grains are quite rich in protein, they contain relatively low amounts of lysine, an essential amino acid, compared to the levels found in animal protein. However, for some cereal species, high lysine cultivars have been developed (see p. 6). There are significant amounts of mineral and trace elements, e.g. iron, magnesium, calcium, manganese, and zinc, but these may not be entirely available in a nutritional sense, being bound to phytate. Some processes, such as malting, leavening, and fermentation, improve this availability. The average content of fat is about 2 per cent. Cereal fat is highly unsaturated. Although the percentage of fat in the grain is low, corn (maize) oil is of great international importance in trade and consumption.

On a world basis, the important cereals in cultivation are wheat, rice, maize (corn), oats, barley, rye, sorghum, and millets. Of these the most widely used are wheat, rice, and maize; in the main, wheat and rice directly for human food, and maize for animal feed. The cereals were probably amongst the first crops brought into cultivation by the Stone Age (Neolithic) people about 8000 BC (see p. 2). Up until that time wild grass seed was gathered for food, as it still is in some parts of the world. Wheat was domesticated in the Fertile Crescent of the Near and Middle East, maize in the tropical highlands of Mexico, and rice in India and China. Ancient people no doubt selected plants for cultivation for various reasons. As far as cereals are concerned, some desirable characters would be compact seeds that are of a nutritious nature and are easy to store, and non-shattering heads, so that the seed is not disseminated naturally but can be harvested by the farmer.

The cereal grain is divided into zones or areas which contain the various food constituents.

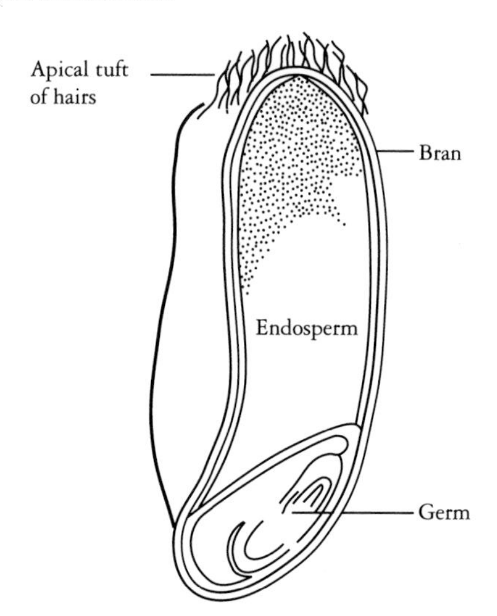

Vertical section of wheat grain

There is a scientific terminology for these zones, but there are also terms which are of common usage in food descriptions—bran (*pericarp*, *testa*, and *aleurone layer*), flour or endosperm (*endosperm*), and germ (*scutellum* and *embryo*). The bran is particularly rich in dietary fibre, the vitamin B complex, and minerals; the flour in starch and protein; and the germ in unsaturated fat, vitamins B and E, and minerals.

Because of this unequal distribution of nutrients, certain cereal processes, such as milling, will affect the nutrient content of the products (see p. 211). The production of white flour from bread wheat grain is a good example. In this instance the object of the process is to remove the bran and germ, leaving the flour. This will remove much of the fibre, vitamins, and minerals, thus affecting the nutritional status of the flour and the resulting bread. White flour retains only 20–50 per cent of such nutrients. There are now many types of bread which, in various ways, have improved nutritionally. Fortification or enrichment with vitamins and minerals, mandatory in certain countries, is one process. Wholemeal and brown bread contain varying amounts of bran. Some breads contain whole cereal grains and, for extra fibre, pea hulls.

The pseudo-cereals are plants grown for the seed (actually fruits) of somewhat similar nutrient composition to the cereals, but not belonging to the grass family. They belong to a diverse group of families—quinoa (Chenopodiaceae), amaranth (Amaranthaceae), and buckwheat (Polygonaceae). In botanical classifications these families are considered closely related. As sources of food and compared to the cereals, pseudo-cereals have decreased in importance over time.

Sugar crops

The main sugars found in plants are glucose, fructose, and sucrose (see p. 200). As far as food is concerned, the term 'sugar' is normally applied to sucrose. The main sources of sugar are sugar cane (*Saccharum officinarum*) and sugar beet (*Beta vulgaris* var. *esculenta*), although various palms in Africa and Asia, the sugar maples of North America, sorghum, carob or locust bean provide a certain amount of dietary sucrose. Sugar cane and sugar beet belong to widely different families, namely the Gramineae (grasses) and Chenopodiaceae.

Sugar cane (a tropical crop) produces about 65 per cent of the world's sugar; the rest comes from sugar beet (a temperate crop). It is the stem (17–20 per cent sucrose) of the sugar cane which is processed for sugar; the root (14–18 per cent sucrose) of the sugar beet is utilized. Sugar is virtually 100 per cent sucrose and is therefore the purest common food. It is used as a sweetener in many types of food and beverage, and also as a preservative, a bulking agent, and an antioxidant. Sugar is the cheapest form of energy-giving food, requiring the lowest unit of cultivated land area per unit of energy production. It is readily converted into energy in the body. On the negative side it can be the cause of dental caries and, if consumed in large quantities, may dilute the nutrient density of the diet, showing what have been described as 'empty calories'—that is sugar produces only energy, no other dietary constituents are present. There is really no evidence that sugar is the direct cause of cardiovascular disease, diabetes, obesity, or cancer.

At one time sugar was a luxury item; in the medieval period honey was used as a sweetener. Cane or beet sugar is now predominant although, since the 1970s, there has been some decline in consumption because of the production of corn (maize) invert sugar, fructose, and glucose syrups. There has also been interest in other sweeteners, some of which are 'artificial' like saccharin or aspartame; others are 'natural'. A number of 'natural' sweeteners have been extracted from plant sources but are relatively little-used. An interesting example is thaumatin, a sweet protein, taken from the berries of the West African plant *katemfe* (*Thaumatococcus daniellii*, family Marantaceae). This is two thousand times as sweet as sucrose. It is used in chewing gum and animal feed. Sweeteners other than sucrose can be useful in calorie-reduced and diabetic diets, although some leave an unacceptable aftertaste.

Sucrose is fermented to produce industrial ethyl alcohol (ethanol). In some countries (e.g. Brazil) the alcohol is used as a motor fuel. This utilization of biofuels helps conserve the fossil fuels.

Oilseeds and nuts

Many plant species have some oil in their seeds but relatively few are utilized as commercial oilseeds. Only about twelve species are of international importance, although about two hundred have been or are utilized on a local scale. In the commercial seed the amount of oil varies (18–50 per cent) according to species and cultivar. To be successful commercially, not only must a seed contain a reasonable amount of oil, but the plant also must have good agronomic characters. The term 'oilseed' is used to describe structures that, in a botanical sense, are truly seeds (e.g. rape, soya) and fruits (e.g. olive, sunflower). In contrast to the cereals, the oilseed species are found in a wide range of plant families.

Not all vegetable oils are employed in food—some are of industrial importance, e.g. castor (*Ricinus communis*). A limited number of oilseed species are cultivated in temperate regions (e.g. rape) but most species are grown in subtropical and tropical lands. Up until relatively recent times, oilseeds were exported from these subtropical and tropical countries to Europe and the USA for processing. Today there is a considerable amount of processing in the countries of origin.

The extraction of oil can take place in facilities of varying sizes, ranging from small village presses to large factories where the oil is extracted by mechanical presses or expellers, and by oil solvents (e.g. light petroleum). After refining, the oil is utilized in a number of food products such as cooking and salad oils, and also margarine; there may also be industrial uses in soap, paints, lubricants, and varnishes. Most vegetable oils employed in food are unsaturated (either polyunsaturated or monounsaturated) (see p. 202). Saturated food oils are coconut, palm kernel, and cocoa. The conversion of vegetable oils into margarine and spreads includes a process of 'hardening' which involves the addition of hydrogen. This leads to an increase in the degree of saturation. The hardening process produces so-called *trans* fatty acids which might, under certain conditions, be deleterious to health (see p. 202).

Following oil extraction, the seed residue by-product contains up to 50 per cent protein, and is consequently utilized as an animal feed, although some residues may have potentially toxic (e.g. peanut) or antinutritional compounds, such as aflatoxin, and glucosinolates (e.g. rape) (see p. 28).

In addition to the usual sources of vegetable oils, other species have been investigated as commercial possibilities. Rarely has success been achieved, but one plant, jojoba (*Simmondsia californica*), although not now of importance in food, has given interesting results. The seed oil is actually a liquid wax and substitutes for sperm whale oil, thus helping whale conservation. An interesting development, at least in UK supermarkets, has been the appearance or reappearance of oils such as hazel (*Corylus avellana* and other hazel species), walnut (*Juglans regia*), and grape seed (*Vitis vinefera*).

In the popular sense the term 'nut' is applied to a seed or fruit with an edible kernel inside a brittle or hard shell; the botanical definition is somewhat more complicated. The kernel is highly nutritious with, according to species, up to 30 per cent protein and up to 70 per cent oil. Dried chestnuts (*Castanea sativa*) are unusual in that the major food reserve is starch (50–60 per cent). Nuts are a good source of minerals, including calcium, phosphorus, iron, sodium, and potassium, and also the vitamin B complex, and vitamin E. No doubt because of their highly nutritious nature, nuts have long been part of the human diet; nut remains have been found in archaeological sites dating back to before 10 000 BC. Some species serve both as popular nuts and commercial sources of oil, e.g. peanut (*Arachis hypogaea*) and almond (*Prunus amygdalus*). Nuts are consumed in their natural form, and in nut butters, confectionery, ice cream, and in various recipes. Immature kernels (green nuts) of some species are eaten, good examples being almond and walnut, which may also be pickled. Green walnuts contain some vitamin C. In some individuals, certain nuts (peanut is notorious) can cause allergies.

Legumes (pulses)

The Leguminosae (Fabaceae) is a large family—its members being popularly known as legumes. On a world basis it is second in importance to the grasses (cereals) as regards food plants. The family can easily be recognized by (a) the flower with its petals comprising a large upper standard, 2 lateral wings, and a boat-shaped keel, (b) the fruit known as the legume or pod containing the seeds or beans. Legume roots bear nodules containing bacterial capable of 'fixing' atmospheric nitrogen. This property has been utilized for a very long time in crop rotations.

Food products of the legumes include dry, edible seeds (beans, pulses), immature green seeds, oilseeds (e.g. soya), green pods, spices (e.g. tamarind), young shoots, leaves, and sprouts (germinated seeds). In addition, some legumes are used as green manures, fodder plants, and cover crops.

As in the case of the cereals, the legumes are amongst the oldest crops cultivated by the human race. Between the cereals and legumes there is a parallel domestication: wheat, barley, pea, lentil, broad bean, and chick-pea in West Asia and Europe; maize and common bean in Central America, ground nut in South America; pearl millet, sorghum, cowpea, and bambara groundnut in Africa; rice and soya bean in China. Very good archaeological remains of pea, lentil, and chick-pea, abound in Near East Neolithic settlements and soon afterwards they were found in vast areas from the Atlantic coast of Europe to the Indian subcontinent. Later, remains of faba beans and grass pea are found, followed much later by fenugreek, and later still by lupins.

Pulses are important in human nutrition in a number of ways. They are very good sources of protein (20–40 per cent), which is relatively rich in essential amino acids, usually including lysine, but deficient in sulphur-containing amino acids, particularly methionine and cystine. A diet combining cereal and pulse protein therefore complements the essential amino acid requirement of the consumer. Pulses are important in populations where animal protein is scarce. The carbohydrate content of pulses is between 13 and 65 per cent with, in most species, starch constituting 40–55 per cent of digestible carbohydrates—soya bean and ground nut being notable exceptions. In general, legume seed fat content is low (1.0–2.5 per cent) except in soya, groundnut, winged bean, lupin, and chick-pea. They are reasonably good sources of some essential minerals, such as calcium, phosphorus, potassium, iron, magnesium, zinc, and copper, but with a low sodium content. Pulses are excellent sources of some B vitamins (thiamin, riboflavin, niacin, folic and pantothenic acids). Immature seeds, pods, leaves, and sprouts also have carotenes, and vitamins C and E. Legumes are quite a good source of dietary fibre.

Although legumes contain an impressive range of nutrients, there may also be antinutritional and toxic substances (see p. 208) in the seed. These include: (a) lectins or haemagglutinins, found in many legume seeds, which in humans can cause nausea, vomiting, diarrhoea, abdominal pain; (b) cyanogenic glycosides—some early reports in Europe associated these substances with cyanide poisoning caused by lima beans imported from the tropics; (c) stachyose and raffinose—these carbohydrates are said to be responsible for flatulence; (d) *Lathyrus sativus* peas under certain circumstances can bring about lathyrism—a neurological disease; (e) digestive enzyme inhibitors—soya and some other legumes contain proteins which can reduce protein digestion and utilization by the consumer; (f) alkaloids, found in lupins; (g) fabism, caused by faba beans; (h) goitrogens—some reports have claimed the presence of these substances, in soya and ground nut, lead to the enlargement of the thyroid gland; (i) saponins.

As has been stated, the legumes are of enormous importance in food in spite of their various antinutritional or toxic constituents. No doubt, over the very long period of time that they have been utilized, human selection has produced more acceptable forms, and proper, often traditional, food processing (soaking in water, cooking, fermentation) can eliminate these undesirable constituents. Dry pulses for sale to the public often carry warnings about proper processing prior to consumption. Problems may arise with seeds transferred to societies unfamiliar with processing needs. Other interesting changes during the evolution of domesticated legumes have been an increase in side size, and pods that do not shatter to release the enclosed seeds, thus enabling a more efficient harvest. This retention of the seed on the plant is a state of affairs rather similar to that existing in the cereals.

A number of pulses can be grown in areas of low rainfall and poor soil.

Fruits and vegetables

In the strict botanical sense, the term 'fruit' applies to the structure, usually containing seeds, which develops from the flower ovary after pollination and fertilization. In the food sense, fruits are succulent structures (with seeds or seedless) exhibiting a pleasant aroma and flavour. There are many fruit species, showing a range of habits (herbaceous, shrubby, and tree-like), and coming from a variety of families, although two families, namely the Rosaceae (apple, pear, plum, strawberry, and others) and the Rutaceae (orange, lemon, grapefruit, and others) are of outstanding importance. It is difficult to define the culinary term 'vegetable'. Vegetables (and salad plants) are edible plant products which can be modified stems (e.g. potato), roots (e.g. cassava), leaves (e.g. cabbage), and even fruits in the scientific sense (e.g. tomato). Normally vegetables are not sweet and are usually eaten with meat, fish, or savoury dishes. There are some exceptions to the above distinction; for instance, rhubarb is prepared as a fruit but is actually a leaf-stalk and bananas may be consumed as a fruit or cooked as a vegetable. Well-known vegetable plant families are Solanaceae (potato, tomato, and others), Cruciferae (Brassicaceae) (cabbage, kale, turnip, swede, and others), Umbelliferae (Apiaceae) (carrot, parsnip, and others). Fruits and vegetables were harvested from the wild long before organized cultivation, as shown by fruit remains which have been found in numerous archaeological sites. However, vegetable remainss are not as well preserved. Although fruits and vegetables were brought into cultivation thousands of years ago, their domestication is not so old as that of cereals and pulses.

Fruits and vegetables form an essential part of a well-balanced diet. In fresh condition they are a major source of vitamin C. Amounts vary according to species (citrus fruits 40–80 mg/100 g; black currant 70–190 mg/100 g). A freshly-dug potato contains 21 mg/100 g vitamin C but this decreases enormously during storage (there is a 75–80 per cent loss over 9 months). Nevertheless, because of the large amounts of potatoes consumed in the UK and North America, the vegetable is an important contributor of vitamin C. The West Indian cherry (*Malpighia glabra* syn. *M. punicifolia*), a fruit of local importance, contains a very large amount of vitamin C (1000–2500 mg/100 g). Carotenes (yellow, orange, orange-red pigments), some of which are vitamin A precursors, are found in useful amounts in sweet potato, pumpkins, *Capsicum* peppers, and tomato. Carrot is an outstanding source. Some crops (e.g. cassava and yam) do not contain appreciable amounts of carotenes, but are important when consumed in large quantities. Various fruits (e.g. mango, papaya, and apricot) are good sources of carotenes. Vitamins E and the B complex are present in fruits and vegetables but these commodities are not major sources.

Also important in a diet is the contribution of minerals from fruits and vegetables. They contain a range of minerals and are important sources of calcium and iron. The amount of potassium present, according to species, is variable but is normally high (e.g. fresh dates, about 400 mg/100 g; apricots, about 300 mg/100 g).

Their flavour and aroma also make fruits and vegetables important in a diet. Sweetness is related to the sugars present, usually glucose, fructose, and sucrose. Generally speaking, fruits and vegetables do not provide much energy, with the notable exceptions of the starchy staples (e.g. potato, cassava, and yams). Organic acids (usually citric and malic) are involved as flavour characteristics. There are also some special situations (e.g. glucosinolates in *Brassica* vegetables; and the bitter naringin of grapefruit). Finally there are numerous volatile compounds (e.g. esters, alcohols, and aldehydes) involved in the flavour and aroma of fruits and vegetables.

Food colour can be important in a diet. Carotenes have already been described. There is also the green chlorophyll as well as some red and purple colours (e.g. strawberry, plum, red cabbage, and red-skinned potato cultivars), which are due to anthocyanin pigments.

The amount of fat is low, generally below 1 per cent (avocado is an exception). Protein content too is low, but this must be balanced with the amount of commodity consumed and some of the staples have a well-balanced amino acid composition.

Fruits and vegetables are most important in providing fibre (both insoluble and soluble) in a diet.

Spices and herbs

The known utilization of spices and herbs extends over some 5000 years, starting with the ancient Egyptian, Chinese, and Indian civilizations. Up until reasonably modern times they were expensive items, sometimes the equivalent of precious metals. Spices and herbs were highly regarded in the Greek and Roman Empires, members of the latter being responsible for carrying these commodities to the rest of Europe. The search for direct trade routes to the spice-producing areas of the East and India produced some great voyages of exploration, e.g. Columbus and Vasco da Gama in the fifteenth century. In the seventeenth and eighteenth centuries the possession of spice areas in the East led to wars among the Portuguese, Dutch, and British.

Spices and herbs have served many functions according to the historical period such as embalming, medical uses, the masking of bad food and body odour, the enhancing of food flavour, and food preservation (because of antioxidant and antimicrobial properties). Modern medicine has led to a decrease in the usage of herbal medicine in western countries, although in many parts of the world (e.g. Africa, India, China, and South America) it is still important and there is now a revived interest in western countries. Modern methods of food preservation, such as refrigeration and canning, have really eliminated the need for herbs and spices to disguise the unpleasant taste of decaying food, although they are still used to enhance food flavour.

The difference between a spice and a herb is not always easy to define, but in general terms, spices can originate from various parts of a plant body (seeds, fruits, bark, roots) and they tend to grow in semi-tropical climates. On the other hand, herbs are the leafy parts of soft-stemmed plants and are found in more temperate regions. The aroma and flavour of a spice or herb relates mainly to its essential (volatile) oil, which is a complex mixture of organic compounds (e.g. alcohols, aldehydes, and esters). In food products, spices or herbs may be used in their entire or powdered states. Much use is also made of the essential oil, produced by distillation, and 'oleoresins' which are organic solvent extracts of the spice or herb, containing certain other constituents in addition to the essential oil.

Spices appear to originate from a number of plant families but the mint family (Labiatae or Lamiaceae), the parsley family (Umbelliferae or Apiaceae), and the tarragon family (Compositae or Asteraceae) contain some well-known herbs.

Palms

A family containing species with various uses in food is that comprising the palms (Palmae or Arecaceae). Some species produce sago starch. Other food products obtained from palms are vegetable oils, 'cabbages' or palm hearts (terminal buds), sugar, palm wine and distilled spirit, desiccated coconut, and animal feeds. Apart from food, the palms have many other uses in building and other industries.

Non-flowering plants

Most food crops are flowering plants (angiosperms) but some other plant groups are also utilized. Seaweeds or marine algae are quite important in various parts of the world. All algae contain the green pigment chlorophyll. This is the only pigment present in the group known as the green algae (Chlorophyceae) but, in other groups, such as the red algae (Rhodophyceae), and the brown algae (Phaeophyceae), there are additional pigments.

The fungi are another important group of non-flowering plants used as food. Probably best-known in this respect are the mushrooms and truffles, where the edible parts are the spore-bearing structures, although the fungi also consist of fine threads or 'hyphae' which grow through the soil. Fungi do not contain chlorophyll and, because of this, together with some other characteristics, they are sometimes regarded as a group distinct from plants and animals. The commercial cultivation of a few mushroom species is important, but in certain parts of the world there is also a deep interest in collecting wild species, although this requires the ability to identify fungi correctly because a number of species are toxic. In more recent times, there has been commercial exploitation of micro-fungi as food.

Ferns comprise another group of non-flowering plants used for food. The young shoots (fiddleheads) and sometimes the rootstocks of various species have been consumed in a number of countries, either from the wild or as a commercial crop. The fiddleheads or bracken fern *Pteridium aquilinum* are popular in Japanese cuisine, but care must be taken in consumption since bracken contains a number of toxic substances including carcinogens.

The seeds of angiosperms (flowering plants) are enclosed in fruits. In members of the other group of plants (the gymnosperms) the seeds are exposed. Of the gymnosperms some pine species provide edible seed kernels, while the stem pith of some cycads has been used as sago.

THE NEW OXFORD BOOK OF FOOD PLANTS

Grain crops: Wheats

(For general information on grain crops, see p. xvi)

*More wheat is produced annually than any other cereal crop and it is probably the world's foremost food plant. The crop is grown throughout the temperate regions of the world but only in the highland areas of the tropics and subtropics. Some 80 per cent of the world's wheat production takes place in Russia, the United States of America, China, India, France, Canada, Australia, Turkey, Pakistan, and Argentina. There are several wild and cultivated wheat species and thousands of cultivars. The two most important species are bread wheat (*Triticum aestivum *syn.* T. vulgare*) (**1, 2**) and durum or macaroni wheat (*Triticum durum*) (**3**). Of the other cultivated species, club wheat (*Triticum clavatum*) is of some importance in the United States, others are grown to a limited extent, for example emmer (*Triticum dicoccum*) (**4**) and rivet, cone or English wheat (*Triticum turgidum*) (**5**). The latter was at one time the main wheat in southern England but was replaced by bread wheat.*

*The origin and domestication of cultivated wheats from wild species has attracted widespread attention. Most of the knowledge gained is based on the archaeological remains of grains but use has also been made, in more recent times, of biochemical techniques. However, the story is far from complete. It is generally accepted that wheat was domesticated from wild species in the Fertile Crescent of south-western Asia about 8000 BC, together with barley and pulses. The cultivated wheats arose from the wild species by processes of mutation and hybridization. Einkorn (*Triticum monococcum*) and the more important emmer were early cultivated species which spread from south-west Asia into Europe, North Africa, and India. The grains of these species retain the hulls after threshing. Today, einkorn and emmer are relic crops, cultivated only to a very limited extent. Bread wheat probably entered cultivation after the domestication of einkorn and emmer and, like the latter species, migrated into Europe, North Africa, and Asia. Its modern form is free threshing to produce naked grains, as is durum wheat. Until near the end of the sixteenth century wheat was confined to Asia, Africa, and Europe but, in the two centuries that followed, it was taken to North and South America, and to South Africa. Somewhat later, cultivation of the crop started in Australia and New Zealand.*

The types and distribution of the important dietary constituents of the wheat grain follow the usual cereal pattern. Protein concentration may be high for the cereals but this will depend on the species, cultivar, or environmental conditions.

BREAD WHEAT (1–2) *Triticum aestivum* syn *T. vulgare*. On a worldwide basis, bread wheat constitutes about 90 per cent of the wheat grown; the remainder is devoted essentially to durum wheat. Bread wheat has awned (glume projections) and awnless forms. There are spring wheats (sown in spring and harvested in late summer) and winter wheats (sown in autumn and harvested in early summer). The colour of the grain varies from yellow to red-brown but cultivars are usually described as white or red. Cultivars may be classified as hard (vitreous endosperm) or soft (mealy endosperm). These are milling characters. Wheat flours are said to be strong (a relatively high protein content) or weak. Hard wheats are used in bread manufacture; soft for the production of cakes, cookies, biscuits, and pastry.

To make leavened or porous bread, the basic ingredients are flour, water, yeast, and salt, which are mixed together to give a dough which rises because of yeast fermentation and is finally baked. Among the cereals, the ability to produce leavened bread is outstanding in bread wheat (other cereals, e.g. rye and durum wheat, produce poorly leavened bread). This extraordinary ability depends on the wheat protein complex known as 'gluten', which is elastic, expands during fermentation, and retains the released carbon dioxide to give a porous bread. Without yeast, wheat flour produces a flat bread (e.g. the chapatis of the Indian subcontinent). The extraction of starch and gluten from the wheat grain or flour are well-known industrial processes. Gluten may be added to bread to increase its protein content.

Some children and adults suffer from coeliac disease, an intolerance to wheat gluten (also to rye). There is some doubt about oats and barley, but maize and rice are tolerated satisfactorily. Wheat germ oil, highly unsaturated and containing vitamin E, is a minor article of commerce. The quantities of hard wheat imported into the United Kingdom from countries such as Canada and the United States for the purpose of bread making have decreased in recent years because of hard cultivars produced in the United Kingdom and better baking technology. Wheat being such an important crop, much breeding is carried out to develop new cultivars. This has been a feature of the so-called Green Revolution which, for example in India, led to a doubling of wheat available in the decade up to 1980.

DURUM WHEAT (3) *Triticum durum*. Compared to bread wheat, far less durum wheat is grown and it has a lower resistance to cold, long winters but is better adapted to drought conditions. In Europe most cultivation takes place in Italy; elsewhere it is grown in Australia, North Africa, Ethiopia, Russia, and North and South America. On average, the grain has a higher protein content than bread wheat. Milling separates the grain into bran, germ, semolina (relatively coarse particles of endosperm), and flour. Pasta (e.g. spaghetti, macaroni, lasagne) is prepared from semolina; the flour is used in noodles and other products. Durum wheat is used to make leavened bread but its gluten does not retain carbon dioxide to the same extent as bread wheat. Other products of durum wheat are couscous and bulgar.

LIFE SIZE WHEAT PLANT × ⅛

1 & 2 **BREAD WHEAT** awnless and awned forms 1A Spikelet and grains
1B Flowering spikelet 1C Wheat plant
3 **DURUM WHEAT** 3A Grains 4 **EMMER** 4A Spikelet and grains
5 **RIVET** or **ENGLISH WHEAT** 5A Grains

Grain crops: Rye, oats, barley, triticale

RYE (1) *Secale cereale.* Rye is an important crop in the cooler parts of northern and central Europe and Russia, cultivated up to the Arctic Circle, and 4000 m above sea-level. The plant is extremely hardy and can grow in sandy soils of low fertility. Rye probably originated from weedy types in eastern Turkey and adjacent Armenia, but it is not as old as wheat. In all rye-producing countries more than 50 per cent of the grain is used in animal feed, but it is also important in human nutrition. Rye is utilized for making so-called black bread (*Schwartzbrot*), including pumpernickel, although the bread can vary in colour because rye flour may be mixed with that of wheat which lightens the colour and adds gluten. Like wheat, rye produces a leavened bread but it is inferior to wheat in bread-making quality because its dough is less elastic and gas retentive. Rye bread has a characteristic stronger flavour compared with wheat bread, and has a lower energy content, is higher in minerals and fibre, and its protein has a higher lysine content. A so-called sour-dough process, involving lactic acid fermentation, may be utilized in bread making. Scandinavian crispbread (*knaeckebrot*) has wide popularity because of its flavour, texture, and excellent storage properties.

Rye is used for making whisky in America, gin in The Netherlands, and beer in Russia. Young plants are used as fodder for livestock. The mature straw is too tough for that purpose but it has its uses for bedding, thatching, paper making, and straw hats.

Ergot (*Claviceps purpurea*), a fungus parasitic on rye, is poisonous to man and animals. Eating rye bread contaminated with ergot may cause gangrene, abortion, hallucinations, or other unpleasant symptoms—St. Anthony's Fire of the Middle Ages. Ergot preparations are sometimes still used in pharmacy.

OATS (2) *Avena sativa* The species most cultivated is *Avena sativa* (white or yellow oats), *A. byzantina* (red oats) comes second. In recent years many planned crosses have been made between the two species, leading to numerous cultivars. It is a cool-season crop and a large proportion of world production is in the northern hemisphere, Russia being the main producer. The plant existed in Europe during the Bronze Age but did not become established in Britain until the Iron Age. It was probably domesticated from weedy types in wheat and barley fields.

Most of the world production of oats is used for livestock feed in the form of grain, pasture, forage, hay, and silage. The grain is also important in human nutrition as porridge, grits, oatmeal, rolled oats in breakfast foods, also in infant foods and cookies. In most cultivars and after threshing, the groat or seed remains enclosed within hulls, removed in milling. In human food products the whole groat is utilized. As consumed, oats contain the largest quantity of protein of any cereal grain (maybe 20 per cent according to cultivar). The fat content (5–9 per cent) is the highest among cereal grains and is highly unsaturated.

Oat bran is rich in water-soluble dietary fibre (see p. 201) and there is good evidence that, if added to a diet, it may significantly reduce serum cholesterol levels of hypercholesterolaemic patients.

Some use is made of oatmeal in the cosmetic industry. Hulls are a by-product of oat milling and from these furfural is prepared which, with its derivatives, is used in making nylon, in oil refining, and in some other industrial processes, although furfural production from oat hulls is now very limited.

BARLEY (3 & 4) *Hordeum vulgare* syn. *H. sativum.* This plant has a wide range of cultivation, from the Arctic, to desert oases. It is more salt tolerant than other cereals. Barley is a cool-season crop but it can tolerate high temperatures if the humidity is low. Major production takes place in most of Europe, the Mediterranean fringe of North Africa, Ethiopia, the Near East, Russia, China, India, Canada, and the United States of America. Two major forms of the plant are cultivated: two-rowed and six-rowed. The former is normally grown in the United Kingdom. After threshing, the grain usually retains hulls, although there are naked cultivars. In order of importance, barley produces animal feed, malts, and human food. As in the case of wheat, barley was domesticated in the Fertile Crescent of south-western Asia about 8000 BC (see p. xiv) and then migrated into Europe, North Africa, and Asia.

A number of barley products and by-products are used in animal feed—harvested grain, hay, straw (after alkali treatment), silage, pearlings and bran, malt culms (sprouts or rootlets), brewers' and distillers' spent grains, and brewers' spent hops and yeast.

Malting is a process whereby barley grains (sometimes other cereals) are germinated to produce enzymes, then dried and lightly cooked (kilned) to give a product, after the culms are removed, known as malt. The crushed malt is mixed with warm water, to convert the grain starch by the enzyme action to fermentable sugars. The removal of spent grains gives the 'wort' which is fermented by yeast to give alcohol. Hops are added to give flavour during beer production; whisky, gin, and vodka are distilled.

The amount of grain protein varies by cultivar; lower quantities and higher starch are preferred for malting. The amount of fat is low, not 2 per cent. In the West, products such as pot (Scotch) and pearl barley are prepared by successively grinding away the outer hull and grain layers, and are then used in soups, stews, and sauces. The nutrient value of these products is less than the original groat. Barley flour may be prepared from pearl barley. Bread made from this flour is flat and heavy because of lack of gluten, but it was widely consumed in Europe before Roman times. In some Eastern countries (e.g. Japan, Korea, and China) large quantities of barley, often the naked forms, are used in human food and drink, such as Japanese barley tea. 'Barley water', made by soaking pot or pearl barley and often flavoured with lemon or orange, is said to be nutritious. Frequently it is given to children and invalids. Malt is used in the production of alcoholic drinks, malt vinegar, in some sweets and confectionery, and as a carrier for vitamin-rich cod-liver oil.

TRITICALE (× *Triticosecale*) Triticale is the first man-made cereal and is a hybrid of wheat (*T. aestivum* or *T. durum*) and rye which combines the quality of wheat with the hardiness of rye. It is cultivated on a limited scale in a number of countries.

LIFE SIZE PLANTS × ⅛

1 **RYE** 1A Spikelet and grains 1B Plant
2 **OATS** 2A Spikelet and grains 2B Plant
3 **TWO-ROWED BARLEY** 3A Plant 4 **SIX-ROWED BARLEY** 4A Spikelet and grains

Grain crops: Maize or corn

This very important cereal crop, third in importance worldwide, is known as maize in many countries, corn in America, and mealies in South and central Africa. So far, the earliest maize material, dating back to 5500 BC, *has been found in the Tehuacán Valley, Puebla, Mexico. The annual teosinte (*Euchlaena mexicana*) was possibly its ancestor. Maize spread across the Americas. Columbus carried the seed to Europe. It is an amazingly adaptable plant in that its cultivars are able to grow in regions from the tropics to temperate areas and from sea-level to 3500 m. Consequently it is cultivated in almost every continent.*

Although maize grain is important in the human diet, most is fed to livestock. Almost 50 per cent of the world production takes place in the United States, mainly in the famous Corn Belt consisting of 12 states (Iowa, Illinois, Nebraska, Minnesota, Indiana, Ohio, Wisconsin, Michigan, Missouri, South Dakota, Kentucky, and Kansas). Other corn belts are in the Danube basin, the Po valley of Italy, northern China, north-eastern Argentina, and south-eastern Brazil. Maize grain is an important staple food in South Africa. It is a major export of some countries, for example the United States, Argentina, South Africa, and Thailand. Relatively little is grown in the United Kingdom, most of it for forage.

MAIZE (1) *Zea mays*. Maize is an annual warm-season crop which gives its best yield when it takes 130–140 days to mature. The male flowers form the 'tassel' at the top of the stem; the female flowers are borne on ears lower down the stem. The ear is covered by modified leaves which form the 'husk'; the long styles of the female flowers, known as the 'silk', protrude from the ear.

In the centre of the ear is the 'corn cob' which bears the grains, 300–1000, in rows. Usually only one or two ears mature on each plant. A popular article of international commerce is the 'baby corn' produced in Thailand and elsewhere. These are harvested before pollination or fertilization. They contain almost 90 per cent water and very little fat, protein, and carbohydrate.

MAIZE TYPES (2–4) The maize types are based mainly on the nature of the endosperm: 'dent' type (**2** and **2A**), hard and soft endosperm, the soft contracts, resulting in a depression at the crown; 'flint' type (**3**), hard endosperm; 'floury' type, soft endosperm; 'popcorn', hard endosperm, kernels burst on heating; 'sweet corn' (**4** and **4A**), endosperm contains a high proportion of sugar (sucrose); 'pod corn', an ornamental type. Kernels are normally yellow or white, although other colours can occur.

Dent is the type cultivated to the greatest extent. The production of 'hybrid' seed, first developed in the United States, is one of the triumphs of agricultural science, not only leading to greater yield and increased resistance to disease, but also to special situations such as higher lysine cultivars (for example, opaque-2), waxy corn, and high-sugar corn.

Maize is utilized in many ways. It is very important in animal feed. Popcorn is a popular traditional snack food in the United States, and also in the United Kingdom. Sweet corn has a wide appeal as a vegetable in the fresh, canned, frozen, or dehydrated states.

On the industrial scale, maize may be milled in the dry or wet condition. Dry milling removes the bran and germ and reduces the endosperm to grits, cornmeal, and cornflour. These can be used in a wide range of foods such as breakfast cereals, pancakes, cookie mixes, snack foods, and animal feed. Wet milling essentially separates the kernel starch and gluten (protein). Starch (natural or chemically modified) has a wide variety of food and industrial uses, also it is processed to make food sweeteners (glucose and fructose syrups). Maize gluten is an important constituent of animal feed. The germ contains a large amount (25–50 per cent) of an oil which is internationally important in food. It is highly polyunsaturated (60 per cent linoleic acid), stable, and does not precipitate under refrigeration. Maize is used to make beer, whisky, and gin.

The grain is employed in a number of traditional dishes throughout the world, for example the porridges (sometimes fermented) of Latin America, Africa, and Asia; *polenta* in Italy, and hominy in North America. If maize is utilized as a staple in the diet, there is the possibility of the deficiency disease pellagra because niacin (see p. 205) is in a bound form in the grain. 'Tortillas' are a most important food in Mexico and Central America. The maize used is lime treated (developed by the native Americans); this makes the niacin available. Yellow corn contains useful quantities of pigments, one of which, β-carotene, can be converted into vitamin A.

After the removal of the seed, great quantities of cobs remain. They have a number of industrial applications.

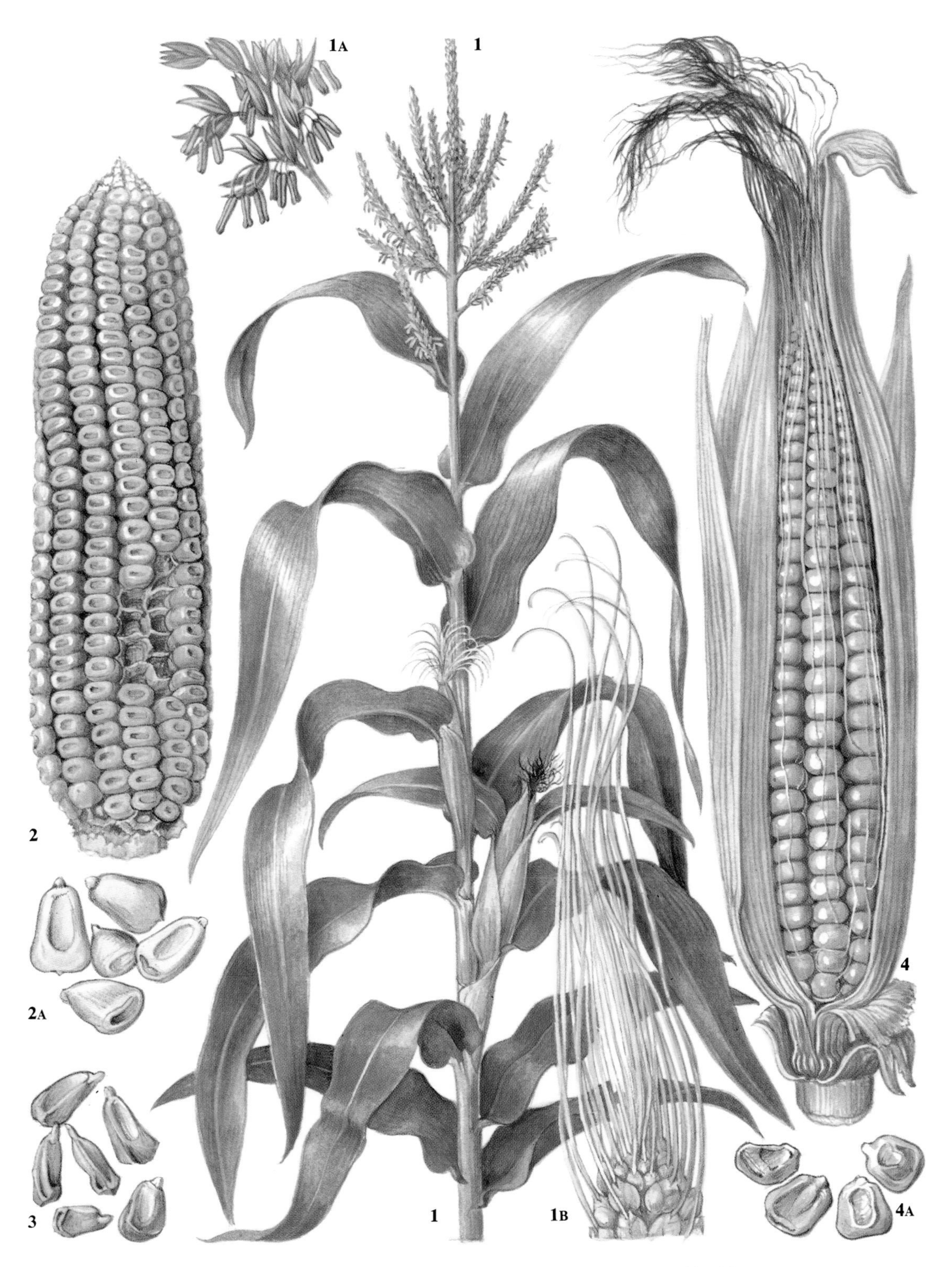

PLANT × ⅛ EARS × ⅔ GRAIN AND FLOWER DETAILS × 1

1 **MAIZE** or **CORN** plant 1A Male flowers detail 1B Female flowers detail
2 **DENT-TYPE MAIZE** ear 2A Grains 3 **FLINT-TYPE MAIZE** grains
4 **SWEET CORN MAIZE** immature ear with husks 4A Grains

Grain crops: Rice

Rice is second to wheat as the world's most important staple food grain; in the humid and subhumid tropics, rice is the primary source of carbohydrate and protein. It is cultivated on 10 per cent of the earth's arable land. Unlike wheat, 95 per cent of rice is grown in the less industrially developed nations, primarily in Asia, although reasonable amounts are cultivated in Africa and Latin America. The crop is also found in the United States and some parts of Europe, for example Spain and Italy. Only 4 per cent of world production moves in international commerce, the main exporters being Thailand, Vietnam, Pakistan, and the United States.

The usual cultivated species is Oryza sativa, *although* Oryza glaberrima *is grown in West Africa, but the latter species is gradually being replaced by cultivars of* O. sativa. *There is no precise knowledge concerning the area or areas of domestication of the crop; however, material dated to 2500* BC *has been found in India, and to 3000* BC *in China.*

RICE (1–5) Rice has many cultivars, some of which possess awns. There are four main methods of cultivation:

1. Irrigated. This is the best known method where the plant is grown in standing water which is under control and is adjusted according to the growth stage.
2. Upland. Plants cultivated in dry land but, because of lack of water and inadequate nutrition, yields are very low.
3. Rainfed lowland. The level of standing water depends on rainfall.
4. Flood-prone. Fields subjected to medium/deep flooding from rivers and tides in river mouth deltas.

Seed may be sown directly or seedlings transplanted. Harvesting and threshing are by hand (most of Asia) or mechanized (as in California). The first phase of milling of the harvested grain, known as rough or paddy rice, removes the hulls to give brown or wholegrain rice. In the second phase, the bran and germ are removed to give white or polished rice. Clearly, on a worldwide basis, enormous quantities of hulls (rich in silica) are produced. These have been utilized in a variety of ways, for example in animal feeds (although they have been regarded sometimes as an adulterant), bedding, litter, fuel, and building materials. Brown rice is now a well-accepted food product. Rice bran is a useful constituent of animal feeds and, in some countries, the source of an oil used in food and for some industrial purposes. There is recent evidence that the bran may reduce blood cholesterol. Although rice flour is used in food products, the complete kernel is the part used mainly in food—in contrast to wheat where the flour is usually utilized.

Rice grains are often classified according to size and shape—short, medium, and long. Compared to other cereals, rice is relatively low in protein but the essential amino acid pattern is good. Because of vitamin and mineral distribution, and also fibre, brown rice is regarded as a healthy dietary constituent. As the result of the removal of the bran and germ, white rice is low in vitamin B_1 (thiamin) and certain other B vitamins. In people heavily dependent on white rice, this may lead to the disease beriberi. The process of 'parboiling' (of ancient origin in India and its modern commercial equivalent) consists of steeping rough rice in hot water, steaming it and then drying it prior to milling. This leads to the movement of vitamins and minerals from the hulls and bran into the endosperm, thus the resulting white rice is nutritionally superior to the usual product.

Rice is associated with many national dishes, for example Chinese cuisine, Indian curries, Italian *risotto*, and Spanish *paella*. Glutinous, sweet, or sticky rice is popular in Japanese, Thai, and other eastern cooking; Basmati rice is scented (aromatic). Rice is found in many food products, such as breakfast cereals, soups, and baby foods. In Japan it produces an alcoholic drink called *sake*; in the rest of the world it may be used as an adjunct to beer manufacture.

Forming part of the Green Revolution, the first of the modern rices, IR8, was released by the International Rice Research Institute (Philippines) in 1966. Since then a number of new cultivars (also from other countries) have appeared. They have increased the rice yield two to three times.

'Rice-paper' can be made from the pith of an oriental tree—*Tetrapanax papyrifer*, but substitutes, such as wheat or potato flour, are also utilized.

AMERICAN WILD RICE *Zizania aquatica.* Wild rice is an aquatic freshwater grass usually found growing naturally in shallow lakes and rivers of the Great Lakes region of North America (north-eastern United States/eastern Canada). It is a large plant with a panicle bearing female spikelets above the male spikelets and extending about 1 m or more over the water surface.

The seed has been harvested from natural populations by the North American Indians since long before recorded history. Harvesting was carried out by bending the panicles over canoes and tapping them gently with a stick to release the grain. Grains that fell into the water became the plants of the following year and, also, provided an important wildfowl food. Wild rice grain, when harvested, is still enclosed within hulls. Traditionally the Indians heated or 'parched' the seed over fires and pounded them to release the hulls. The grain, thus treated, was an important winter food.

Wild rice is the only grass grain species domesticated (during the 1970s) in historical times because the wild populations did not meet demand. Present-day commercial production involves mechanized seeding, harvesting, and processing, the development of non-shattering cultivars, and some cultivation in paddy fields.

The grain, when presented for sale, is elongated and shiny black/brown. Processing involves heating and this gives the grain a nutty flavour and chewy texture, the so-called 'caviar' of grains.

Nutritionally, its protein content (12–15 per cent) is quite high for a cereal and so are the concentrations of the amino acids lysine and methionine. Total amounts of B vitamins (thiamin, riboflavin, and niacin) exceed those of other cereals; the fat is unsaturated. There is some international trade involving wild rice.

PLANT × ¼ PANICLES LIFE SIZE DETAILS × 2

1 **RICE** plant 2 Flowering spikelet detail
3 Panicle of rice grain 4 Panicle of awned variety
5 Details of spikelets and polished grains

Grain crops: Sorghum and millets

Millets are small-seeded cereals; sorghum has larger seeds. The different species can be cultivated in varying degrees in arid and semi-arid zones, and in areas of uncertain rainfall and poor soil, where larger-seeded cereals cannot be grown. Their grains have good storage properties (apart from sorghum) and may have useful mineral contents. They form the staple diet of a considerable proportion of the world's population, mainly in Asia and Africa. The grains may be consumed like rice or the flour converted into gruel, porridge, or unleavened bread. Another common product is beer. In addition, seeds and other parts are used as animal feed. Relatively little improvement work through plant breeding has been applied to the millets.

SORGHUM (1–2) *Sorghum bicolor* syn. *S. vulgare*. Sorghum is an important human food plant in Africa, South-East Asia, India, Central America, and China; in the United States of America and Australia it is of importance as animal feed. It exhibits a great deal of variation in the form of a number of races, for example *guinea*, *kafir*, and *durra*. There is general agreement that sorghum was domesticated over 5000 years ago in Africa and then spread through South-East Asia to China. Sorghum was introduced into the United States in the 1850s but did not become important until the 1930s. It was later improved through hybridization, as in the case of maize. In general appearance, sorghum is somewhat similar to maize and may grow into a large plant. Human food products produced from the grain are porridges (thin and thick; fermented and non-fermented), *roti* or *chapati* of India (but usually made from wheat), 'tortilla' of South and Central America (but usually made from maize), *kisra* (fermented flat bread of the Sudan), and African beer. Seeds with dark brown/red seed coats have a high tannin (polyphenol) content, greater than white/yellow seed. In dark seed, tannins are antinutritional in that they are bitter and decrease palatability, also they decrease protein digestibility by binding with seed proteins and/or digestive enzymes. White/yellow seeds are therefore better for food products, although red/brown seeds on the plant are less attacked by birds, also they are preferred for beer manufacture. Other minor uses are starch and alcohol from grain; syrup from sweet-stemmed cultivars; brooms from broomcorn cultivars; plant bases and stems as fuel and thatching.

FINGER MILLET (3–4) *Eleusine coracana*. Finger millet is an important staple food in parts of East and central Africa, also in India. The English name aptly describes the seed head arrangement. Its grain provides porridge, gruel, and beer. The seed heads may be stored for long periods, up to 10 years or more, without deterioration or weevil damage. This property, together with its ability to tolerate adverse conditions, makes it an excellent crop for arid zones. Domestication of the crop probably took place in Africa, followed by migration to India.

BULRUSH OR PEARL MILLET (5) *Pennisetum typhoideum*. Bulrush millet is cultivated as a staple in the drier parts of tropical Africa (particularly the northern territories of West Africa) and in India. In size and general appearance, it is like maize or sorghum but bears a seed head which resembles a bulrush. The seed is normally brown; however, the colour may vary between near white and black. It is the most widely grown of all the millets and very tolerant of drought. This millet is not so well developed as sorghum, but semi-dwarf hybrids have been adopted in India. The crop probably originated in West Africa some 5000 years ago, while at least 2000 years ago it migrated to East and central Africa, also to India. It was first known in Europe in the sixteenth century and introduced into the United States in the 1850s, but is now relatively unimportant there. In West Africa, *degue* is a food product of the plant. A high incidence of goitre has been reported among eaters of this grain. The goitrogen has been identified as a thioamide chemical.

OTHER MILLETS There are several millets of minor importance:

(1) teff (*Eragrostis tef*), widely grown in Ethiopia, from which is produced *injera*—a traditional flat bread;
(2) 'hungry rice' or fonio (*Digitaria exilis*, *D. iburua*), found in the savannah areas of West Africa;
(3) adlay or Job's tears (*Coix lachryma-jobi*) which is hardly a millet because of its large seeds but is of local importance as a cereal in the Philippines and some other countries;
(4) kodo millet (*Paspalum scrobiculatum*), a minor grain crop throughout India;
(5) brown-top millet (*Brachiaria ramosa*), a native of India. It has been grown in Georgia, Florida, and Alabama for hay and pasture.

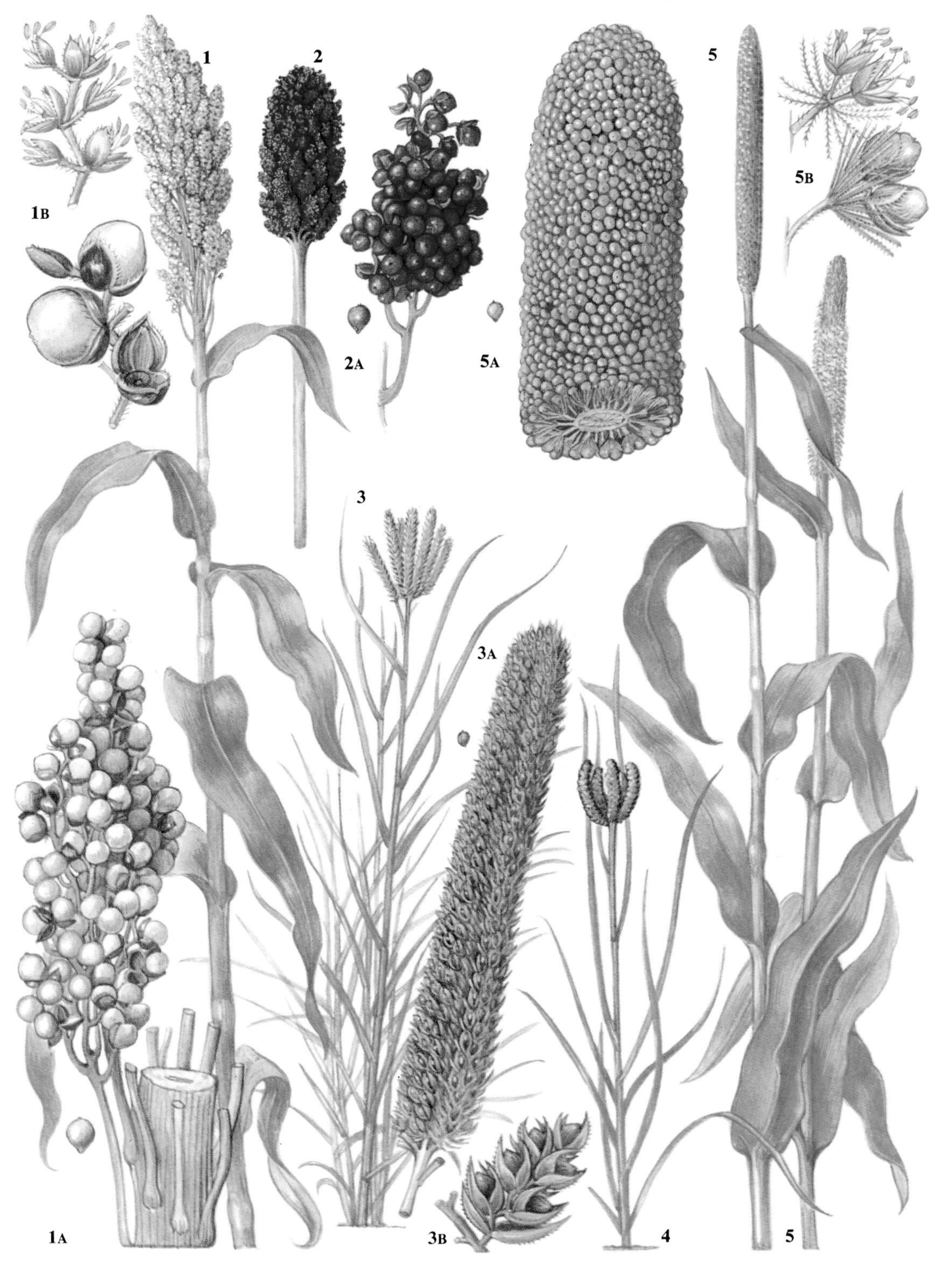

PLANTS × ⅛ DETAILS LIFE SIZE SPIKELETS × 3

1 **SORGHUM** white-grained type 1A Detail of ear and seed 1B Spikelets
2 **RED-GRAINED SORGHUM** 2A Detail of ear and seed
3 **FINGER MILLET (INDIAN)** 3A Spike and seed details 3B Spikelet 4 **FINGER MILLET (AFRICAN)**
5 **BULRUSH MILLET** 5A Detail of ear and seed 5B Spikelets

Grain crops: Millets

COMMON MILLET (1) *Panicum miliaceum.* This is also known as 'proso', 'hog', 'broomcorn', 'Russian', and 'Indian' millet. As with other millets, it is utilized as a human food and animal feed, the grain having a slightly nutty flavour. In the United Kingdom, and no doubt other countries, it is well known as a bird-seed. It is of ancient origin (possibly domesticated in central and eastern Asia) and has been cultivated in Europe since early times, having been grown by the early Lake Dwellers. This millet was the *milium* of the Romans, the *dokhan* of the Hebrews, and one of the millets of the Old Testament. Cultivation takes place mainly in eastern Asia (for example Mongolia, Manchuria, and Japan), India, eastern and central Russia, the Middle East, and the United States. In the US it is the largest millet crop. It grows further north than most other millets and is one of the most popular cereals in northern China, commanding a price equal to that of wheat. The crop is said to have the lowest water requirements of any cereal.

The seed head of the plant is a compact panicle and resembles an old-fashioned broom. Its grains may be whitish, straw-coloured, or reddish-brown.

LITTLE MILLET (2) *Panicum miliare.* Little millet is a minor crop grown throughout India but is of no great importance elsewhere. It is really a smaller version of *Panicum miliaceum* and can be cultivated on soils that produce little or nothing else, giving a crop, however small, even in drought years.

FOXTAIL MILLET (3) *Setaria italica.* This plant is also known as 'Italian', 'German', 'Hungarian', or 'Siberian' millet and is cultivated in Asia, south-eastern Europe, and North Africa; being the most important millet in Japan, and widely grown in India. As other millets, its grain may be used in human food. In Russia, beer may be manufactured from this millet; in the United Kingdom it is used as a bird-seed; and in the United States it is grown for hay and silage.

It is considered to have been domesticated from *Setaria viridis*, with an origin in eastern Asia. The millet was recorded as one of the five sacred plants of China as early as 2700 BC. Because of its short duration of growth, it was a suitable crop for nomads and was probably brought in this way to Europe during the Stone Age. Seeds have been found in the Lake Dwellings of Europe.

Foxtail millet has a terminal characteristic spike-like panicle, the spikelets being surrounded by bristles. The seed is of various colours: white, yellow, red, brown, or black.

JAPANESE MILLET (4) *Echinochloa frumentacea.* It is the quickest growing of all millets, with a crop available in 6 weeks. The millet may be utilized as human food in the Orient and India; in the United States it is grown as fodder. In Japan and China it may be cultivated as a substitute when a rice crop fails; in Egypt it is frequently grown as a reclamation crop on land too saline for rice.

The inflorescence is densely branched and usually purple-tinged; the grain is light brown to purple.

PLANTS × ⅛ EARS AND SEEDS LIFE SIZE SPIKELETS × 3

1 **COMMON MILLET** plant	1A Ripe ear and seed	1B Spikelets
2 **LITTLE MILLET** plant	2A Ripe ear and seed	2B Spikelets
3 **FOXTAIL MILLET** plant	3A Ripe ear and seed	3B Spikelets
4 **JAPANESE MILLET** plant	4A Ripe ear and seed	4B Spikelets

Pseudo-cereals: Quinoa and buckwheat

BUCKWHEAT (**1**) *Fagopyrum esculentum.* Buckwheat is an annual plant growing to a height of some 20–50 cm (**1**) with white or pink flowers and producing black or grey triangular seeds (**1A**) about 6 mm long, which botanically are fruits. It is cultivated in Russia, China, Japan, Poland, Canada, Brazil, the United States of America, South Africa, Australia, and some other countries, but probably not now in the United Kingdom. The common buckwheat is *Fagopyrum esculentum.* It grows quickly and is tolerant of low soil fertility. Two other species are tartary buckwheat (*F. tartaricum*) and perennial buckwheat (*F. cymosum*). The common buckwheat probably originated in Central and north-eastern Asia, spreading through China, Japan, India, and later, in medieval times, to Europe. It was possibly cultivated in China during the fifth and sixth centuries AD. The plant first appeared in France and Germany during the fifteenth century, and Italy at the beginning of the sixteenth century. Various pathways have been suggested for the movement of buckwheat from Asia to Europe, including a pathway from Asia Minor. The crop has sometimes been described as saracen in Europe (for example *sarrasin* in French), which would indicate a Moslem connection.

The grain contains about 75 per cent starch, 11 per cent protein, which is particularly rich in lysine, and about 2 per cent fat with a composition similar to that of cereals. From the leaves, stems, flowers, and fruit of the plant can be extracted a phenolic compound known as rutin, a drug for vascular disorders.

Its grain flour is used in the preparation of a number of food items—pancakes, soups, porridge, bread, pasta, dumplings, and biscuits, and is often mixed with cereal or soya-bean flour. In Japan, noodles known as *soba* are made from it.

QUINOA (**2**) *Chenopodium quinoa* To the ancient Incas of South America, quinoa was a vital and sacred food. The Inca emperor planted the first seed of the season. It was cultivated at the high altitudes of the Andes, replacing maize grown at lower levels. Other pseudo-cereals of this region were kaniwa (*Chenopodium pallidicaule*) and kiwicha (*Amaranthus caudatus*). Quinoa was probably domesticated at several locations in the Andes some 3000 to 5000 years ago. After the Spanish Conquest, cereals such as wheat and barley were introduced. This led to a decline in the amount of quinoa cultivated, perhaps also through active discouragement by the Spaniards because of the plant's position in Inca culture and religion. Today it is grown as a staple in the highlands of Argentina, Bolivia, Chile, Colombia, Ecuador, and Peru. Some very limited cultivation takes place outside South America, for example in the United States and United Kingdom.

Quinoa is an annual plant (**2**), growing to a height of between 1 and 3 m and bearing grain (**2A**), some 2 mm in diameter, of various colours—white, yellow, pink, orange, red, brown, or black (**2B**). The grain contains about 48 per cent starch; an average of 18 per cent protein, which is high in lysine and other essential amino acids; 4–9 per cent of unsaturated fat; and good quantities of calcium, phosphorus, and iron. The information about vitamin content is scarce and variable; however, the grain is said to be a good source of vitamin E, but low levels of carotenes and vitamin B have been reported. This nutrient composition compares very favourably with that of the common cereals. The seeds of most cultivars contain bitter-tasting saponins which need to be milled or washed out during food processing. These saponins constitute a chemical defence against birds.

The grain has been utilized in a number of food products (either in the entire state or as flour), for example breakfast cereals, biscuits, soups, beer, animal feed, and bread. As quinoa grain contains no gluten, wheat flour must be added to it to make leavened bread.

PLANTS × ½ HEADS AND GRAINS × ½ but QUINOA GRAIN × 1

1 **BUCKWHEAT** plant 1A Seeds
2 **QUINOA** plant 2A Grain 2B Heads of grain

Sugar crops

(For general information on sugar crops, see p. xvii.)

SUGAR-CANE (1) *Saccharum officinarum.* This is a large (up to 6 m high) perennial grass cultivated in about 70 countries, mainly in the tropics but also the subtropics. Brazil and India grow large quantities of sugar-cane. The main demand for white refined sugar has traditionally been in the developed world so that, even if sugar-beet is grown in the countries concerned, imports of raw sugar take place from countries such as Cuba, Hawaii, Barbados, and Puerto Rico. *Saccharum officinarum* has thick stems (5 cm in diameter) and is the 'noble' cane; *S. sinense* (China) and *S. barberi* (India) have thin canes (2 cm in diameter). Almost all common cultivars grown today are hybrids of the various species.

Saccharum officinarum was used from ancient times for chewing and is thought to have evolved from the wild *S. robustum* in the South Pacific area, probably New Guinea. Three paths of migration have been suggested for *S. officinarum*:

(1) about 8000 BC to the Solomon Islands, New Hebrides, and New Caledonia;
(2) about 6000 BC through South-East Asia to India; and
(3) between AD 500 and AD 1100 from Fiji through various countries to Hawaii.

The sugar-cane stem is easily transported. There is some doubt about the origin of *S. sinense* and *S. barberi.* These crops could be naturally occurring or evolved as hybrids between *S. officinarum* and *S. spontaneum.* The Islamic conquests (AD 600–800) stimulated the westward movement of sugar-cane from India through the Middle East to several Mediterranean countries. It was then taken to parts of Africa and across the ocean to the New World, including Brazil and the West Indies. The late seventeenth and eighteenth centuries saw a rapid expansion of the sugar industry in the Caribbean area. This was associated with the notorious slave trade.

For successful growth, sugar-cane requires a rich soil and good rainfall. The plant is propagated by cuttings; inflorescences, or 'arrows', are not often produced. Harvesting, namely the cutting of the canes, is still widely carried out by hand although mechanical means are of increasing importance.

Cane sugar was first produced in India, probably about 1000 BC, simply by evaporating the juice, squeezed from the cane, over an open fire to give a dark product, containing varying amounts of sucrose, which deteriorates rapidly. The method is still used in some parts of the world and the resulting sugar is given various names, for example *gur* in India, *jaggery* in Africa, and *panela* in Latin America. These are 'non-centrifugal' sugars.

The vast majority of cane sugar produced is known as 'centrifugal'. In the country of origin, the harvested canes are taken to the factory where they are crushed, the sugar is extracted with water to give an impure solution which is purified, concentrated by evaporation, and the sugar crystallized. Sugar crystals are separated from the molasses, or brown syrup, by centrifugation. The sugar produced is raw (96–99 per cent sucrose) and brown specialities are demerara and muscovado. The cane residue is known as 'bagasse' and is used in paper making, animal feed, and as a fuel. Molasses has a number of uses, including the manufacture of rum. The raw sugar may be exported to countries where it is refined to give the white crystalline substance, which is virtually 100 per cent sucrose.

SUGAR-BEET (2) *Beta vulgaris* var. *esculenta.* This is the most important source of sugar in temperate countries and is closely related to the various beet root and leafy vegetables (see p. 181). All are thought to have evolved from wild sea-beet (*B. vulgaris* var. *maritima*), a common seashore plant of Europe and western Asia. It has been used as a vegetable since the first century AD. In the middle of the nineteenth century it was noted that the roots of the Silesian sweet fodder beet contained about 6 per cent sucrose (modern cultivars contain up to 18 per cent sucrose). The first factory for the extraction of beet sugar was erected in Silesia in 1801 and the industry spread throughout continental Europe. Napoleon encouraged the study and cultivation of the crop because of the Royal Navy's blockade of imports of cane sugar from the West Indies. In the United Kingdom, cultivation of sugar-beet started in earnest in the 1920s. Today, the crop is grown throughout Europe and in North America.

Sugar-beet has whitish conical roots, almost half a metre in length. It is a biennial, in the first year producing the rosette of leaves and the root, which is harvested in autumn and early winter. If allowed to remain in the ground, flowering would take place in the second year. The leafy tops are a good animal feed.

The extraction of sugar from the root is carried out in essentially the same way as extraction of cane sugar, except that the process proceeds directly to white refined sugar. Both the root residue and molasses are used in animal feed. Molasses may also produce industrial alcohol. Filter cake, the residue left behind after the purification of the sugar-beet juice, is utilized as a manure.

QUARTER LIFE SIZE PLANT × 1/20

1 **SUGAR-CANE** 1A Flowering plant
2 **SUGAR-BEET** 2A Inflorescence

Other sugar crops

A number of palms in certain tropical Asian and African countries provide sugar (sucrose) by the non-centrifugal method. The soft apical part of the stem (terminal bud) or inflorescence is tapped for the sweet sap, often containing more than 10 per cent sucrose. This sap may be boiled down to give a brown, sticky sugar. Alternatively, the sap can be fermented to give alcoholic 'toddy' or 'palm wine' which may be distilled to produce 'arrack'. Generally speaking, palm sugar production is a village industry, although in some countries (Burma, India, and Cambodia) it is quite significant.

WILD DATE PALM (1) *Phoenix sylvestris*. This relative of the true date palm (*Phoenix dactylifera*) (see p. 112) is planted on quite a considerable scale in India as a source of sugar. To obtain the sugar, some of the leaves covering the tender top portion of the stem are removed by the climber and a cut is made into the stem. As it exudes, the sap is collected in a receptacle suspended on a bamboo cane. As described previously, this sap may be boiled down to give a brown, sticky sugar. About 3.5 litres of sap are required to produce 0.5 kg of sugar. Toddy and arrack may be prepared from the sap. Palms are fit to be tapped from about 8–10 years old, and a plantation may yield as much as 18 tonnes of crude sugar/hectare/year.

SUGAR OR GOMUTI PALM (2) *Arenga pinnata* syn. *A. saccharifera*. (see also p. 21, 2, 2A) This palm grows wild from Annam through South-East Asia to the Philippines. It is sometimes cultivated, commercial plantations being established in Indonesia. The palm has been introduced into the Pacific Islands and a few places in Africa. Inflorescences are male or female, flowering takes place when the plant is 7–10 years old. Both male and female inflorescences can be tapped for sap. Flowering continues for about 2 years, which can be extended to 10 years, then the palm dies. The inflorescences are normally beaten with wooden mallets to stimulate the sap flow. Sugar (about 75 per cent sucrose) is produced from the sap. A village stand may produce 25 tonnes of sugar/hectare/year. Other food materials produced from the palm are starch, young leaves, and immature fruits. Grubs of the palm beetle are reared on fallen stems and eaten raw, fried, or cooked. There is an international trade in these food products.

PALMYRA OR BORASSUS PALM (3) *Borassus flabellifer*. Considered here only as a source of sugar, its other uses are described on p. 112. It grows in the drier areas of tropical Asia and is particularly abundant in India where it is planted frequently. The plant does not grow well in the humid tropics. The palms are either male or female, the male producing more sugar than the female. Methods of sugar production from the sap are as described for other palms. In some Indian states, prohibition of alcoholic drinks has meant that more trees are used for sugar production than for toddy manufacture. Flowers form on the trees about 15 years after sowing. Tapping then commences and may continue for 30–40 years. The annual tapping period is usually 4–5 months, during the hot season. In this period, a single tree can yield between 200 and 350 litres of sap.

OTHER SUGAR-PRODUCING PALMS Other palms may be tapped for sap leading to sugar or toddy production. The nipa palm (*Nipa fruticans*), which lives in the brackish water of estuaries and extends from Sri Lanka to the Philippines, contains a sap with about 17 per cent sugar.

Some palms are not primarily utilized as sugar producers but are occasionally tapped. These include the coconut, the buri palm (*Corypha elata*), and the fishtail or toddy palm (*Caryota urens*).

SUGAR MAPLE (4) *Acer saccharum*. This is a North American tree with a natural range extending from south-eastern Canada to north-eastern America. It is a large tree, up to 40 m high. Its sap contains 1–3 per cent sucrose. The optimal flow of sap from the trees occurs in late winter and early spring. Traditionally, the native Americans tapped the tree by making cuts in its bark and channelling the sap into containers. To obtain the sugar and syrup, they boiled the sap by dropping hot stones into it, then froze the resulting concentrate, afterwards removing the ice which formed at the surface. Up to a relatively short time ago, a small number of tap holes were drilled into the tree and the sap collected in buckets. The sap was concentrated by boiling in kettles. Since about 1970, networks of plastic tubing have been used to transport sap to factories where it is concentrated to produce maple-sugar and syrup used in confectionery, puddings, and ice-cream.

Black maple (*Acer nigrum*) is also a source of sugar and syrup.

TREES, SMALL SCALE SPADICES AND MAPLE LEAVES × 1/8

1 **WILD DATE PALM** tree 2 **SUGAR PALM** male spadix
3 **PALMYRA PALM** tree 3A Male spadix 3B Female spadix
4 **SUGAR MAPLE** tree 4A Leaves

Sago and palm hearts

SAGO PALM (1) *Metroxylon sagu.* This is a tree, up to 10 m in height. The palm probably originated in New Guinea and the Moluccas but has now spread to other parts of South-East Asia and neighbouring Pacific Islands. It grows on swampy land under hot and humid conditions, although it can tolerate drier situations. Most sago palms are wild or semi-wild but there are some plantations. Propagation can be carried out by suckers. The palm only flowers (**1A**) once in its life, at 10–15 years. Just before this there is a build-up of starch reserves in the pith of the trunk, and it is at this stage that the trees are felled. The trunk has its bark removed and is cut into lengths of about 1 m. The exposed pith is pounded and pulverized, or rasped to loosen the starch which is then washed out. In New Guinea, the wet starch is a staple, being boiled, fried, or roasted. It can be dried to give a flour. Pearl sago, known in Western countries for making puddings and sweet dishes, is prepared by pressing wet starchy paste through a fine sieve and drying it on a hot surface which will cause gelatinization of the starch. A trunk may produce 110–136 kg starch. Nutritionally the sago product provides only starch. Malaysia and Indonesia are important exporting countries. The young apical shoot, or cabbage, is widely eaten as a vegetable.

OTHER SAGO-PRODUCING PLANTS A number of other palms have a starchy pith and produce sago. These include *Caryota urens* (India), *Metroxylon rumphii* (Indonesia), *Arenga saccharifera* (see p. 18), *Oreodoxa oleracea* (American tropics), *Eugeissona utilis* (Borneo), *Phoenix acaulis* (India and Burma). Starch is also produced from the pith of some cycads—*Cycas circinalis* (Sri Lanka, India), *Cycas revoluta* (Japan). These are gymnosperms, plants with seeds which are exposed and not enclosed in fruits.

PALM HEARTS These are young apical shoots, or cabbages, often utilized as vegetables. There is some international trade in canned palm hearts. Palms processed for canning include *Euterpa oleracea, Cocos nucifera, Bactris gasipaes,* and *Daemonorops schmidtiana.* There is export of this commodity from some countries of Central and South America, also from Thailand. Palm hearts contain less than 1 per cent total sugars, about 2 per cent protein, and 1.5 mg vitamin C per 100 g.

TREES SMALL SCALE   SPADICES × ⅛

1 **SAGO PALM**   1A Part of flowering spadix
2 **SUGAR PALM**   2A Flowering female spadix

Oilseeds and fruits: Coconut palm

(For general information on oil crops and nuts, see p. xvii.)

COCONUT *Cocos nucifera*. This palm, being tolerant of salty, sandy soils, is often found growing at the top of beaches in tropical lands, although plantations are also to be found inland. It is essentially a smallholders' crop. The plant also grows in some subtropical regions, for example Florida. Its origin and domestication have been the subject of much discussion and there is no universal agreement, but a commonly held view is that it originated in the Melanesian area of the Pacific, then was taken in prehistoric times to Asia. Later, it was carried to East Africa, Panama, and the Atlantic coasts of the Americas and Africa. No doubt humans were responsible for much of this dispersal but a number of botanists are of the opinion that the coconut fruits could float in sea water and be dispersed by ocean currents over considerable distances without losing their ability to germinate if a suitable site is reached.

Tall palms (1) are the most commonly planted and may attain a height of 20–30 m with an unbranched stem and a crown of 25–35 leaves. Flowers are male or female with flowering commencing at 6–12 years. There are also dwarf palms, 8–10 m high. The young fruits (2) on the tree are initially green but become, on maturity, yellow, orange, red, or brown. A mature fruit (3, 4) is 20–30 cm long with a weight of 1.2–2.0 kg. The fruit consists of an outer skin (epicarp), a fibrous region (mesocarp), and a hard shell (endocarp) which encloses the well-known nut of commerce. At one end of the shell are three soft areas or 'eyes', through one of which the young shoot and root emerge. Inside the shell is the thin brown seed coat covering the important white endosperm, or 'meat', containing a tiny embryo. The centre of the endosperm is occupied by a cavity with, in the mature state, some water. Fruit may appear on the tree at 6 years and fruiting can carry on for about 80 years. A tree will bear some fruit at all times of the year, a distinct advantage over many crops.

The fruit is harvested in a number of ways. Skilled climbers use a rope passed around the tree trunk either as a belt or looped into stirrups for their feet. Harvesters may cut the fruit off the tree with a knife attached to a long bamboo pole. Often the fruit is allowed to fall naturally on the ground and collected at intervals. In some parts of South-East Asia, pig-tailed monkeys are trained to climb the palm and throw down the coconuts. Propagation of the tree is carried out by seed germination (5).

Coconut provides an important vegetable oil. Fruits are dehusked, that is the epicarp and fibrous region are removed. The resulting nuts are split open and the endosperm, or meat, extracted. This is dried to give 'copra', containing about 65 per cent oil, which is saturated (contains a large percentage of lauric acid). After oil extraction, the residue, known as 'poonac' in the East, with almost 20 per cent protein, is utilized as an animal feed. The oil is used in the manufacture of margarine, soap, cosmetics, and confectionery. It is a useful substitute for cocoa butter. Also, it can be used directly to fuel unmodified diesel engines. Desiccated coconut (70 per cent oil, 6 per cent protein) is prepared from fresh endosperm by shredding and drying, after first removing the seed coat. This product may be found in confectionery and baked goods. In the East and the Pacific region, fresh endosperm is eaten. Coconut 'milk', used in curries and sweets, is prepared by squeezing freshly grated endosperm through a sieve. This is not to be confused with coconut 'water' (with 5 per cent sugar, amino acids, minerals, and vitamin C) which is found in the centre of unripe fruits and is a popular tropical drink. The young endosperm is jelly-like and may be eaten. This immature state is retained permanently in the cultivar 'makapuno'. The coconut 'apple', part of the germinating embryo, is eaten in some regions. Coconut pollen may be sold in health food shops. Coconut palms are tapped at the unopened spathe for toddy (containing some ascorbic acid; the yeast provides vitamin B), arrack, and sugar. Toddy is sometimes used to produce vinegar. Palm cabbage is the delicate terminal bud, eaten raw, cooked, or pickled—it is sometimes canned. Naturally, the removal of the bud from the tree destroys the palm.

Apart from food, the coconut palm has many other uses. The fibrous region (mesocarp), known as 'coir', is made into products such as mats, ropes, brushes, and brooms. At present in the United Kingdom coir forms part of some soil composts—a useful substitute for peat which needs to be conserved. The wood is used for building and carvings; the leaves for baskets and thatching. Fruit shells have many uses, such as fuel, drinking receptacles and, when finely ground, as a filler in plastics.

The coconut palm is grown throughout the tropics. Major producers include the Philippines, Indonesia, India, Sri Lanka, the Pacific region, Malaysia, and Mexico.

TREE SMALL SCALE DETAILS × ⅛ OPENED NUT × ⅔

1 **COCONUT PALM**
2 Immature fruits 3 Opened ripe fruit and nut
4 Opened nut 5 Young plant

Oil palm

OIL PALM *Elaeis guineensis*. This oilseed plant is unique in that its fruit contains two types of oil—palm oil and palm kernel oil. On a worldwide basis, it produces more edible oil than any other species except soya bean. Its level of oil productivity per unit soil area per year is the highest of all vegetable oil crops. It is a native of tropical West Africa where it is still of local importance. In the present century the plant was developed as a plantation crop in West Africa, Zaire, South-East Asia, and Latin America. The most rapid expansion since 1960 has been in South-East Asia (Malaysia and Indonesia). Some 200 years ago it was taken by slaves to Brazil. The crop grows best in lowland regions with a high rainfall and close to the equator.

The oil palm is a tree (**1**) growing to a height of 20–30 m, which may live for up to 200 years. Its flowers (**2**, **3**) are male or female and borne in separate inflorescences on the same plant. In contrast to coconut, the fruits (**4**) are smaller (2–5 cm long) and in large bunches, with up to 200 fruits in each bunch. A tree may produce 2–6 bunches a year. The ripe fruits are usually orange in colour, although violet and black forms may occur. As in coconut, the fruit shows three regions: the outer skin, or 'epicarp'; a fibrous pulp, or 'mesocarp'; and a black shell, or 'endocarp', covering the endosperm containing a minute embryo. However, in contrast to coconut, the fibrous pulp contains a large amount of oil (about 50 per cent) which is known in commerce as palm oil. As in coconut, the endosperm contains oil (about 50 per cent). This is known in commerce as palm kernel oil. The two oils have different characteristics. Palm oil contains unsaturated and saturated fatty acids in roughly equal proportions; palm kernel oil is saturated (a large percentage of lauric acid). Fruits are harvested in essentially the same way as coconut.

Propagation is by seed germination. Successful germination usually requires heat treatment, such as in fermenting vegetable matter or germinators heated by open fires, hot water, or electricity. The seedlings are raised in nurseries and then planted out into the fields. A number of oil palm forms are to be found:

1. 'Dura'. The fruits have a thick shell, a medium pulp, and a large kernel. This form is common in West Africa. The Deli oil palm, which performs better in South-East Asia than in West Africa, is in this group.
2. 'Tenera'. The fruits have a thin shell, a medium to high pulp, and a small kernel.
3. 'Pisifera'. The fruits are without shells and have tiny kernels. This form is of little commercial value but is good for breeding.
4. 'Dumpy'. This oil palm was originally discovered in Malaysia and is of interest because of its restricted height.

Because of the great popularity of palm oil in West Africa, methods of oil extraction vary from the simple to the mechanized. One of the simple methods is to cut the fruit bunches into sections, sprinkle with water, ferment for 2–4 days, then pick the fruits off the sections. Consequently, the fruits are boiled, pounded in a wooden mortar, and the pulp is soaked in water so that the oil rises to the surface. It is then skimmed off and boiled to remove the last traces of water. The nuts are cracked one by one to release the kernels. More advanced processing methods involve hand-operated presses and small mills. In Malaysia large mills process the fruits.

Palm oil is an article of international commerce and is used in the manufacture of soap, candles, margarine, shortenings, domestic frying oil, and snack foods. When freshly extracted the oil is orange or yellow due to the presence of carotenes, including β-carotene—the precursor of vitamin A. As such, the oil is important, nutritionally speaking, as a rich source of β-carotene, in West Africa and elsewhere. However, industrial refinement destroys β-carotene. Palm kernels and palm kernel oil are also articles of international commerce, although palm kernel processing is now often a feature of the country of origin. Palm kernel oil is used in confectionery, ice-cream, margarine, and as a cocoa-butter replacer. The residue, with a protein content of about 20 per cent after oil extraction, is employed as an animal feed.

As with the coconut palm, the oil palm is tapped for toddy (palm wine). The oil palm can also provide an edible 'cabbage'.

MALE SPADIX × ¼ FRUITING SPADIX × ½ FLOWER AND FRUIT DETAILS × 1

1 **OIL PALM** (small scale)
2 Male spadix 2A Detail of male flowers
3 Branch of female flowers 3A Details of female flower
4 Fruiting spadix 4A Details of fruits and nut

Oilseeds and fruits: Olive, sesame, peanuts

OLIVE (1) *Olea europaea.* The olive (1) is a small evergreen tree which grows best in a Mediterranean climate. Spain, Italy, and Greece are very important producers but it is also cultivated outside the region in countries such as the United States of America and Argentina. The fruit (**1A**) has a skin, a fleshy pulp, and a stony kernel (basically the same pattern as oil palm). As the fruit matures its colour changes from green to black. On some archaeological evidence it is possible that olive was domesticated in eastern Mediterranean countries 10 000 years ago, and for thousands of years it has been closely associated with human religious, cultural, medical, and food requirements—there are many Biblical references.

As regards food, the fruit provides oil and table olives. The pulp contains up to 40 per cent or more oil, the kernel only a small percentage. The oil is monounsaturated with a high percentage of oleic acid and is used for cooking, in salad dressings, and for food preservation. In the United Kingdom, at least, it is used in a spread which therefore has a large amount of monounsaturates. The oil is also utilized in cosmetics, the pharmaceutical industry, and for certain medical purposes. Olives are cold pressed (rare among oilseeds) and the first pressings, which require no further treatment, are described as 'virgin', of which there are various types, such as 'extra virgin'. The residue (pomace) after oil extraction is used in animal feed.

Both green (immature) and black olives are pickled in brine. They contain less oil than those used for oil extraction. Prior to pickling, the bitter glycoside (oleuropein) is commonly neutralized with caustic soda or another lye solution. The olives are sometimes pitted (stone removed) and stuffed with pimentos, onions, almonds, or other food materials.

SESAME, BENISEED, SIMSIM, GINGELLY, TIL (2) *Sesamum indicum.* This crop, one of the most ancient of oilseeds, was domesticated in Africa and, early on, was taken to India. Today it is cultivated in China, India, Africa, the United States of America, Central and South America. The plant (**2**, **2A**, **2B**) is an annual, growing to a height of 1–2 m and with white, pink, or purplish flowers. Its fruits are capsules containing white, yellow, grey, red, brown, or black seeds. In harvesting, the whole plants are cut and stacked in an upright position, often against a rack; as they dry, the capsules split open at the apex (hence the expression 'open sesame') and the seeds shaken out on to a cloth. In some cultivars, the capsules do not split open; mechanical harvesting is therefore possible. The seeds contain about 50 per cent of a highly unsaturated oil (oleic and linoleic acids predominate) and 20–25 per cent protein. The oil, which rarely becomes rancid because of the presence of phenolic material, is used in the manufacture of margarine, cooking fats, soaps, paints, and as a lubricant and illuminant. In India it is employed as a ghee substitute and the basis of scented oils used in perfumery. The residue, left after oil extraction, is a valuable animal feed. Throughout the world, sesame seeds are used in food in various ways: in soup, porridge, sweetmeats, nut snacks, and sprinkled on top of cakes, bread, and pastries.

GROUNDNUT, PEANUT, MONKEY NUT (3) *Arachis hypogaea.* This crop belongs to the important food family Leguminosae (Fabaceae) and supplies both oil and nuts. It is of South American origin, being grown there since at least 3000–2000 BC, but it is now cultivated in many tropical and subtropical countries. Important producing countries include India, China, the United States, Argentina, Brazil, Nigeria, Indonesia, Burma, Mexico, and Australia. Some countries, for example the United States, grow groundnut primarily for consumption of whole nut products; others, for example India and China, for oil.

The plant is an annual, growing to a height of 15–60 cm and with yellow flowers (3). Two forms are available: erect (**3B**) or prostrate (creeping). The flowers are self-pollinated and, after pollination, the flower stalk elongates, forcing the young pod (fruit) into the ground where it matures. Harvesting is by hand, particularly of the prostrate forms; erect forms can be harvested by mechanical means.

The groundnut pod (**3A**), with its wrinkled surface network, usually contains two nuts (seeds) in the Spanish or Virginia types, three to four (possibly up to six) in the Valencia types. Nuts are extracted from the pod either by hand or by mechanical means.

These nuts are rich in nutrients. There is 50–55 per cent oil, with a high proportion of the unsaturated oleic and linoleic acids. The seeds contain about 30 per cent protein. Also, they are good sources of the essential minerals. The nuts have a good supply of vitamins E and the B complex. Oil extracted from the seed is used for cooking, as a salad oil, in margarine, in India as vegetable ghee, and for fish preservation. The seed residue, with 50–55 per cent protein, can, under certain circumstances, be a useful animal feed.

Around the world, the nuts have many uses. In the East, they are found in soups, stews, and curries; in West Africa, there is a common dish known as 'groundnut chop or stew'. The nuts may be present in confectionery, snack foods, sweetmeats, and consumed in various forms (salted, unsalted, roasted). They are available commercially in the shell or shelled. Low-energy peanuts, with 50 per cent of the oil removed, are available. High-protein peanut flour has been used to supplement milk beverages in India and to raise protein levels in bread and biscuits. Peanut butter is made by removing the skin (seed coat) and germ (embryo) and grinding the roasted nuts.

Under certain conditions, the nuts may become infected with the moulds (fungi) *Aspergillus flavus* and *A. parasiticus.* These moulds can produce chemicals (mycotoxins) known as 'aflatoxins' (see p. 210) which are carcinogenic. Under no circumstances should infected nuts or nut residues be introduced into food or animal feed chains. Groundnut oil is not affected.

Clearly, enormous amounts of shell (pod wall) are available every year. These are used as fuel and in industry. Stems and leaves of the plant can be consumed as forage.

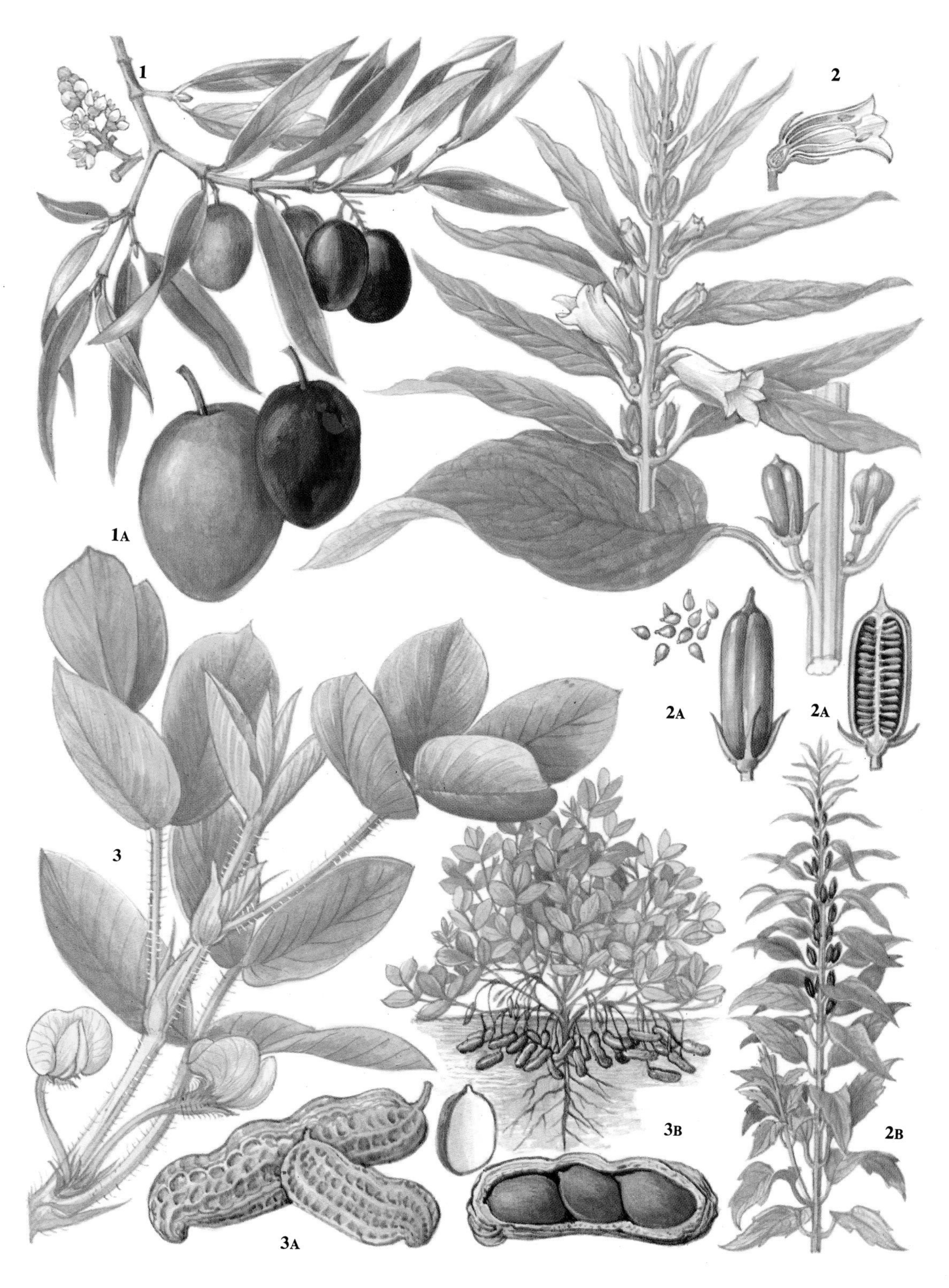

LEAFY SHOOTS × ⅔ FRUITS LIFE SIZE PLANTS × ⅛

1 **OLIVE** flowers and fruiting branch 1A Fruits
2 **SESAME** flowering stem 2A Seed pods and seeds 2B Part of plant
3 **PEANUT (GROUND NUT)** flowering shoot 3A Pods and nuts 3B Plant

Other oil producing plants

SOYA BEAN OR SOY BEAN (1) *Glycine max*. Soya has been an important food plant in the Far East for thousands of years. It was cultivated in China in at least 3000 BC, from which country it spread to Korea, Manchuria, Japan, and Indonesia. In the West it has only been of importance in the past few decades. The major world producers of seed are the United States of America, Brazil, China, and Argentina, with the United States in the lead, although the plant was only introduced into that country at the beginning of the twentieth century as a hay or pasture crop. Soya cultivation has not been successful in the United Kingdom.

Although the whole plant can be utilized as food or fertilizer (pasture, fodder, hay, silage, or green manure), it is the seed or bean which is of major food importance. Many of the Far East uses have now been adopted in the West. The seed contains 30–50 per cent protein (higher than other legumes) with a good balance of essential amino acids and 14–24 per cent of highly unsaturated oil (with considerable proportions of linoleic and oleic acids). The B vitamins, thiamin, riboflavin, nicotinic acid, and folic acid, are well represented and there are good amounts of minerals. Soya is the world's most important oilseed, the extracted oil being used for cooking, as a salad oil, in margarine, and shortenings; there are also some industrial uses, such as in paint and soap. The seed residue is a common constituent of animal feed.

Soya seeds are eaten whole, split, or germinated to give bean sprouts. They may be processed to give soya milk and cheese, or fermented to give soya sauce. Roasted seeds and soynuts may be included in cakes and candies. 'Tofu', 'tempeh', 'miso', and others are Oriental cheese-like soya products. Soya flour, grits, protein concentrates, and isolates are used in bakery products and meat extenders, also in 'analogues' which are imitations of food products such as meat. Soya is the main commercial source of 'lecithin', a substance with numerous food uses, such as an emulsifier in margarine. The bean's protein may lower blood cholesterol and its isoflavones may be effective against certain types of cancer (see p. 206).

The crop is an annual belonging to the family Leguminosae (Fabaceae), grows to a height of 20–180 cm, has white or lilac flowers, and pods with 2–3 seeds (yellow, green, brown, or black).

SUNFLOWER (2) *Helianthus annuus*. This plant belongs to the family Compositae (Asteraceae) and grows to a height of 0.7–3.5 m. Its flower head (10–40 cm in diameter) consists of outer yellow ray florets for attraction only, and inner brownish disk florets which are fertile. Bees are the most effective pollinators, consequently bee colonies are frequently placed in sunflower fields, giving rise to honey as a by-product. The 'seeds' (botanically speaking, fruits known as achenes) are white, brown, black, or striped.

Sunflower probably originated in the south-western part of North America and its seed has been utilized as food by North American Indians for thousands of years. There is evidence of cultivation as early as 900 BC. The plant was introduced into Europe in the sixteenth century. Russia developed sunflower as a commercial oilseed in the early nineteenth century. The crop is still important in that country, together with the United States and Argentina. Its seed contains 27–40 per cent of polyunsaturated oil (a high percentage of linoleic acid) and 13–20 per cent protein of high biological value. In Russia, the oil content has been raised to 50 per cent. Sunflower seed with a high percentage of the monounsaturated oleic acid has been developed. The oil is used in salad and cooking oils, margarine, and shortenings, together with some industrial uses, including satisfactory results when added to diesel fuel. Its seed residue is good animal feed.

Roasted whole sunflower seeds have long been a popular snack item in Russia. This usage has now spread to other countries where the seeds (sometimes with the hulls removed, sometimes salted) are found as a snack food or in confectionery. The seeds are widely fed to birds and other small animals.

RAPE (3) There are two species: *Brassica napus* (oilseed rape, swede rape) and *B. campestris* syn. *B. rapa* (turnip rape, 'sarson', 'toria'). The crop is annual or biennial and is an important oilseed in countries with a temperate climate or as a winter crop in subtropical climates. Canada, China, India, and Europe are significant producing areas. In the United Kingdom, only *B. napus* is cultivated. There is evidence that rape (probably *B. campestris*) was cultivated at least 2000 years ago in China and India; *B. napus* is of more recent origin, possibly in Europe. Rape belongs to the family Cruciferae (Brassicaceae) and shows the characteristic feature of four petals, in this case yellow, in the form of a cross. Its black, brown, or sometimes yellow seeds are found in a pod-like fruit (siliqua). Many important vegetables are also found in the Cruciferae.

Rape seed contains about 40 per cent oil which is highly unsaturated. Originally the oil contained a high percentage of erucic acid. Investigations on experimental animals indicated that this acid had a deleterious effect on the heart. Consequently plant breeding produced cultivars with under 5 per cent erucic acid. High erucic acid cultivars are good for certain industrial purposes. Low erucic acid oil is used for cooking and in margarine and other edible substances; some use of the oil has been made in motor fuel. The seed residue has about 36 per cent protein but, originally, contained quantities of substances known as 'glucosinolates' which are goitrogenic and potentially toxic (see p. 207). Cultivars have been produced with very low quantities of glucosinolates which can be combined with low erucic acid cultivars to give 'double zeros', known in Canada as 'canola'. Modern rape seed residues are used widely in animal feeds. Rape pollen can be a powerful allergen. The crop is also a forage plant.

OTHER OIL PLANTS. Those deserving mention include cotton (*Gossypium* spp.). Its main product is seed fibre (lint) but the unsaturated oil is widely used in edible materials. The seed residue contains a phenolic substance, 'gossypol', which is potentially toxic to some animals, but the manufacturing process normally eliminates any toxicity. Another oil plant which has attracted attention in recent years is safflower (*Carthamus tinctorius*), no doubt because its seed oil is highly unsaturated with a large percentage of linoleic acid.

PLANTS × 1/8 SEEDS LIFE SIZE DETAILS × 2/3

1 **SOYA BEAN** plant 1A Seeds 1B Details of flower, leaves, and pods
2 **SUNFLOWER** plant 2A Seeds
3 **RAPE** plant 3A Seeds 3B Detail of ripe fruits

Nut trees of temperate climates

HAZEL OR COB (1) *Corylus avellana.* Although a number of *Corylus* species are found throughout the world, *C. avellana* and *C. maxima,* also their hybrids, are most important as regards nut production. Hazel is a tree or bush which may grow to 6 m high. In the United Kingdom in the early spring, its pendulous male catkins are conspicuous because of their yellow colour; the female flowers are less conspicuous, occurring in short buds with the crimson styles exposed. The nuts, in clusters of one to four, are globose or ovoid, 1.5–2 cm long, with a hard brown shell partially enclosed by a deeply lobed involucre or husk. Hazel grows wild in Europe and West Asia.

FILBERT (2) *Corylus maxima.* This grows wild in south-eastern Europe. It is a larger tree than *C. avellana,* also its ovoid to subcylindrical nuts are somewhat larger, but the most important distinction between the two species is that, in *C. maxima,* the involucre extends beyond the nut and is constricted at the apex although, because of hybridization, this distinction is not always clear.

Archaeological shell remains indicate that hazel kernels have been an important constituent of the human diet since prehistoric times. The nut was described by Pliny, Theophrastus, and Dioscorides. The place and time of domestication is not clear but it was already cultivated by the Romans. A good deal of hybridization leading to the production of many cultivars was carried out in the nineteenth century. Hazels are cultivated in many temperate countries; major producing countries include Spain, Italy, and Turkey. Apart from being a dessert nut, the kernels are employed in confectionery and sweetmeats. The kernel contains about 18 per cent protein and up to 68 per cent oil (lower quantities have been recorded), sometimes available commercially, with a chemical constitution similar to olive oil. In Turkey, hazel shells have been used as fuel, to colour wine, and to make vegetable carbon.

SWEET OR SPANISH CHESTNUT (3) *Castanea sativa.* It is not certain, but the plant could have been domesticated in northern Turkey and the Caucasus and then taken into southern Europe. The Romans probably introduced it into France and Britain. Most commercial nuts contain relatively large amounts of protein and fat, chestnut is an exception with only 2–4 per cent protein and 2–5 per cent fat. However, it contains a very large amount of starch—up to 70 per cent. The nuts are consumed roasted or boiled and have been processed to produce flour, bread, porridge, poultry stuffing, fritters, animal feed, and sweetmeats, such as the famous *marrons glacé* of France.

The sweet chestnut is a large tree, up to 35 m high, with a broad crown. The leaves are 8–25 cm long, oblong lanceolate, and coarsely toothed. Its male and female flowers are borne in separate inflorescences in July. The male catkins, 8–16 cm long, are conspicuous because of their yellow anthers. The female flowers are usually borne in threes, each with seven to nine red styles. The glossy, brown nuts, 2–4 cm wide, are enclosed in a green cupule (bur) densely covered with long, branched spines.

There are a number of *Castanea* species. *C. dentata* was an important North American species but, since the beginning of the twentieth century, has largely been destroyed by fungal diseases. The Chinese chestnut, *C. mollissima,* is grown commercially to some extent in the United States, but most chestnuts (*C. sativa*) are imported from Italy.

ALMOND (4) *Prunus dulcis* syns *P. amygdalus, Amygdalus communis.* This was one of the earliest nut trees domesticated in Old World agriculture. It was probably taken into cultivation in the eastern part of the Mediterranean basin about the same time as olive, grape vine, date palm, and not later than 3000 BC. It is now the most important world tree-nut crop and is cultivated commercially in several Mediterranean countries (e.g. Spain, Italy, France, and Portugal), also California (the largest producer), South Africa, and Australia.

The nut kernel contains 40–60 per cent unsaturated oil (a large percentage of oleic acid) and about 20 per cent protein. It is a very important dessert nut and is utilized in confectionery, almond butter, almond paste, macaroons, marzipan, health foods, and ice-cream. Sugared or candied almonds are sometimes given as a traditional gift to guests at weddings in certain European countries. Two botanical varieties have been recognized: sweet almond (*P. dulcis* var. *dulcis*) and bitter almond (*P. dulcis* var. *amara*). Bitter almond kernels contain a glycoside substance known as 'amygdalin'; there is none, or only a trace, in sweet almond. Amygdalin, under certain circumstances, produces the toxic prussic acid (hydrogen cyanide), although the bitterness of the kernels should deter anyone from eating enough to be poisoned. The triglyceride oil extracted from the kernels is known as 'sweet almond oil'—used in the cosmetic, confectionery, and baking trades. 'Sweet almond oil' is to be contrasted with 'bitter almond oil' (benzaldehyde) produced from amygdalin and used in pharmacy. Prussic acid is eliminated during the production of bitter almond oil.

Almond is a small tree, up to about 30 m high, and belonging to the Rosaceae family which contains well-known fruit trees (e.g. plum, apricot, apple, and pear). There are many cultivars, most of which, to produce fruit, must be cross-pollinated by other cultivars, for which purpose hives of honey bees are placed in the orchards.

The fruit consists of a hull (epicarp and mesocarp), shell (endocarp), which may be thin (paper), soft, or thick, and the kernel (seed). At maturity, the hull splits open. This type of fruit is known as a 'drupe'. Such a fruit is found in stone fruit (e.g. peach, plum, and apricot) but in these the mesocarp becomes fleshy and edible. Palm fruits (e.g. coconut, oil palm, and date) are also drupes.

NUTS LIFE SIZE BRANCHES × ¼

1 **WILD HAZEL-NUT** 1A Flowering and fruiting branches 1B Cultivated **COBNUTS**
2 **FILBERT NUTS** 2A Fruiting branch 3 **SWEET CHESTNUTS** 3A Flowering and fruiting branches
4 **ALMOND-NUT** 4A Flowering and fruiting branches

Nut trees of temperate climates

COMMON OR PERSIAN WALNUT (1) *Juglans regia*. This tree grows wild from the Balkans through central Asia to some parts of China. It was domesticated thousands of years ago, probably in north-eastern Turkey, Caucasus, and northern Iran. Today it is cultivated commercially in many countries. California is the largest producer but there is considerable production in other countries, for example Turkey, China, Russia, Greece, Italy, France, and Rumania. There are about 15 *Juglans* species, all with edible nuts; however, the one being described is the most important. The nut kernels contain about 15 per cent protein and almost 70 per cent unsaturated oil (considerable percentages of linoleic and oleic acids). They are employed as dessert nuts and in confectionery, cakes, and ice-cream. The extracted oil has been used since early times in paints and also has various food uses, being particularly popular in France. The kernel residue can be employed as an animal feed; the shell flour has been utilized in the manufacture of plastics. In the United Kingdom, young fruits pickled in vinegar are popular. Ripe walnuts, like other nuts, contain vitamins E and B, except B_{12}; in addition, young fruits contain vitamin C. In France, an alcoholic drink is made from the young fruit.

The tree grows to a height of some 30 m. Its leaves are alternate, pinnate, with 5–11 (rarely 13) obovate or elliptic leaflets, 8–16 cm long, the terminal leaflet largest. The male flowers, with a small, lobed perianth and 3–40 stamens, are borne in long, pendulous catkins. The female flowers are solitary or few in number. The fruit is a green drupe, like almond, with a hull containing the walnut. Walnut timber is highly valued.

BLACK WALNUT (2) *Juglans nigra*. This is one of the best known and most widely distributed of North American trees. Its shell is usually thicker than that of the common walnut and its kernel has a richer flavour. Cultivars have been produced with thinner shells. In the United States the food uses of this nut are essentially the same as those of the common walnut.

The tree grows up to 45 m in height. Its dark-green leaves, 30–60 cm in length, have 11–23 serrate leaflets, each 5–12 cm long.

WHITE WALNUT OR BUTTERNUT (3) *Juglans cinerea*. This is a smaller North American tree. Its pinnate leaves have 7–17 irregularly serrate leaflets, glandular, hairy, 5–12 cm long. The glandular hairy fruits are in groups of two to five. The shell, though hard, is generally not difficult to crack and the food uses of the kernel are similar to the two previous species.

Some other species of *Juglans* are of limited local importance as sources of edible nuts—for example, the Japanese walnut (*J. sieboldiana*).

PISTACHIO (4) *Pistacia vera*. This tree grows wild in the Middle East and central Asia, as far east as Pakistan and India. The nuts of the wild pistachio are an important food of migratory nomads in northern Iran and Afghanistan. Major producing areas of the cultivated plant are Iran, Turkey, and California; there is also cultivation in other countries, for example Syria, Italy, India, and Greece. Domestication probably took place in central Asia.

The green kernels have a unique flavour and pistachios are one of the most desirable nut types. They contain about 18 per cent protein and 55 per cent unsaturated oil. Because of the value of the commodity as a dessert nut, it is unlikely that oil extraction will ever become a commercial proposition. For eating, the nuts are presented raw or salted in their shells, also the kernels are employed in confectionery, ice-cream, candies, and bakery goods. The shells split longitudinally prior to harvest; this is desirable for the consumer because the kernels can be marketed in the shell which is easy to remove.

Pistachio is a deciduous tree, up to 10 m in height. Its leaves have three to seven ovate leaflets, 5–10 cm long. Male and female flowers are borne on different trees, in axillary racemes. The fruit is an ovoid drupe.

PECAN (5) *Carya pecan* syn. *C. illinoensis*. This is the most important nut tree native to North America, being found in many parts of the United States and Mexico. The nuts constituted an important food for the North American Indians. Its many cultivars are grown in some 20 states of the USA, also in Australia, Brazil, and South Africa. The kernel contains up to 70 per cent monounsaturated oil (almost 80 per cent oleic acid) and up to 18 per cent protein. Oil is not really extracted on a commercial basis but any available is sold mostly for specialized purposes in the pharmaceutical and cosmetic industries. The kernels may be eaten raw or salted, in ice-cream, cake, bread, candies, confectionery, vegetarian croquettes, sandwiches, and pecan pie. Pecan nuts are exported, sometimes with the shells polished and dyed. The shell flour has the same uses as that of walnut.

The pecan is a large tree, up to 50 m high. Its leaves are pinnate, with 7–17 leaflets, 10–18 cm long. The male flowers are borne in three-branched, pendulous catkins; the female in 2–10 flowered spikes. The fruit is a drupe, 4–9 cm long, which at maturity splits into four valves to reveal the nut. Both pecan and the walnuts belong to the same family—Juglandaceae.

NUTS AND WALNUT FLOWERS LIFE SIZE BRANCHES × ¼

1 **EUROPEAN WALNUT** 1A Fruiting shoot 1B Female flowers and male catkin
2 **BLACK WALNUT** 2A Leaf 3 **BUTTERNUT** 3A Leaf
4 **PISTACHIO** 4A Fruiting branch 5 **PECAN** 5A Flowering branch 5B Fruiting branch

Nut trees of warmer climates

BRAZIL-NUTS (1) *Bertholletia excelsa.* These are the seeds of one of the tallest trees (40–50 m) of the Amazonian forest and they are harvested entirely from wild trees. The nuts are contained in large, woody, round fruits (0.5–2.5 kg in weight). The fruits are allowed to fall to the ground, are split open by the collectors, and the nuts, 10–25 in a fruit and packed like orange sections, are extracted. There have been instances of the heavy fruits falling from the tree on to the heads of collectors and causing injury, even death. The kernel contains up to 17 per cent protein and 65–70 per cent of a monounsaturated (almost 50 per cent oleic acid) oil. There is some local oil extraction from broken seeds. Most nuts originate from Brazil, but Bolivia, Peru, Colombia, the Guianas, and Venezuela are also involved. A small proportion of the nuts is consumed locally, most are exported to the United States and Europe where they are eaten raw, roasted, salted, or in ice-cream, bakery, and confectionery products.

CASHEW-NUTS (2) *Anacardium occidentale.* These are the products of a medium-sized tropical tree (up to 12 m in height) which probably originated in north-eastern Brazil. The Portuguese had taken the plant to East Africa and India by 1590. It then spread to Sri Lanka, Malaysia, and India. Today the main producer of the nuts is Mozambique. It is an easy plant to cultivate. Its fruit is peculiar. The enlarged fruit stalk (yellow, red, or scarlet) is actually pear-shaped but is known as the 'apple', 5–10 cm in length. It contains carotenes, some of the B complex vitamins, also vitamin C. A number of food products have been derived from the 'apple', including fruit juice, jams, wine, and liquor. The greyish-brown, kidney-shaped nut, 2–4 cm in length, is found at the apex of the 'apple'. Its shell contains an unpleasant oily liquid known as cashew-nut shell liquid (CNSL) which can produce swellings and blisters on the human skin. Great care must therefore be observed when the nuts are shelled to produce the kernels. CNSL has a number of industrial uses, including the production of resins. Prior to marketing, the brown seed coat is removed from the kernel, which contains about 17 per cent protein and 45 per cent monounsaturated (70–80 per cent oleic acid) oil. The kernels are consumed as dessert nuts and in bakery and confectionery products.

PINE KERNELS (3) *Pinus* spp. Most food plants belong to the group known as angiosperms, where the seeds are contained in fruits in the true botanical sense. Pines are gymnosperms, there being no fruits and the seeds are exposed, in the case of the pines, on the woody scales of the female cones. The pines are trees and the seed kernels of various species are utilized as food in many parts of the world: *P. pinea,* Stone pine or *pignolia,* Spain, Portugal, Italy, cultivated to some extent; *P. edulis,* piñon, Colorado, New Mexico, Arizona, Utah, seeds collected in the wild; *P. monophylla,* single-leaf piñon, western USA; *P. cembroides,* Mexican piñon; *P. cembra,* central Europe; *P. gerardiana,* Afghanistan, Pakistan, Himalayas; *P. sibirica,* Russia; *P. pumila,* Japan; *P. koraiensis,* Korea, Japan, China, now exported from the latter country to Europe and the United States.

According to species, the kernel contains 47–68 per cent unsaturated fat and 12–31 per cent protein. They are utilized as raw or roasted dessert nuts, in many meat, fish, and game dishes, sauces, soups, sweetmeats, cakes, and puddings. Pine kernels are said to have a distinctly different flavour compared to other dessert nuts. For the North American Indians they were an important food source, shell remains have been found in a Nevada site some 6000 years old. Roman soldiers carried the nut to Britain.

Other gymnosperm seeds consumed as food include Chile pine or monkey-puzzle (*Araucaria araucana*), Chile; Parana pine (*A. angustifolia*), Brazil; bunya-bunya pine (*A. bidwilli*), Australia; cycads (*Cycas* spp.), Asia; gnetum (*Gnetum gnemon*), Asia; ginkgo (*Ginkgo biloba*), China and Japan.

QUEENSLAND OR MACADAMIA NUTS (4) *Macadamia integrifolia,* smooth shelled; *M. tetraphylla,* rough shelled. The tree is an evergreen growing up to 15 m in height. Its fruit, with a fleshy husk, contains the nut. The plant is native to the coastal subtropical rain forests of south-eastern Queensland and northern New South Wales. Its nut has been utilized by Aboriginal tribes since ancient times and was domesticated for the first time in Australia in 1858, the only native Australian plant developed as a commercial food crop. In 1882 it was introduced into Hawaii and has developed there into an important crop, third after sugar-cane and pineapple. It is considered one of the world's finest gourmet nuts with a unique delicate flavour. The kernel contains about 70 per cent monounsaturated fat and 8 per cent protein. It is utilized as a roasted and salted dessert nut, also in confectionery, bakery products, and ice-cream.

AUSTRALIAN OR MORETON BAY CHESTNUT (5) *Castanospermum australe.* This is of local importance in Australia where the seeds are consumed by Aborigines. The seeds are normally roasted before being eaten because in the fresh state they may be harmful. In recent years a substance known as 'castanospermine', extracted from the seeds, has been investigated as a possible antiviral agent in respect of the AIDS virus.

The plant is a large evergreen tree, belonging to the pea family (Leguminosae/Fabaceae). Its leaves are compound, with 11–15 leaflets. Its yellow, orange, or reddish flowers are succeeded by pods up to 20 cm long, each containing several large, brown seeds.

NUTS AND SEEDS LIFE SIZE   FLOWERS AND FRUITS × ¼

1 **BRAZIL-NUT** and kernel   1A Fruit   2 **CASHEW-NUT** and kernel   2A Cashew apples
3 **PINE KERNELS**   3A Pine cone   3B Stone pine tree (small scale)
4 **QUEENSLAND NUT**   4A Fruit and flowers
5 **MORETON BAY CHESTNUT**   5A Flowering branch and pods

Exotic water plants used as food

LOTUS (1) *Nelumbo nucifera.* This plant has been held sacred to Buddhists in the Near and Far East for over 5000 years. It is indigenous in Asia from Iran to China and Japan, also in north-eastern Australia. The plant belongs to the water-lily family—Nymphaeaceae. Its 'root', which is botanically speaking an underground stem known as a rhizome (**1B**), grows in the mud at the bottom of shallow ponds, lagoons, marshes, and flooded fields. The large, bell-shaped leaves (up to 1 m across) and the attractive white, pink, or red flowers grow above the surface of the water.

Lotus has been cultivated in China for at least 3000 years, the 'root' and seeds being the important food products in countries such as China, Japan, and India. They are exported to many other countries. As food, the 'roots' can be roasted, sliced and fried as chips, pickled, or candied. They contain about 2 per cent protein, 0.1 per cent fat, and 6 per cent starch; a good deal of sodium, and vitamins B, C, and E. In China, a highly digestible starch, like arrowroot, is prepared from them. The seeds can be eaten raw, roasted, boiled, or candied, although the bitter green embryo must be removed first. They contain 60 per cent starch, 17 per cent protein, 2.5 per cent fat, and are rich in vitamin C.

WATER-CHESTNUTS (2) *Trapa* spp. These are aquatic plants with floating diamond-shaped leaves and submerged, finely divided, feathery leaves (sometimes regarded as roots). The small white flowers are succeeded by dark-brown woody fruits, 2.5–5 cm across. These fruits bear two or four horns, according to species. The plant belongs to the family Trapaceae. *Trapa bicornis* (two fruit horns) is 'ling', grown in China, Japan, and Korea. Its kernels are eaten boiled, in various regional dishes, or preserved in honey and sugar. It was a very important grain in China before the twentieth century. *Trapa bispinosa* (two fruit horns) is singhara nut, grown in Kashmir, India, and Pakistan. *Trapa natans* (four fruit horns) is the water caltrops or Jesuit's nut. It was a common food of ancient Europeans, its use dating back to neolithic times. The kernels contain 16 per cent starch and 2 per cent protein. Because of possible toxicity, the kernels should be boiled for an hour before consumption.

CHINESE WATER-CHESTNUT OR MATAI (3) *Eleocharis dulcis* syn. *E. tuberosa.* This belongs to the sedge family or Cyperaceae (contrast with *Trapa*). It is cultivated in China, Taiwan, and Thailand in shallow marshes, lakes, and flooded fields. The 'nut' is actually a tuber or corm from which grow tubular leaves, 1–2 m in height. New corms are formed at the ends of horizontal rhizomes. The corm contains 1.4 per cent protein, 0.2 per cent fat, and 5.6 per cent starch, with an equal amount of sugars (sucrose, glucose, and fructose), also vitamins B, C, and E, and good amounts of phosphorus and potassium. It is an ingredient of oriental soups, chop suey, salads, meat and fish dishes, and puddings. The corm has a delicious flavour and a unique crisp texture. It is exported fresh and canned from China, Hong Kong, and Taiwan to the United States and Europe.

1B **LOTUS** root × ½

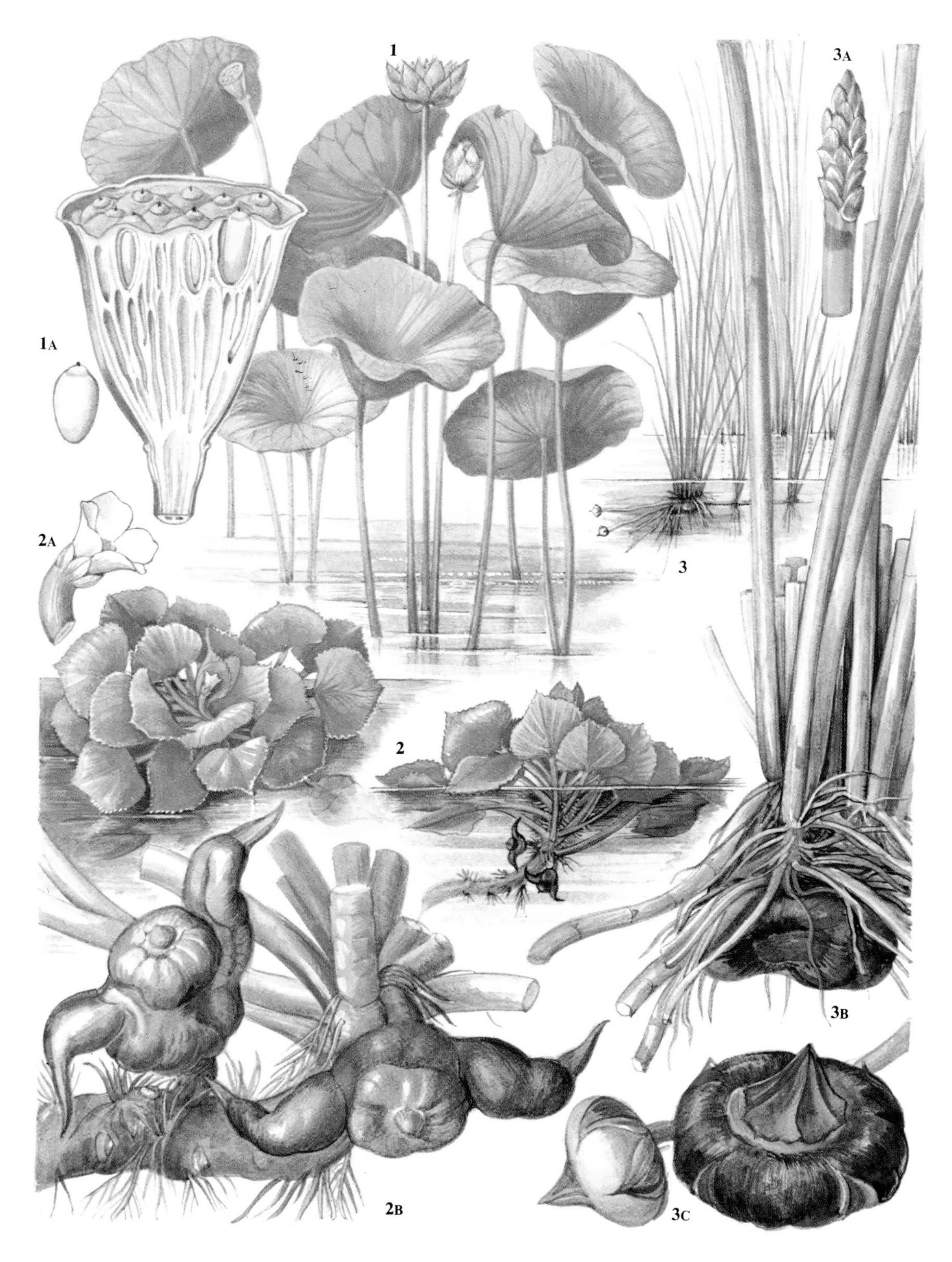

PLANTS × ⅛ DETAILS LIFE SIZE

1 **LOTUS** 1A Opened seed head and seed
2 **WATER-CHESTNUT** 2A Flower 2B Fruits
3 **CHINESE WATER-CHESTNUT** 3A Flowers 3B Base of plant 3C Corms

Exotic legumes

(for general information on legumes, see p. xviii.)

PIGEON-PEA OR RED GRAM (1) *Cajanus cajan*. This is the world's fifth most important pulse crop, cultivated in India, Africa, South-East Asia, and also the West Indies where it is an important cash crop. It is unusual for a pulse crop (normally annuals) as it is a woody, short-lived (up to 5 years) perennial, from 1 to 4 m in height, although it is sometimes cultivated as an annual. It is drought resistant and less suitable for the wet tropics. The flowers are yellow and the pods contain three to four seeds which may be white, greyish, red, brown, purplish, or speckled, with a white hilum (point of attachment). This is an ancient crop. Both Africa and India have been suggested as the original regions of domestication. It was cultivated in Egypt before 2000 BC.

Young green seeds are eaten as a vegetable in many countries and are canned in some parts of the West Indies. The mature seed is utilized as a pulse and, when split, constitutes the 'dhal' of India, second in importance to chick-pea. The pulse contains about 20 per cent protein and almost 60 per cent carbohydrate, mainly starch, and little fat. The immature green seed has 7 per cent protein, 20 per cent carbohydrate, and almost 70 per cent water. Pigeon-pea may replace soya to manufacture 'tempeh' and the seeds may be germinated to give sprouts. Its green pods are sometimes used as vegetables while the tops of plants with fruits make excellent fodder, hay, and silage.

BAMBARA GROUNDNUT (2) *Voandzeia subterranea* syn. *Vigna subterranea*. This is mainly cultivated in the drier areas of tropical Africa but also in America, Australia, central Asia, Indonesia, Malaysia, the Philippines. It was probably was domesticated in northern Nigeria and the Cameroon. The plant is a promising crop for semi-arid areas, being able to tolerate drought and poor soil better than many other crops. The stems are very short and prostrate, from which the leaves with long stalks arise thickly giving the appearance of a close bunch of leaves rising from one point on the ground. The flowers are pale-yellow, and give rise to pods (1–3 cm in length) which are buried in the earth, like groundnut (*Arachis hypogaea*). They have a wrinkled surface and contain one or two seeds (7–15 mm in diameter) which are white, yellow, brown, red, black, or mottled. The seed contains about 18 per cent protein, 60 per cent carbohydrate (mainly starch), and 6 per cent fat (compare with *Arachis hypogaea*). As they are very hard, prolonged boiling is required to make them palatable. For this reason, immature seeds are often consumed, also young pods.

WINGED BEAN, GOA BEAN, OR ASPARAGUS-PEA (3) *Psophocarpus tetragonolobus*. This interesting legume is a twining vine which grows to over 3 m when supported. Its pods (see illustration below), up to 40 cm in length, are square in section with four smooth or serrated wings. It is grown in quantity only in Papua New Guinea and South-East Asia. All parts of the plant are eaten: seed (33 per cent protein, 16 per cent largely unsaturated fat, 32 per cent carbohydrate); root tubers (unusual for a legume—yam-bean, *Pachyrrhizus*, is another example), with 8–10 per cent protein and up to 30 per cent starch; immature green pods (1–3 per cent protein); and leaves (5–7 per cent protein). At present it is only of local importance but it does have potential in the wet tropics. This legume should not be confused with *Lotus tetragonolobus* which also has winged pods and is also known as asparagus-pea, occasionally grown in temperate regions.

OTHER EXOTIC LEGUMES Other tropical legumes are only of minor or local importance as sources of food, such as certain species of *Phaseolus*, in addition to those figured on pages 41 and 43, but which are also pulses with similar characteristics: *P. aconitifolius* is the mat- or moth-bean of India, where it is grown for food, in the south-western United States it has been grown for fodder and green manure; *P. trilobus*, known as pillepesara, is a perennial from India; *P. angularis*, the adzuki bean, is probably a native of Japan, Korea, Manchuria, India, and neighbouring areas of southern Asia, also introduced to some other countries; *P. acutifolius* is the tepary bean, native to the south-western United States and Mexico where cultivation is most important, although it has been introduced to a number of other countries; *P. calcaratus*, the rice-bean, is most widely cultivated in China, Korea, Japan, India, and some other eastern countries, the beans are usually boiled and eaten with rice or instead of rice.

Hausa or Kersting's groundnut (*Kerstingiella geocarpa*) is similar to bambara groundnut but has a restricted range of cultivation in the drier parts of West Africa.

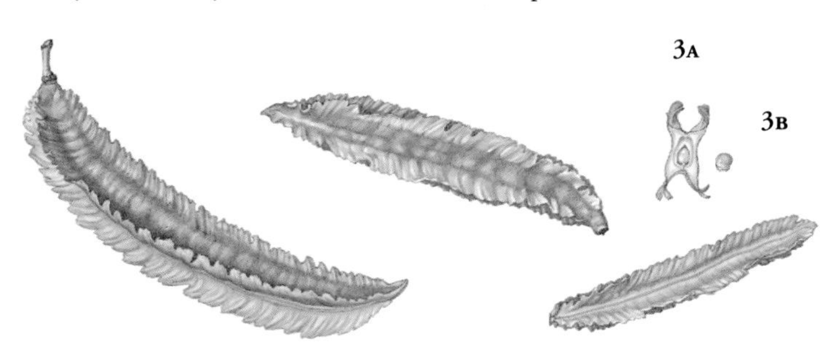

3 **WINGED BEAN** pods 3A Pod section 3B Seed × 1/3

TWO-THIRDS LIFE SIZE SEEDS LIFE SIZE

1 **PIGEON-PEA** flowers 1A Seed pods 1B Seeds
2 **BAMBARA GROUNDNUT** young plant 2A Pods and seeds

Runner beans and French beans

SCARLET RUNNER BEAN (1) *Phaseolus coccineus* syn. *P. multiflorus.* This bean is a native of the uplands, at an altitude of about 2000 m, of Central America (Mexico, Guatemala, and possibly some other countries) and was domesticated in that area. Material of the plant, about 2000 years old, has been found in the Tehuacán valley (Mexico). It was introduced into Europe in the sixteenth century.

In its original area it is cultivated as a perennial and is grown for its tender pods, green and dry seeds, tuberous starchy roots. Both dry seeds and roots require boiling in water prior to consumption because of toxic constituents (see p. 208). In temperate lands it is cultivated as an annual (it is popular in the United Kingdom) for the tender green pods. The mature seeds contain 17 per cent protein, 65 per cent carbohydrate, 2 per cent fat; the pod contains about 2 per cent protein, 0.5 per cent fat, 3 per cent carbohydrate, and a range of vitamins B, C, and E, and carotenes.

The plant with its twining stem can grow to a height of 4 m or more but, in cultivation, it is usual to restrict its height to that of the support by 'pinching out' the tip. Its flowers are scarlet, white, or variegated, giving rise to green pods, usually 20–30 cm in length, which contain seeds that vary in colour—pink to purple, dark mottled. Plant breeding has been directed at producing longer pods which are less 'stringy' (fibrous). The plant is sometimes grown as an ornamental.

COMMON, FRENCH, KIDNEY, HARICOT, SNAP, OR STRING-BEANS, FRIJOLES (2–8) *Phaseolus vulgaris.* This is the best known and most widely cultivated bean in the world and it possesses a host of common or local names. Archaeological remains, dated to about 5000 BC, have been found in the Tehuacán valley, also in Peru. It was introduced to Europe in the sixteenth century by the Spaniards and Portuguese; they also carried it to Africa and other parts of the Old World. The bean grows wild in the mountains, 500 to 2000 m above sea-level, in parts of Central and South America. Today, it is widely cultivated in the tropics, subtropics, and temperate areas. The bean is the main pulse crop throughout tropical America (Brazil produces the most) and many parts of tropical Africa—it is a minor crop in India and most of tropical Asia. In temperate areas, the bean is grown mainly for the young pods.

There are many cultivars. The crop may have a twining stem (pole type) up to 3 m in height, or it may be erect or bush (dwarf) with no support; there are also intermediate types. Its petals may be white, yellowish, pink, or violet; the pods are usually green but yellow (wax), purple, and green streaked with red kinds are also known. 'String' pods must be picked before they become fibrous; 'stringless' can be picked at a later stage. Seeds vary in shape (kidney-shaped to globose), size (7–18 mm long), and colour (black, white, red, buff, brown, or various combinations of these colours).

Pods contain a large amount of water, about 2 per cent protein, 0.5 per cent fat, 3 per cent carbohydrate with carotenes, vitamins B, C, and E. They are marketed fresh, canned, or frozen. The dried seeds (pulses) have 22 per cent protein, 1.6 per cent fat, and 50 per cent carbohydrate, with vitamins B and E only. These seeds have been given a variety of names, for example berlotto, haricot, cannellino, pinto, pea-bean, navy, red kidney, marrow, black bean, and flageolet. They are associated with a number of famous food products and dishes, for example canned baked beans in tomato sauce; France's famous *cassoulet,* beans cooked with meat pieces; and red kidney beans in Mexican *chilli con carne.* As with other legumes, these pulses must be well soaked in water and cooked prior to consumption because of the presence of antinutritional substances, particularly lectins (see p. 208). An interesting type of *P. vulgaris* known as *nuñas* grows above 2500 m altitude in parts of South America. These beans are produced essentially for home consumption and 'pop' when cooked, rather like popcorn.

PLANTS × ⅛ FLOWERS AND PODS × ⅔ SEEDS × 1 FLOWER SECTION × 2

1 **SCARLET RUNNER** 1A Flowers and pod 1B Seed 2 '**CLIMBING PURPLE-PODDED KIDNEY BEAN**' 2A Seeds
3 '**PEA-BEAN**' pod 3A Seeds 4 '**CANADIAN WONDER**' 4A Flower section 4B Pod 4C Seed
5 '**DEUIL FIN PRÉCOCE**' flowers and pod 5A Seeds 6 **BROWN HARICOT** seeds
7 **WHITE HARICOT** seeds 8 '**MEXICAN BLACK**' flowers and pod 8A Seeds

Tropical pulses

BUTTER-BEAN, LIMA BEAN, OR MADAGASCAR BEAN (1) *Phaseolus lunatus*. This was probably domesticated in both Central and South America—archaeological specimens have been discovered dating back to 5000 BC. It is now also found in many tropical, subtropical, and warm temperate areas of North America, Africa, and Asia. According to cultivar, the plant may be a small annual bush (30–90 cm in height) or a large climber (2–4 m tall). It is cultivated as an annual or perennial. The seeds are very variable in size (1–3 cm in length), shape, and colour (white, cream, red, purple, brown, black, or mottled). The dried beans are utilized as pulses, also the green immature beans. In the United States of America, which is the world's largest producer of butter-beans, the immature seed is canned or frozen. The pulse gives a protein-rich flour which is added to bread and noodles in the Philippines; it is used in bean paste in Japan. The pods and leaves may also be utilized as food. The pulse contains 20 per cent protein, 1.3 per cent fat, and 60 per cent carbohydrate; the green seed will contain less of these constituents but more water. The mature seed contains the glycosides 'phaseolunatin' and 'linamarin' (see p. 208) which can produce toxic hydrocyanic acid (prussic acid), although the amount produced relates to the cultivar. United States' legislation allows a maximum content of 20 mg hydrocyanic acid per 100 g seeds. This hazard can be eliminated by soaking and boiling the seeds in water, which should be changed during the process.

CHICK-PEA OR BENGAL GRAM (2) *Cicer arietinum*. This is one of the world's three most important pulses (the other two are *Phaseolus vulgaris* and *Pisum sativum*). It was domesticated in the Fertile Crescent (see p. xiii) of the Middle East together with wheat, barley, and other pulses, and possibly evolved from the wild *C. reticulatum*. The crop spread, probably reaching the Mediterranean area by 4000 BC and India by 2000 BC. In the sixteenth century it was taken to the New World by the Spaniards and Portuguese. Today the greatest production of the plant takes place in India, where it is the most important pulse, but there is also considerable production in the Middle East and Mediterranean countries. Production has decreased to some extent in India because of the introduction in the 1960s of new wheat cultivars and improved irrigation, part of the Green Revolution.

The plant has well-divided leaves, giving it a feathery appearance. Its pods are oblong (2–3 × 1–2 cm) and contain one or two beaked seeds which may be white, yellow, red, brown, or nearly black. Chick-pea likes a cool, dry climate and is grown in India as a winter crop. Compared to other pulses (other than groundnut and soya bean) it contains somewhat less protein (maybe as low as 17 per cent) but more fat (5 per cent). It is used in India to make 'dhal' (see *Cajanus cajan*), also the seed flour is a constituent of many forms of Indian confectionery. In the Mediterranean area the cooked seeds plus sesame oil and other flavouring form a well-known side-dish—'hummus'.

BLACK GRAM, URD, OR WOOLLY PYROL (3) *Phaseolus mungo* syn. *Vigna mungo*. This is one of the most highly prized pulses of India, although it is also cultivated to a lesser extent in South-East Asia, East Africa, the West Indies, and the United States. The crop is drought resistant. Its seeds are boiled, eaten whole, or used in 'dhal'; the seed flour is utilized in porridge, bread, and biscuits, and the pods, containing in the mature state 4–10 black or olive-green seeds, can be eaten as a vegetable when green. The plant is sometimes grown as a green manure, cover, and forage crop. In Japan it is favoured for bean sprouts. The pulse contains about 25 per cent protein, 1 per cent fat, and 40 per cent carbohydrate.

GREEN GRAM, GOLDEN GRAM, OR MUNG BEAN (4) *Phaseolus aureus* syn. *Vigna radiata*. This is an important pulse crop in India and, as black gram, is also cultivated elsewhere in the world. It was probably derived from *P. radiatus*. The pulse is utilized in food in the same way, and has virtually the same chemical composition, as black gram. All pulses can be germinated to give bean sprouts, but green gram is the pulse most widely used for this purpose in North America, Asia, and Europe (see below). Bean sprouts, which may be eaten raw or cooked, are popular in salads, and also in oriental cooking. Compared with the pulse, the sprout has less protein, fat, and carbohydrate, but there is an increase in B vitamins (thiamin, riboflavin, and niacin) and vitamin C. The seed is said to cause less flatulence than other pulses. The plant is rather similar to black gram but is less hairy. It has purplish-yellow flowers, giving rise to pods which contain 10–15 seeds, usually green, but sometimes yellow or black.

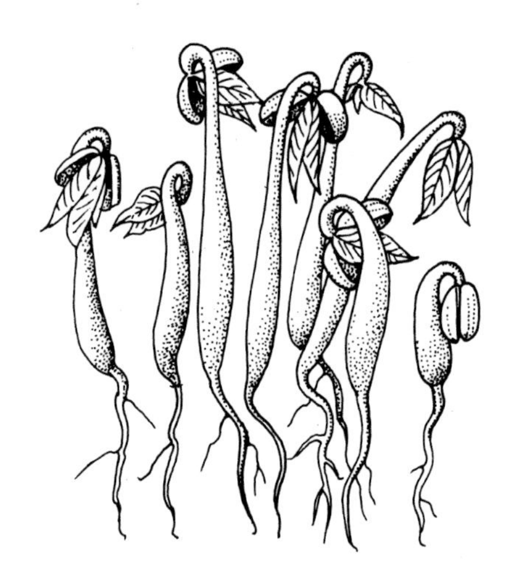

PLANTS × 1/8 FLOWERS AND FRUITS × 2/3 SEEDS × 1

1 **BUTTER-BEAN (LIMA BEAN)** plant 1A Flowering stem 1B Ripe pod 1C Seeds
2 **CHICK-PEA** plant 2A Flowering stem 2B Fruiting stem 2C Seeds
3 **BLACK GRAM** flowers 3A Pods 3B Seeds 4 **GREEN GRAM** shoot 4A Pods 4B Seeds

Large-podded beans

FABA, BROAD, HORSE-, FIELD-, TICK-, OR WINDSOR BEAN (1) *Vicia faba*. This is a temperate crop which originated in the Mediterranean region or south-western Asia, together with the other pulses—pea, lentil, chick-pea—although the time of its origin was probably later. The earliest archaeological remains of faba beans were of the neolithic period (6800–6500 BC) and found in Israel. Dated about 3000 BC, numerous remains appeared rather suddenly in the Mediterranean region and central Europe. Until the introduction of *Phaseolus* beans from the New World in post-Columbian times, it was a common food for many Mediterranean and Near East civilizations, including the ancient Egyptians, Greeks, and Romans. It spread along the Nile valley to Ethiopia, also to northern India and China. The plant's wild progenitor is not known.

Today it is cultivated in over 50 countries, with China producing about 65 per cent of the world's crop. It is an erect, hardy annual, easily recognizable by a four-ribbed stem, single or sparsely branched. The compound leaves are composed of a few large leaflets and bear large stipules at their base. The white, black-blotched flowers are borne in axillary clusters. Faba beans can be divided into 'longpods' with up to eight seeds, and 'Windsors' with shorter pods containing up to four seeds. The seeds are very variable in shape, colour (white, green, buff, brown, purple, or black), and size (in length, 6–20 mm). Seed size has sometimes been used to form varieties within the species.

The immature green seeds can be cooked as a vegetable, or canned or frozen. The dry mature seeds can be utilized as a human food or animal feed. Seed flour has been incorporated into bread. Whole immature pods have been utilized as food. In Egypt faba beans are an important part of the diet of many people in a food known as *foul*. The dried seed contains 25 per cent protein, 1.5 per cent fat, 49 per cent carbohydrate; the young bean has about 80 per cent water and, of course, much less of the other constituents, but it does contain carotenes and vitamin C.

Faba beans in the Mediterranean are responsible for a form of haemolytic anaemia known as 'favism' (see p. 208). It occurs in individuals with a genetic deficiency of an enzyme. The Greek philosopher, Pythagoras, was said to suffer from the disease.

JACK BEAN OR HORSE-BEAN (2) *Canavalia ensiformis*. This legume has its origin in Central America and the West Indies—archaeological material, dated 3000 BC, has been found in Mexico. It is now widely distributed throughout the tropics. The plant is utilized for fodder and as a green manure or cover crop; its young pods, immature and mature seeds can be eaten, although, because of toxic constituents, the mature seeds must be well boiled in water first. The mature seed contains about 24 per cent protein, 1 per cent fat, and 55 per cent carbohydrate. It is the source of commercial preparations of the enzyme 'urease'. The plant is a bushy annual with pods 10–35 cm in length, and containing 3–18 white seeds.

Canavalia gladiata is the sword-bean. It is used for essentially the same purposes as *C. ensiformis* but, in contrast, its origin lies in the Old World and it is a climbing perennial with pods 15–40 cm in length, containing 5–10 dark-red seeds.

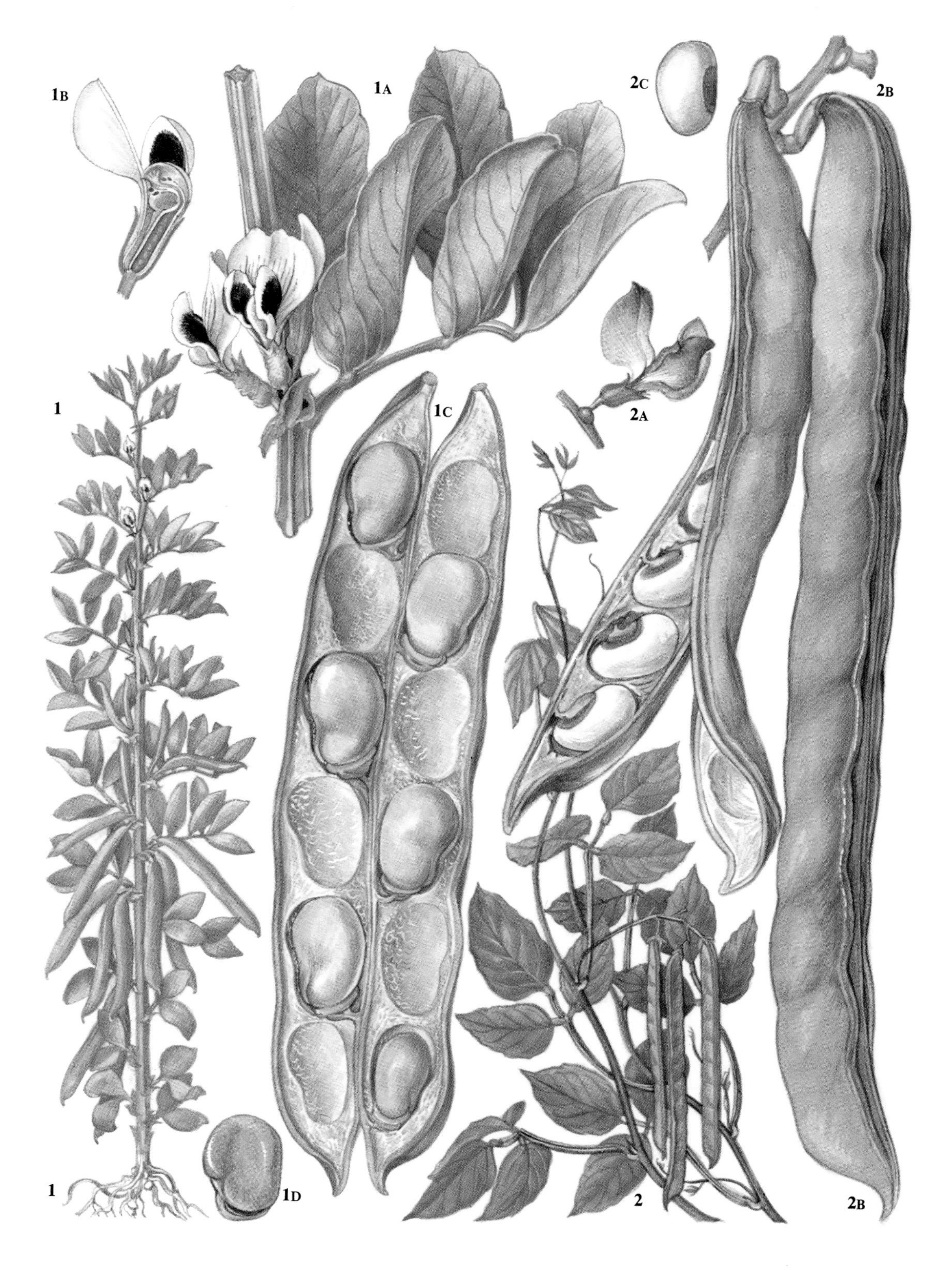

PLANTS × 1/8 PODS, FLOWERS, AND SEEDS × 2/3 SECTION × 1

1 **BROAD BEAN** plant 1A Flowers 1B Flower section 1C Opened pod 1D Seed
2 **JACK BEAN** plant 2A Flower 2B Ripe pod 2C Seed

Peas and lentils

PEA (1) *Pisum sativum.* This is cultivated in many temperate countries and as a cool-season crop in the subtropics, also at higher altitudes in the tropics. The plant was domesticated about 8000 BC in south-western Asia and the eastern Mediterranean region, together with wheat, barley, and some other legumes. Its cultivation spread eastwards to India and reached China about AD 1000. Pea was probably grown in England by the Romans. Dry peas were utilized as food in Europe from very early days, although green peas were not used until the sixteenth century. The plant reached Africa before the advent of Europeans. It is today the world's second most important pulse, Russia and China together produce almost 80 per cent of the world's production of dry peas; the United States and the United Kingdom are the largest producers of green peas (only one-tenth of dry-pea production).

The dry or mature seeds may be cooked whole, split, or ground into flour and used in soups, pease pudding, convenience foods, or rehydrated and canned ('processed' peas). The seed coats (hulls) have been utilized as a fibre additive in bread or health foods and the protein added to increase both the amount and quality of protein in such foods. Dry seeds have been incorporated into animal feed. They contain about 23 per cent protein, 1 per cent fat, and 59 per cent carbohydrate. Green or immature peas are cooked as a vegetable, much of the production is now canned or frozen. This must be done within a few hours of picking, before which the seed quality is assessed in terms of sugar content, texture, and colour. Green seeds contain much more water than mature seeds but less protein, fat, and carbohydrate. Some pea cultivars ('mangetout', sugar-peas, snow-peas) (2) have pods lacking the stiff papery inner parchment and are consumed whole. Peas are particularly free of toxic constituents. In some parts of the world, the young green tops of the plant are eaten, also the plants are suitable as forage, hay, silage, and green manure.

Pisum sativum is a glaucous green, climbing annual (30 cm to 3.5 m in height) with large, leaf-like stipules. Its leaves consist of one to three pairs of leaflets. The tendrils, by which the plant climbs, are thread-like modified leaflets towards the tip of the leaf. In recent years, 'leafless' cultivars have been developed where the leaflets, and possibly the stipules, have been converted into tendrils. The flowers are white, purple, or pink. The seeds are very variable as regards surface features (round or wrinkled), and colour (green, brownish, white, or blue). In the United Kingdom, well-known dry peas are green 'marrowfats' and 'split yellow' (white seed coat). *Petit pois* (3) are small and very acceptable peas. Wrinkled peas contain more sugar than the smooth types.

ASPARAGUS-PEA (4) *Lotus tetragonolobus.* A native of southern Europe which is occasionally grown for its young edible pods which can be steamed and served with butter. It is a hairy annual with trailing stems 15–40 cm long. The greyish-green leaves have three broadly ovate leaflets. The flowers, 3.5—4 cm long, are of a beautiful brownish red colour. Its mature pods are 5–8 cm long with four prominent longitudinal ribs and contain smooth, brown seeds. This plant should not be confused with *Psophocarpus tetragonolobus* (see p. 38) because of the winged pods.

LENTIL (5) *Lens culinaris* syn. *L. esculenta.* This is one of the oldest crops cultivated by humans. It was domesticated about 8000 BC together with pea and chick-pea in the Fertile Crescent. The crop spread northwards into Europe, eastwards to India and China, and into Egypt. In the Old Testament, the red pottage for which Esau sold his birthright was made of lentils. Today, the Indian subcontinent is the largest producer, but lentils are cultivated in most subtropical and warm temperate countries, including Ethiopia, Syria, Turkey, and Spain.

Lentil seeds, entire or split, are used in soups and 'dhal'. In some areas they are fried, seasoned, and consumed as snack food. Flour made from the pulse can be mixed with cereals in cakes, invalid, and infant food. The seed has few antinutritional factors. It contains about 25 per cent protein, 1 per cent fat, and 56 per cent carbohydrate. In India the young pods are eaten as a vegetable. The seeds are a source of commercial starch for the textile and printing industries. The residues (straw, pods, and leaves) left after seed threshing constitute a valuable livestock feed—in the Middle East sometimes fetching a better price than the seed.

It is a much-branched annual, 25–40 cm tall, with slender, angular stems. Its leaves are pinnate, with four to seven pairs of more-or-less oval leaflets, about 1.3 cm long. The flowers are pale blue, white, or pink. The pods, rarely more than 1.3 cm long, are flattened and contain one or two biconvex or lens-shaped seeds, which vary in colour from grey to light red speckled with black. Lentils on sale as pulses can be green, yellow, orange, red, or brown.

TWO-THIRDS LIFE SIZE   PLANTS × ⅛

1 **GARDEN PEA**   1A Flower   1B Plant
2 **'DWARF SUGAR PEA'**   2A **'MANGETOUT'** flower   3 **'PETIT POIS'**   3A Flower
4 **ASPARAGUS PEA**   4A Flower   5 **LENTIL**   5A Flowers   5B Plant

Some other pea-like plants

COWPEA (1) *Vigna unguiculata.* This annual legume was probably domesticated in West Africa about 3000 BC. It spread elsewhere in Africa and reached the Indian subcontinent some time after 1500 BC. Also, the plant was taken to South-East Asia. It was carried to the New World by the Spanish and Portuguese in the seventeenth century. Cowpea shows much variation in form—there are bushy, erect types and prostrate, spreading, twining, climbing forms (0.3–4 m in height or length). The flowers are white, pale mauve, pink, or dark purple, giving rise to pods which vary in colour (tan or pink through red and purple to almost black) and size (12–30 cm in length). Longer pods are found in yard-long beans, where pods may attain a length of 1 m. Seeds can be buff, brown, red, black, or white in colour. White seeds have pigment confined to a narrow 'eye'—these are often described as black-eyed beans, which have become identified with the 'soul food' of the American Deep South.

Mature seeds, young pods, and leaves are consumed. The seed contains about 22 per cent protein, 2 per cent fat, and 60 per cent carbohydrate; the young pod has 3 per cent protein, 0.2 per cent fat, and 8 per cent carbohydrate. The plant is a forage or cover crop. Cowpeas are sometimes divided into three groups known as (1) common, (2) catjang, and (3) yard-long or asparagus-peas. The greatest part of world seed production takes place in West Africa although, in recent times, Brazil has produced a considerable quantity of the pulse. Cowpeas are grown in Texas, Georgia, and California, where the immature seeds may be canned or frozen.

A related species, *Vigna vexillata,* is occasionally cultivated in Ethiopia and the Sudan for its starchy root.

LABLAB, HYACINTH-BEAN, OR BONAVIST BEAN (2) *Lablab niger* syn. *Dolichos lablab.* This legume is probably of Asian origin and has been cultivated in India since earliest times. It is also widely grown in South-East Asia, Egypt, and the Sudan. Young pods and young and mature seeds are utilized as food. The pulse contains about 25 per cent protein, 0.8 per cent fat, and 60 per cent carbohydrate; the young pod has 5 per cent protein, 0.1 per cent fat, and 10 per cent carbohydrate.

Although the plant is a perennial, it is often grown as an annual. It is normally a twining plant (1.5–6 m tall) but bushy forms also occur. The flowers are white or purple, giving rise to pods (5–15 cm in length) containing three to six seeds (white, cream, buff, reddish, brown, or black).

Dolichos biflorus is the 'horse-gram'. It is the 'poor man's' pulse crop in southern India. The seeds are a feed for cattle and horses. In Burma, the dry seeds are boiled in water and fermented to give a product similar to soya sauce.

PLANTS × 1/8 DETAILS × 1

1 **COWPEA** plant 1A Flowers 1B Ripe pod 1C Seeds
2 **LABLAB** plant 2A Flower 2B Pods 2C Seeds

More legumes

TAMARIND (1) *Tamarindus indica*. This is a semi-evergreen tree, up to 20 m in height, with leaves consisting of 10–20 leaflets. The flowers are pale yellow streaked with red, giving rise to usually curved pods, 5–10 cm in length, with 1–10 seeds embedded in a brown, sticky pulp. Tamarind grows wild in the drier areas of tropical Africa but was introduced into India early on. The seeds (63 per cent starch, 16 per cent protein, 5.5 per cent fat) can be eaten as a pulse, but the pod pulp, constituting about 40 per cent of the pod, is better known. The pulp (containing tartaric acid and sugars) has a sweet–sour flavour and is included in sweetmeats, curries, and chutneys. It is an article of international commerce.

GRASS-PEA OR CHICKLING VETCH (2) *Lathyrus sativus*. This is a minor pulse crop cultivated in traditional agriculture in the Mediterranean basin, south-western Asia, and the Indian subcontinent. Archaeological specimens, dated 6000–5000 BC, have been found in the Near East. India is now the main producer of the crop, where it can grow in dry places and poor soils. In India it is the cheapest pulse available. The seeds may be boiled in water and eaten or split to make 'dhal'. Seed flour can be made into chapatis, paste balls, and curries. In India, the pulse is consumed by the very poor and in times of famine. Seeds eaten over a prolonged period of time can cause 'lathyrism' (see p. 208)—a paralysis of the lower limbs. The main causative agent is a 'non-protein amino acid' (β-oxalyl-diamino-propionic acid). Seeds contain about 25 per cent protein, 1 per cent fat, and 61 per cent carbohydrate.

Grass-pea is a straggling or climbing plant not unlike the ordinary pea (*Pisum sativum*) but has much narrower and more elongated leaflets. The flowers are blue (sometimes violet or white) and giving rise to small, flat pods (2.5–4 cm in length) containing three to five white, brown, or mottled seeds with flattened sides.

LUPINS (3) *Lupinus* spp. Several lupin species have been grown in Mediterranean lands for over 3000 years as pulse and forage plants, also as green manure. They are: white lupin (*L. albus* syn. *L. termis*), yellow lupin (*L. luteus*), and narrow-leaved lupin (*L. angustifolius*). 'Tarwi' (*L. mutabilis*) was used by the Incas in South America and it is still utilized as human food in the Andes. From the Mediterranean area, lupin species were taken to New Zealand, South Africa, Australia, and elsewhere in the world. On a worldwide basis, lupins are minor pulse crops.

The protein concentration in the seed varies according to the species, but it could be as much as 44 per cent; oil concentration could reach 17 per cent. Lupin seeds have been used as a coffee substitute and its seed flour has been suggested as an alternative to soya, although there is relatively little utilization of the seed as a human food except in subsistence agriculture in various parts of the world. One difficulty is the presence of toxic alkaloids (see p. 207). Forms of the plant with relatively high concentrations of alkaloids are described as 'bitter', those with much lower concentrations are said to be 'sweet'. The alkaloid concentration of seeds may be reduced by washing in water.

White lupin grows to a height of 1.5 m, with palmate leaves of seven to nine leaflets. Its flowers are white and the pods (up to 13 cm in length) contain large, off-white seeds.

CLUSTER BEAN OR GUAR (4) *Cyamopsis tetragonolobus*. This is indigenous to India. Its immature pods are eaten and the plant is used as fodder or green manure. The seed is rich in mucilaginous galactomannans which are extracted commercially to give 'guar gum', of value in food as a thickener and stabilizer. Guar is cultivated for its gum in India and the south-western United States. There is some evidence that guar gum included in bread might reduce blood cholesterol and sugar levels in the consumer.

It is an annual bushy plant (1–3 m in height) with small purplish, pink, or white flowers in dense clusters. The pods are 4–10 cm long with 5–12 white, grey, or black seeds. The seed contains 30 per cent protein, 46 per cent carbohydrate, and 2 per cent fat.

PLANTS, FLOWERS, PODS, SEEDS × ½

1 **TAMARIND** foliage with pod 1A Flower 1B Pod 1C Seeds
2 **GRASS-PEA** plant with flowers 2A Pods 2B Seeds
3 **WHITE LUPIN** foliage with flowers 3A Pods 3B Seeds
4 **GUAR** (see above and opposite page) foliage and flowers 4A Developing pods 4B Mature pods with seeds
4C Young pods and seeds

Apples (1): Crab apples and apple origin

(For general information on fruits and vegetables, see p. xviii.)

*The cultivated apple (*Malus x domestica*) is probably the most widely distributed of tree fruit crops in the world, being cultivated in many temperate countries. Its fruit, botanically speaking, is a pome. The cultivated apple was selected from the wild apple of which some 25–30 species are distributed over the temperate belt of Europe, Asia, and North America. Crab apple is a slightly ambiguous term and might refer to (1)* Malus *species generally, (2) wild* M. sylvestris, *(3) ornamental* Malus, *or (4) seedlings arising in the wild from discarded seeds. Wild apples have small fruit and reproduction is entirely by seed, giving rise to seedlings which show a great deal of variation. Apples for food were collected from the wild long before domestication; numerous archaeological remains have been found in European neolithic sites. It would appear that the cultivation of olive, grape vine, and date palm developed to a great extent between 3000 and 2000 BC, but there has been no sign of parallel development in apple.* Malus sieversii, *of central Asia, is now often regarded as the main ancestor of the domestic apple, with input from other species, notably* M. orientalis. Malus sylvestris *of Europe is thought to be a minor contributor. In Europe, apple cultivation seems to have commenced in classical times and is possibly associated with the technique of grafting, known to the ancient Greeks, which allows the propagation of desirable clones—seed propagation gives very variable results.*

Cultivated apples are utilized for cooking (culinary) and dessert purposes, also cider and juice manufacture. Crab apples may be used to make jelly and the plants can be very attractive ornamental trees, many with purple leaves.

MALUS SYLVESTRIS (1) A wild apple of Europe.

'TRANSCENDENT' (2)

'JOHN DOWNIE' (3)

MALUS BACCATA (4) This species, found through eastern Asia to northern China, is very hardy and resistant to apple diseases, and has interbred freely with other crab apples. The brilliant-red, cherry-like fruit, with no trace of a calyx when ripe, makes excellent jelly. The round-headed tree is usually 12–16 m high. *Malus manchurica* is a related form with larger fruit, and is the one most cultivated. Several other eastern Asiatic species, including *M. hupehensis*, are also popular in gardens for their beautiful flowers.

'GOLDEN HORNET' (5) There are many other *Malus* species and ornamentals. In the United Kingdom, there is a collection at the National Fruit Collections, Brogdale, Kent and the national collection is at Hyde Hall in Essex.

Classification of apples

The largest collection of domestic apples (some 2000) is at Brogdale and there are many more throughout the world. The main characters used for cultivar identification are those of the fruit and include season, size, shape, colour, nature of 'eye' (remains of calyx and stamens), appearance in vertical and horizontal cross-sections, flavour, and use. Certain features of the tree, such as habit and blossom, are extra clues but not the main ones.

The scheme of classification used at Brogdale is by John Bultitude, a development from the Bunyard scheme (*The Oxford book of food plants* (1st edn)). There are eight groups based on fruit characters (for details see Morgan and Richard 1993).

TWO-THIRDS LIFE SIZE SECTIONS × 1

1 **MALUS SYLVESTRIS** 1A Blossom 1B Sections of flower and immature fruit
2 **'TRANSCENDENT'** 3 **'JOHN DOWNIE'**
4 **MALUS BACCATA** 5 **'GOLDEN HORNET'**

Apples (2): Historical cultivars

Apple trees have been valued for a very considerable time as a source of fruit. One reason is the long span of availability of fruits of different cultivars. The Romans possessed named late cultivars. In more modern times, apples have been stored under refrigerated and controlled atmospheric conditions.

UP TO SIXTEENTH CENTURY (THE MONTHS GIVEN REFER TO THE FRUITING SEASON) (2) 'Court Pendu Plat', described about 1613 but known for 400 years, and possibly since Roman times, is a richly flavoured little dessert apple, unusual in its very late flowering season (December–April). 'Court Pendu Gris', mentioned in 1420.

'London Pippin' or 'Five Crown', possibly a continental sort, has been documented in Somerset since 1580. It is a very late green dessert or cooking apple, now widely grown in Australia (February–March).

'Royal Russet' ('Leathercoat Russet') was known before 1597. This large, late, cooking russet keeps well and has a good flavour (until March).

'Golden Pippin' was the most highly prized dessert apple in the sixteenth and seventeenth centuries. It has a golden yellow skin with an orange flush and a remarkably crisp and juicy flesh. Many different cultivars were raised from it (November–March).

'Nonpareil', known before 1600, is a small, conical apple with greenish rather soft but crisp flesh. It is mentioned in 1870 as 'so well known as to need no description' (until March).

'White Joaneting'. This very old sort, known before 1600, is still the earliest of all apples to ripen. It is a greasy, yellow-skinned fruit, sometimes red-flushed, of pleasant flavour (July).

Famous sixteenth–early seventeenth century cultivars omitted are 'Api', 'Catshead', and 'Genet Moyle'.

SEVENTEENTH CENTURY 'Autumn Pearmain', mentioned in 1629, is a typical example of the hardy, disease-resistant apples of the time. It has a golden yellow fruit with some russet tinge (September–October).

'Golden Reinette'. Known before 1650, these continental dessert apples of good flavour, are usually russetted and red-yellow. They are widely grown on the continent (November–March).

'Boston Russet'. Raised before 1650 in the United States, this delicious apple probably comes from pips taken to America by emigrants (January–March).

'Devonshire Quarrenden'. Known before 1650, it was possibly originally French. It has a deep crimson fruit with white juicy flesh (August–September).

'Flower of Kent'. Popular for several hundred years. Like many others now almost forgotten, this large green fruit is said to be the apple that Sir Isaac Newton observed falling and thus formulated the theory of gravity (November–December).

EIGHTEENTH CENTURY 'Ashmead's Kernel'. This apple was raised in 1720 by a Dr Ashmead in Gloucester (December–March).

'Newtown Pippin'. Known before 1760, this crisp juicy dessert apple is yellow or greenish yellow. It is widely grown in the United States and Canada, but is not always successfully cultivated in England (until March).

'Hawthornden'. Raised before 1790 at Hawthornden in Scotland, this is an excellent cooking apple with an almost white skin (October–December).

'Wagener'. Raised in New York State before 1800, this is an example of many good apples raised in the United States from European stock. The rather hard fruit is golden with a carmine flesh. It is remarkably prolific and disease free (November–February).

'Cornish Gilliflower'. Known before 1800, this delicious late apple needs a warm climate (December–May).

Other famous cultivars raised in the eighteenth century were 'Ribston Pippin', 'Blenheim Orange', and 'Dumelow's Seedling'.

NINETEENTH CENTURY 'Blenheim'. Raised at Woodstock before 1818, this was for 100 years the most prized winter apple. It is large, yellow-skinned with a dull red flesh, and rather acid (November–January).

'Pitmaston Pine Apple' (3) Raised at Pitmaston before 1820, this remarkable little fruit is crisp and juicy, but more aromatic than its supposed parent 'Golden Pippin' (December–January).

'Coe's Golden Drop' (4) Known before 1820, this apple is greenish-yellow, with a very long stalk.

'Rosemary Russet'. Recorded by 1830, this perfect russet probably existed long before (until February).

'Tom Putt'. Known before 1840, this large crisp acid fruit (primarily a cider apple), with a vivid red-striped skin, is disease free (August–September).

'White Transparent'. Introduced into the United Kingdom before 1850, this apple is characteristic of Russian and Scandinavian types and still widely grown in northern Europe. The smooth, shiny, white-yellow fruit is digestible, crisp, and juicy ('Yellow Transparent' of the United States) (August).

'American Mother'. Sent before 1850 to England from America, this delicious apple is striped red all over, conical, with a richly flavoured, juicy flesh (October–November).

'May Queen' (5) Known before 1890 (until June).

'Bismark'. About 1890, this vivid crimson cooking apple was first sent to England from Tasmania (November–February).

'Christmas Pearmain' (6) Introduced into the United Kingdom about 1896, this apple is rather dry and hard, but keeps well (November–January).

In the nineteenth century a huge number of new cultivars were raised and, in addition to the above, these included, 'Cox's Orange Pippin', 'Sturmer Pippin', 'Worcester Pearmain', 'Gladstone', 'Bramley's Seedling', 'Newton Wonder', 'Peasgood Nonesuch', 'McIntosh' (Canada), 'Golden Delicious', 'Delicious', 'Rome Beauty' (United States of America), and 'Granny Smith' (Australia).

TWO-THIRDS LIFE SIZE

1 **APPLE BLOSSOM ('MAY QUEEN')**
APPLE CULTIVARS
2 **'COURT PENDU PLAT'** 3 **'PITMASTON PINE APPLE'**
4 **'COE'S GOLDEN DROP'** 5 **'MAY QUEEN'** 6 **'CHRISTMAS PEARMAIN'**

Apples (3): Cultivars through the season

In their wild state apples ripen in late autumn before the onset of winter. Many cultivars originating in northern countries, where the summer is short, ripen within 10 weeks of flowering, but there are many keeping, hardy cultivars. In other countries there is a 6 month period between blossom time in March and leaf fall in November. By selection among native and foreign kinds we now have apples which ripen from July until April and May the following year.

JULY–AUGUST The first are often those of Russian or Scandinavian origin and probably all descend from a northern cultivar. They are shiny-skinned fruits with soft, juicy flesh, acid but often delicious in their season, as much used for dessert as for cooking. They are usually white or palest yellow with a brilliant crimson or scarlet flush or few stripes:

'Akero'
'Biela Borodowka'
'Duchess of Oldenburg'
'Red Astrachan'
'Scarlet Pimpernel'
'White Transparent'

Later in August there are other dessert varieties. These are mostly brightly coloured, shiny- or greasy-skinned apples which do not keep for long:

'Beauty of Bath'
'Feltham'
'George Cave'
'Iris Peach'
'Lady Sudeley'
'Laxton's Advance'
'Miller's Seedling'
'Red Melba'

AUGUST–OCTOBER The Codlins ripen in August and September. They are soft, acid, white-fleshed cooking apples with pale whitish or yellow greasy skins, usually oblong or oval in shape with a distinct nose:

'Emneth Early' (1)
'Keswick Codlin'
'Lord Grosvenor'
'Lord Suffield'

By September and early October there is a wider range of dessert apples. Among these there are good sharp-flavoured striped apples:

'James Grieve'
'Michaelmas Red'
'Tydeman's Early Worcester'
'Worcester Pearmain'

Ready at this time are many derivatives of 'Cox's Orange':

'Ellison's Orange' (2)
'Laxton's Fortune'
'Merton Worcester'
'Sunset'

In September and October the cooking apples that derive from the 'Codlins' ripen. These possess mainly a pale green or golden greasy skin, with a flush:

'Golden Spire' (3)
'Arthur Turner'
'Charles Eyre'
'Grenadier'
'Revd W. Wilks'
'Stirling Castle'

LATE OCTOBER–NOVEMBER Now many of the Reinettes, Pearmains, and Pippins ripen—mostly smaller, drier, more richly flavoured dessert apples—and some Russets, all harder, less perishable fruits than the early sorts:

'Autumn Pearmain'
'Egrement Russet'
'Golden Pippin'
'Herring's Pippin'
'King of the Pippins'
'Laxton's Reward'
'Mother'
'St. Edmund's Pippin'

DECEMBER–JANUARY There are innumerable sorts of dry-skinned, russetted apples with hard, juicy, aromatic flesh:

'Braddick's Nonpariel' (4)
'Adams' Pearmain'
'Blenheim Orange'
'Claygate Pearmain'
'Cox's Orange'
'Ribston Pippin'
'Tydeman's Late Orange'
'Winston'

Less usual at this season are the shiny, smooth-skinned fruits:

'Coe's Golden Drop'
'Golden Noble'
'Lambart's Calville'
'Spartan'

The late cooking apples are mainly shiny, green apples with striped, hard, acid fruits cooking to a froth, keeping until the spring if in good condition and by then quite eatable raw:

'Annie Elizabeth'
'Bramley'
'Crawley Beauty'
'Howgate Wonder'
'Lane's Prince Albert'
'Mère de Ménage'
'Newton Wonder'

JANUARY–APRIL A few later varieties stay green until March and April. These are all green or flushed crimson, shiny when ripe:

'Edward VII' (5)
'Monarch' (6)
'French Crab'
'Lemon Pippin' (bright yellow)
'Gooseberry'
'Hormead Pearmain'
'London Pippin'

A few late cooking apples are russetted:

'Diamond Jubilee'
'Reinette de Canada'
'Royal Russet'
'Woolbrook Russet'

The very late-ripening dessert apples belong mainly to the Reinettes group, small, green or green-yellow apples with hard, dry, but well-flavoured flesh, sometimes flushed with crimson red or russetted:

'Allen's Everlasting'
'D'Arcy Spice'
'Heusgen's Golden Reinette'
'King's Acre Pippin'
'Laxton's Pearmain'
'Laxton's Royalty'
'Orleans Reinette'
'Rosemary Russet'
'Sturmer Pippin'
'Wagener'

A few are green with golden or orange tint:

'Easter Orange'
'Grange's Pearmain'
'Sanspareil'
'Granny Smith'
'Ontario'

'May Queen' is a vivid crimson.

These very late apples need a long ripening season and may, in a cold year, lack flavour and prove difficult to store. They are essentially fruits for a good soil and a warm climate. 'Sturmer Pippin', for instance, was much more successful in New Zealand and Tasmania than in England.

TWO-THIRDS LIFE SIZE

APPLE CULTIVARS
1 '**EMNETH EARLY**' 2 '**ELLISON'S ORANGE**'
3 '**GOLDEN SPIRE**' 4 '**BRADDICK'S NONPAREIL**' 5 '**EDWARD VII**' 6 '**MONARCH**'

Apples (4): Cultivars of flavour and quality

Apple flavour is a combination of acidity, sweetness, bitterness, and scent. Malic constitutes 90 per cent of the acid present, the rest is citric. According to cultivar, malic acid ranges from 0.4 to 1 per cent of the fruit. As regards sweetness, the sugars (9–12 per cent of the fruit) present and at harvest are sucrose and fructose, with somewhat less glucose. The bitter or astringent compounds are tannins (average 0.2 per cent of the fruit). Apple scent or aroma is a blend of some 250 traces of various chemical substances, such as volatile esters, alcohols, and aldehydes. Potassium is the main mineral constituent. The amount of vitamin C varies between 3 and 14 mg/100 g, but declines during storage.

The apple flesh may be soft, crisp, or hard. Its proportion of acidity to sweetness is a most noticeable feature. Cooking (culinary) apples with a high acid content cook to a froth and, if sweetened, have a good flavour. Many dessert apples have as much acid but also more sugar, and this balance makes for a richly flavoured fruit which also cooks well. Cooking and dessert apples are often described separately in the United Kingdom but not so in many other countries. Bitterness is chiefly prized in cider fruit but, in a small degree, imparts a nut-like flavour to cooking and dessert apples.

'COX'S ORANGE PIPPIN' (1) This is a favourite British dessert apple, probably from a seedling raised about 1850 from the still highly prized 'Ribston'. It ripens well in Britain and remains firm when cooked, although it is very prone to disease. Its acidity is nicely balanced by sweetness, the skin has a strong, delicious scent, and the flesh has a crisp texture. Such a perfection has led many raisers to use it as a parent for new and more vigorous sorts—for it is difficult to manage. The most modern offspring is 'Fiesta' which is being planted commercially in the United Kingdom, Europe, and New Zealand.

'EGREMONT RUSSET' (2) Of unknown origin, this is the best of the Russets, a group with a sweet, strong taste, crisp and firm, with little juice though never tough or hard, less acid than 'Cox's Orange', and usually without much scent in the skin. It crops well as a small tree. 'St. Edmund's Pippin', 'Sam Young', 'Ross Nonpareil', and 'Boston Russet' all have the same pleasant qualities, and their roughened skin seems resistant to scab.

'GOLDEN DELICIOUS' (3) This is characteristic of an entirely different group, the shiny, thin-skinned, warm-climate apples which possess a refreshingly light flavour, a crisp flesh, and plentiful juice. Though the climate is often not good enough for cultivation in the United Kingdom, it has become a leading apple in most warmer countries. It arose in West Virginia.

'ORLEANS REINETTE' (4) Like 'Cox' and 'Ribston Pippin' this apple can be susceptible to disease in cold areas and on different soils. The best flavoured of the rough-skinned Reinette group, it is sweeter and less juicy than 'Cox's Orange'. It has been known since before 1776.

'CALVILLE BLANCHE D'HIVER' (5) This very old French cultivar is often mentioned as being the most delicious of all apples. It can only be grown to perfection in warm soils or in a cool greenhouse. Its large, golden fruit possesses a light, sweet scent of surprising strength and a delicious, sweet, juicy flesh.

'GRAVENSTEIN' This is widely grown in Russia and northern Europe. It was formerly the major apple of Canada and is still grown in California. It is a fairly large fruit, its strong sweetness balanced by a marked acidity and a penetrating scent.

'NEWTOWN PIPPIN' This variety was raised in the United States of America some 250 years ago and was being shipped to London as a delicacy as early as the mid-nineteenth century. The apples are crisp and firm with a strong flavour, somewhat reminiscent of the pineapple. It is no longer important commercially in the United States, only being grown in Oregon.

'AMERICAN MOTHER' This variety and a few other sorts have a peculiar, and not at all unpleasing, scent or flavour of anise.

'JAMES GRIEVE' This is one of the most acid dessert varieties but it is also accepted as one of the most delicious and refreshing. Its scent is almost as acid as a lemon. Others of this type are 'Upton Pyne' and 'Herring's Pippin'.

Cultivation

In the garden, apples are most successful if grown as dwarf bushes, pyramids, or cordons. For this purpose they are budded on a dwarfing rootstock. Trees are available on a number of rootstocks, the most common is 'Malling 9', then 'M26'. Apples thrive in a wide range of soils, but not in poor acid soils or heavy clay.

TWO-THIRDS LIFE SIZE

APPLE CULTIVARS
1 **'COX'S ORANGE PIPPIN'** 2 **'EGREMONT RUSSET'**
3 **'GOLDEN DELICIOUS'** 4 **'ORLEANS REINETTE'** 5 **'CALVILLE BLANCHE D'HIVER'**

Apples (5): Modern cultivars

For commercial purposes, apples are grown in every continent under temperate conditions. In a number of countries active breeding programmes continually produce new cultivars. An interesting modern development is the appearance of columnar varieties with single stems bearing heavy crops of apples on short spurs.

The cultivars illustrated on page 61 were modern when *The Oxford book of food plants* was first published in 1969. Their current popularities are noted below.

'LAXTON'S FORTUNE' (1) Also 'Laxton Superb', grown in gardens.

'TYDEMAN'S EARLY WORCESTER' (2) Still seen in gardens.

'TYDEMAN'S LATE ORANGE' (3) Not often planted now.

'MERTON CHARM' (4) Favoured by gardeners, also 'Merton Worcester', which was planted commercially.

'SPARTAN' (5) (CANADA), **'IDA RED'** (UNITED STATES), **'MUTSU' OR 'CRISPIN'** (JAPAN) Seen in gardens and to a small extent commercially.

'DISCOVERY' is now the main early apple.

Of the thousands of apple cultivars available, only about 30 account for most of the world production of dessert and culinary types. Some examples are: United Kingdom, 'Discovery', 'Cox's Orange Pippin', and 'Bramley's Seedling'; Europe, the Americas, and the southern hemisphere, 'Red Delicious' (6), 'McIntosh', 'Empire', 'Jonagold', 'Elstar', 'Cox's Orange Pippin', 'Golden Delicious', 'Braeburn', 'Fuji', 'Granny Smith', and 'Gala' (7).

6 **'RED DELICIOUS'** × ½
7 **'GALA'** × ½

TWO-THIRDS LIFE SIZE

APPLE CULTIVARS
1 'LAXTON'S FORTUNE' 2 'TYDEMAN'S EARLY WORCESTER'
3 'TYDEMAN'S LATE ORANGE' 4 'MERTON CHARM' 5 'SPARTAN'

Pears (1)

*Like apple, the common or European pear (*Pyrus communis*) (1) is a most important tree fruit crop of the temperate regions of both hemispheres. Also, like apple, its fruit is a pome. There are about 20 pear species. It is normally accepted that the pear, as apple, originated in the region which includes Asia Minor, the Caucasus, central Asia, and western China. No doubt a number of species contributed to the cultivated forms. Pears have been cultivated in Europe and western Asia since the earliest historical times—Homer and Pliny recorded the names of many pears. In China, pears have been cultivated for over 4000 years. Considerable breeding of pears took place in France, Germany, Belgium, and England, starting in the seventeenth century.*

*Leading producers of the fruit include France, Germany, Italy, Japan, Spain, Turkey, the United States of America, and China. There are over 5000 recorded pear cultivars. The oriental pear (*P. pyrifolia*) is cultivated in China and Japan, also in other eastern countries, the United States, Australia, and New Zealand. Traditional oriental pears with gritty flesh are sometimes described as 'sand pears', the more modern high-quality types as 'nashi pears' (see 7, p. 64). In the nineteenth century, in the United States, hybrids were produced between* P. communis *and* P. pyrifolia, *some of these showed resistance to the notorious bacterial disease of pears—fireblight.*

'HAZEL' OR 'HESSLE' (2) This is characteristic of many old hardy pears and produces immense crops of small, sweet pears.

'FERTILITY' (3) Once grown in large quantities for the market, this has now lost its popularity. Raised about 1875.

'CONFERENCE' (4) A good dessert, canning, and bottling pear. It is widely grown in the United Kingdom and will succeed almost anywhere. Commercial growers appreciate the cultivar. It is partially self-fertile but also shows parthenogenesis (fruit will develop without fertilization).

'WILLIAMS' BON CHRÉTIEN', OR 'BARTLETT' (5) This is well known and of high quality for dessert, canning, and bottling. Raised in Aldermaston (United Kingdom) about 1770. On introduction to America it was named after Bartlett, the importer.

'GLOW RED WILLIAMS' (6) A mutant or bud sport of 'Williams' Bon Chrétien'. It shows considerable red colouring in the fruit skin, leaves, and shoots.

'DOYENNE D'ETÉ' One of the first cultivars to ripen but only lasts a few days. The tree is small and of weak growth. Raised about 1700 by Capuchin monks at Mons.

'JARGONELLE' (ENGLISH JARGONELLE) Known in France before 1600 but not to be confused with 'French Jargonelle'—a distinct cultivar. A very useful early dessert pear.

'CLAPP'S FAVOURITE', 'BEURRÉ D'AMANLIS', AND 'DR JULES GUYOT' These are three good prolific pears which ripen in early September.

'MARGUERITE MARILLAT' Raised in France. A very upright tree with fruit borne on small, upright branches.

Classification

Cultivated pears vary a great deal in their habit of growth (erect or pendulous), in size, in flowering time, and in the amount of fruit they bear. They may live 200 or even 300 years. Fruits may be classified according to maturity date (summer, autumn, or winter), or according to shape: round or flattened; bergamot or top-shaped; conical; pyriform (general pear shape) with a distinct waist; oval; calebasse (long pears).

TWO-THIRDS LIFE SIZE SECTIONS × 1

1 **PEAR BLOSSOM** 1A Sections of flower and immature fruit

PEAR CULTIVARS

2 **'HAZEL'** 3 **'FERTILITY'** 4 **'CONFERENCE'** 5 **'WILLIAMS' BON CHRÉTIEN'** 6 **'GLOW RED WILLIAMS'**

Pears (2)

Pears may be consumed as dessert fruit or used for stewing if the fruit is hard. The fruits can be canned, puréed for baby food, included in jams, and nectar (a pulpy liquid). Pear concentrates have been used as natural sweetening agents; pears may be dried and the alcoholic perry is prepared from the fruit. The criteria of a good dessert pear are juicy or buttery flesh, acid yet sweet, with a marked aroma. The grittiness of pear flesh is related to the presence of stone cells—in modern, high-quality types this grittiness is reduced. Nutritionally speaking, pears are a fairly good source of dietary fibre, substantial amounts of potassium (150 mg/100 g) and reasonable quantities of vitamin C (6 mg/100 g). Sugars (10 per cent of the fruit) consist, in descending order, of fructose, glucose, and sucrose. The main acid present is malic but there is also some citric acid. The chemical composition of the European pear is virtually the same as that of the nashi pear.

'BERGAMOTTE D'ESPÉREN' (1) Raised in Belgium around 1830. An excellent dessert pear, ready for use during December but keeps well until March.

'PACKHAM'S TRIUMPH' (2) A popular market cultivar.

'BEURRÉ HARDY' Raised in France about 1820 and a good dessert pear.

'DOYENNÉ DE COMICE' Introduced to England in 1858 and the best-flavoured and most widely grown dessert pear in the United Kingdom.

'DURONDEAU' (3) Raised in Belgium in 1811 and characteristic of the russet-coated pears.

'THOMPSON'S' Raised in Belgium about 1810 and one of the best garden dessert pears.

'JOSÉPHINE DE MALINES' and **'GLOU MORCEAU'** These are the two latest ripening pears in the United Kingdom.

'OLIVIER DE SERRES' (4) Late ripening in February to April. Does best in the long summers of southern France and Italy.

'PASSE CRASANNE' (5) An old French russet cultivar. Much grown in Italy.

'EASTER BEURRÉ' The last pear to ripen.

Cultivation

Wild pear rootstocks have been used to propagate cultivated pears, particularly in North America. In Europe, quince (*Cydonia vulgaris*) has been mainly used. Pears can be grown as bushes, pyramid trees, fan-trained plants, or cordon trees. Cross-pollination is normally necessary for fruit production. Pears to not keep as well as apples. Cultivars in commercial trade include, 'Packham', 'Beurré Bosc', 'Williams', 'Comice' (**6**), 'Conference', or 'Dr Jules Guyot'.

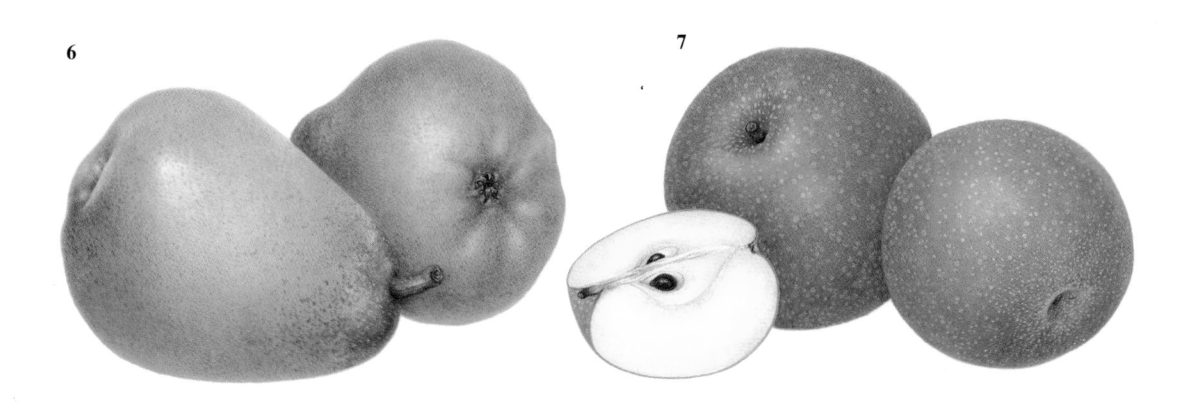

6 **COMICE** × ½ 7 **NASHI PEAR** × ½

TWO-THIRDS LIFE SIZE

PEAR CULTIVARS
1 '**BERGAMOTTE D'ESPÉREN**' 2 '**PACKHAM'S TRIUMPH**'
3 '**DURONDEAU**' 4 '**OLIVIER DE SERRES**' 5 '**PASSE CRASANNE**'

Cider apples and perry pears

The cultivars of these fruits are descended from the same wild stock as the dessert and culinary cultivars. Fermented apple juice produces cider (known in North America as 'hard cider', unfermented apple juice is 'soft cider'). Fermented pear juice gives perry. Generally speaking, these fruits have a higher tannin content than dessert and culinary types and this is responsible for their stringency or bitterness. The production of cider has taken place for over 2000 years. It was a common drink at the time of the Roman invasion of Britain in 55 BC *and was more popular than beer in the eleventh and twelfth centuries in Europe. Cider is now produced in temperate countries throughout the world where apples are cultivated. Traditionally cider apples are of four main types: bitter sweet (low acid, <0.45 per cent; high tannin, >0.2 per cent), bitter sharp (high acid, >0.45 per cent; high tannin, >0.2 per cent), sharp (high acid, >0.45 per cent; low tannin, <0.2 per cent), and sweet (low acid, <0.2 per cent; low tannin, <0.45 per cent). 'Apple brandy' (known in France as 'calvados' and North America as 'applejack') is distilled from cider. Perry pears may not have been introduced into Britain until the Norman Conquest. Pear brandy can be distilled from perry. After juice extraction, the apple residue ('pomace') can be utilized as a source of pectin and an animal feed or fertilizer. Perry pears are smaller than modern dessert cultivars. They are also more gritty (higher stone cell content) which allows an easier extraction of the juice.*

'SWEET COPPIN' APPLE (1)
This is a 'sweet' apple, also a vintage type. Many ciders are made from blends of juices from different cultivars. A vintage cider is made from one cultivar. 'Sweet Coppin' originated in Devon, probably in the early eighteenth century.

'TREMLITT'S BITTER' APPLE (2) A bitter sweet apple which arose in Devon, probably in the late nineteenth century.

'KINGSTON BLACK' APPLE (3) A bitter sharp and vintage apple, originating in Somerset probably in the late nineteenth century. It was popularized in the early 1900s and widely planted in the West Country of England, but declined to some extent due to its susceptibility to canker, scab, and poor crops.

'YELLOW HUFFCAP' PEAR (4) A very large tree producing from its fruit a medium- to high-acid, low-tannin perry.

'THORN' PEAR (5) This old cultivar, known in 1676, is unusually upright in growth.

'TAYNTON SQUASH' PEAR (6) An old cultivar.

'RED PEAR' (7) Another old cultivar, known in England since Tudor times.

Cultivation

Most cider apples and perry pears have no recorded history and little can be learned of their origins except through their names, which may have originated from farm or village names, or that of their raiser. Farm orchards used to be planted in grass with seedlings raised from local trees but the demand from scientifically controlled factories for fruit of uniformly good quality has, in the twentieth century, brought into being a new type of orchard with large numbers of trees carefully arranged to cross-pollinate and to produce a succession of well-balanced fruit.

TWO-THIRDS LIFE SIZE

CIDER APPLES 1 **'SWEET COPPIN'** 2 **'TREMLITT'S BITTER'** 3 **'KINGSTON BLACK'**
PERRY PEARS 4 **'YELLOW HUFFCAP'** 5 **'THORN'** 6 **'TAYNTON SQUASH'** 7 **'RED PEAR'**

Quince, medlar, rose, azarole

These trees of temperate climates are all members of the Rosaceae family which bear fruits of some local interest but of no great commercial importance.

QUINCE (1) *Cydonia vulgaris.* This plant grows wild in Caucasia, northern Iran, and the Kopet Dagh range. It is likely that the quince reached the Mediterranean region only in classical times—it was used by the Romans. The tree is small (5–7 m), sometimes thickly branched and bent as if deformed by wind and the weight of its foliage, but often upright with strong branches. It has rounded oval leaves which are very woolly underneath, and solitary pink and white flowers at the ends of short young shoots in May. Its fruit is similar to that of the pear and the apple, but has many ovules in each carpel or section—up to 20 instead of two. The raw fruit is hard and unpalatable but, when cooked, the flesh turns a brownish pink with a fine flavour. It contains about 6 per cent sugar, is a good source of potassium, and has a vitamin C content of 15 mg/100 g. The malic acid percentage is high (0.8 per cent). It is used to make jams, jellies, and paste, and as a flavouring to be added to cooked apples and pears. In France and Spain, quince jelly is made into a candy—*cotignac.* Similar products are *marmelo* (Spain) and *marmelada* (Portugal)—these might be the origin of the English term 'marmalade', although this is normally made from citrus products. Quince is a common pear rootstock (see p. 64). Some quince cultivars grown in the United Kingdom are 'Portugal', 'Vranja', and 'Maliformis'.

The familiar garden plants known as 'japonicas' are species of the genus *Chaenomeles* but can be distinguished from *Cydonia* by their toothed leaves and united, not free, styles. Their fruits are more acid than those of *Cydonia* but can be utilized for the same purposes. Species include *C. japonica* and *C. speciosa* (syn. *C. lagenaria*).

MEDLAR (2) *Mespilus germanica.* This tree grows wild in several countries of Asia Minor and may have migrated from that region to Europe. The fruit is remarkable in that the five seed vessels are visible in the eye of the fruit, for the fruit is set in the receptacle as in a gaping cup, around the rim of which stand the five conspicuous calyx lobes. It is quite rich in sugars (11 per cent), a good source of potassium, but quite low in vitamin C (2 mg/100 g). Usually the fruit must be allowed to become half rotten or 'bletted' to make it palatable; it becomes soft and brown. At one time these fruits were consumed with port wine at the end of a meal. They can be used to make jam. Cultivars include 'Nottingham' and 'Dutch'.

The medlar is a spreading tree, apt to be deformed by the wind. The wild tree has thorns but the cultivated kinds are thornless. Its flowers are borne at the end of short young shoots in late May or June.

DOG ROSE *Rosa canina* and *Rosa rugosa* (3–4) The fruit, or hip, is an urn-shaped receptacle with numerous achenes (containing the seeds) inside. The hips can be used to make jellies, preserves, and sauces. Rose-hip syrup is very rich in vitamin C (300 mg/100 g).

AZAROLE (5) *Crataegus azarolus.* This member of the hawthorn genus, a reputed native of Crete, is grown in a small way in southern Europe (Spain, France, and Italy) for its fruit (red or yellowish orange). The fruit may be eaten fresh, used in confectionery, jam, and jellies, or fermented to give an alcoholic drink.

SERVICE TREE *Sorbus domestica.* A native of Asia Minor, this has spread over Europe. According to cultivar, its fruit may be apple-shaped or pear-shaped but is only edible when bletted. The fruit has been used to make jam and an alcoholic drink.

ROWAN OR MOUNTAIN ASH *S. aucuparia.* This is a familiar tree in the United Kingdom and is found throughout Europe. Its scarlet fruit has been utilized in the same way as that of *S. domestica.*

TWO-THIRDS LIFE SIZE

1 **QUINCE** 1A Blossom 2 **MEDLAR** 2A Blossom
3 **DOG ROSE** 3A Section of flower 4 **ROSA RUGOSA** 5 **AZAROLE** 5A Flowers

Cherries

Cherries belong to the genus Prunus *which also includes plums (see p. 72), peaches, and apricots (see p. 78). These fruits have a thin skin, a middle fleshy region, and a central stone containing the seed. Such fruits are described as 'one–seeded drupes'.*

*Two species are cultivated—sour cherry (*Prunus cerasus*) and sweet cherry (*P. avium*). The sweet cherry is closely related to wild forms which are distributed over temperate Europe, northern Turkey, Caucasia, and Transcaucasia. Sour cherry probably evolved as a hybrid between* P. avium *and ground cherry (*P. fruticosa*)—a wild shrub found in central and eastern Europe. Cherry cultivation was first reported in classical times but, as no doubt with other plants, fruits were collected from the wild long before.*

The sugars (10 per cent) in cherry flesh consist of glucose and fructose in roughly equal amounts. Cherry is a very good source of potassium but not particularly of vitamin C (11 mg/100 g). The acid (malic) content varies between 0.5 and 2.0 per cent, the higher levels being found in sour cherry.

SOUR CHERRY (1) *Prunus cerasus.* This is cultivated in a major way in various regions, including the United States of America, Russia, Germany, and other East European countries. The fruit is suitable for cooking, and is canned, frozen, processed into pies, dried fruit products, juice, liqueurs (for example, Kirsch), and jam. Sour cherry types are morello (dark fruit) and amarelle (light- to medium-red fruit). The plant is more of a bush than a tree, is self-fertile, and the fruit can be harvested mechanically. Sour cherry is much more resistant to diseases than sweet cherry.

SWEET CHERRY, GEAN, OR MAZZARD (2–3) *Prunus avium.* Germany, Russia, the United States, Italy, Switzerland, France, and Spain, amongst others, produce considerable quantities of cultivated sweet cherries. They are utilized as a dessert fruit and in a number of the ways described for sour cherry. The wild cherry is a large tree, up to 25 m in height; cultivated sweet cherry plants are somewhat smaller. Because of intermediate forms, the old distinction of cultivars into 'Bigarreau' and 'Geans' ('Guignes') is no longer considered valid, but a classification can be based on fruiting times (early, mid-season, or late), or fruit—'white' (yellow flushed with red) or 'black' (dark red to blackish). Sweet cherries are highly susceptible to disease. They require cross-pollination and the fruit is normally hand harvested. For successful fruit production, sweet cherries are fastidious about climate, situation, and soil. Good progress has now been made in the development of dwarfing rootstocks.

Some varieties grown in the United Kingdom are:

Early white	'Frogmore Early'
Early black	'Early Rivers'(2)
	'Bigarreau de Schrecken'
	'Merton Favourite'
	'Merton Heart'
Midseason white	'Kent Bigarreau' ('Amber Heart')
	'Napoleon'(3)
Midseason black	'Roundel Heart'
	'Waterloo'
	'Merton Bigarreau'
	'Merton Bounty'
	'Merton Premier'
Late white	'Florence'
Late black	'Hedelfingen'

HYBRIDS Hybrids between *P. avium* and *P. cerasus* are a group valued for their hardiness and cooking qualities. These may be either black-fruited or red-fruited and they are generally known as 'Dukes' or 'Royal Cherries'.

BIRD CHERRIES *P. padus.* These are hardy northern trees with white flowers in racemes which appear on the young shoots in May. In the United States, one species, *P. serotina*, at least, provides fruit which is used for flavouring rum and brandy, for which purpose it is said to equal the morello cherry.

TWO-THIRDS LIFE SIZE SECTIONS LIFE SIZE

1 **MORELLO CHERRY** 1A Blossom
SWEET CHERRY CULTIVARS
2 '**EARLY RIVERS**' 2A Blossom 2B Sections of flower and immature fruit
3 '**NAPOLEON**'

Plums (1): Sloe, bullace, damson, gage

In many countries in the cooler and temperate parts of both hemispheres of the world, plums, as tree fruit, are next in importance to apples and pears. European plum, bullace, damson, and gage (all Prunus *species) are well known in the West, although there is no universal agreement as regards Latin names. The Japanese plum (*P. salicina *syn.* P. triflora*), said to have originated in China but domesticated in Japan, was introduced into the United States of America around 1870. It has been suggested that the European plum (*P. domestica*) evolved as a hybrid between the wild blackthorn or sloe (*P. spinosa*) and the cherry plum (*P. cerasifera*), possibly in Asia Minor, but it has also been suggested recently that the European plum arose directly from* P. cerasifera. *Like apples and pears, plum fruits were no doubt collected from the wild before cultivation; plum stones have been found in some European neolithic and Bronze Age sites, but the earliest records of cultivation are from Roman times.*

*Hybrids have been created between European and Japanese plums; other plum species, for example the American plum (*P. americana*), have also been utilized in hybridization programmes. Other North American species are chickasaw (*P. angustifolia*), Oregon plum (*P. subcordata*), Texan plum (*P. orthosepala*), and American sloe (*P. alleghaniensis*). China, the United States, Turkey, Bulgaria, France, Germany, Hungary, and Italy produce large quantities of plums.*

Plums and their allies (sloe, etc.) contain about 10 per cent sugar, half of which is glucose, the remainder consisting, in roughly equal halves, of fructose and sucrose. There is about 1 per cent malic acid; both sugar and acid contents vary with dessert and culinary types. The vitamin C content is about 6 mg/100 g in plum, but lower in damson and greengage; in prune it is about 3 mg/100 g. The potassium content is high. Plums and their allies are utilized as dessert, culinary, and dried fruit (prunes); are canned and included in jams and jellies. Prune juice is a laxative and plum purée may be a baby-food item. Liqueurs and spirits can be manufactured from plums, for example Slivovitch.

BLACKTHORN OR SLOE (1) *P. spinosa*. This is a wild plum of Europe and western Asia. It is a shrub up to 4 m in height with thorny shoots, which are dark purple and downy when young, and black bark. The blue-black fruit (1–2 cm in diameter) has a marked bloom and green flesh which is too acid for dessert purposes but is used for sloe wine and gin.

BULLACE (2) *P. insititia*. This is a larger plant with larger fruits than sloe. Fruit colour varies according to cultivar—purple (black bullace) and greenish yellow (shepherd's bullace). This is a useful cooking plum when other plums have finished but it should be left on the tree until touched by the first frosts to improve acceptability.

DAMSONS (3–4) *P. damascena*. The damsons are closely related to the bullaces but the fruit is more oval in shape. They are used for cooking and jam making. The species is said to have originated in Syria. 'Farleigh', 'Prune', 'Merryweather', and 'Frogmore' damsons are well-known varieties in the United Kingdom.

GAGES (5–6) The greengage (sometimes called *P. italica*) could be a distinct species with an origin in western Asia and is often considered the best of all plums, possessing a round, yellowish-green fruit with a red flush and russet dots. 'Green Gage' is very old and is said to have been named originally after Queen Claude (Reine Claude) and introduced from the continent into England by Sir Thomas Gage in about 1720 and then renamed, although it has also been claimed that it was known in England much earlier.

Because of its fine flavour, 'Green Gage' has been crossed with other plum cultivars to give larger fruit. Some authorities do not regard these cultivars as true gages. 'Cambridge Gage' is almost indistinguishable from 'Green Gage' and is excellent for dessert, cooking, preserving, and canning purposes. Other hybrids are 'Bryanston', 'Reine-Claude de Bavay' (introduced into the United Kingdom about 1843), 'Denniston's Superb' (of American origin and raised about 1836), and 'Ontario'.

The 'transparent' gages have transparent fruit skins which, when held to the light, show the stones as shadows within the flesh. 'Transparent Gage' is an excellent dessert plum with golden yellow fruit. Its origin is unknown but probably French. 'Early Transparent' and 'Late Transparent' are other cultivars.

'Mirabelles' are little grown in the United Kingdom but are prized in France for the manufacture of apricot-like jam and an alcoholic spirit.

TWO-THIRDS LIFE SIZE

1 **BLACKTHORN (SLOE)** 1A Blossom 2 **SHEPHERD'S BULLACE** 2A Blossom
3 **'FARLEIGH DAMSON'** 4 **'PRUNE DAMSON'**
5 **'GREEN GAGE'** 6 **'TRANSPARENT GAGE'**

Plums (2): Cooking cultivars

CHERRY PLUM (1) *Prunus cerasifera.* This is not much grown in the United Kingdom for fruit but is used extensively as a rootstock for the European plums and damsons, but not gages. The round fruit (about 1.9 cm in diameter) is usually red with juicy and sweet flesh.

EUROPEAN PLUMS (2–6) *P. domestica.* The origin of the European plum has already been discussed (see p. 72). Trees may grow to a height of 9 m and are erect, spreading, or pendulous. Fruits vary in shape (round, oval) and colour (green, yellow, red, purple, or black).

'RIVERS EARLY PROLIFIC' AND 'CZAR' (2) These are plums produced in the nineteenth century with dark-blue/black fruit ripening early in the season (July, August). Both are essentially culinary types.

'PERSHORE EGG' (3) This has greenish-yellow plums, excellent for jam and canning.

'POND'S SEEDLING' (4) With rose-crimson fruits, this is a good cooking plum but not suitable for canning.

'MONARCH' (5) This was a popular late cooking plum but has now been superseded by 'Marjorie's Seedling'.

'PRUNE D'AGEN', 'PRINCE ENGELBERT', AND 'FELLEMBERG' (6) These are characteristic of plums that are dried to give prunes. Drying may take place on the tree, although this requires a hotter climate than that found in the United Kingdom, but drying can also be carried out artificially.

TWO-THIRDS LIFE SIZE

1 **CHERRY PLUM 'MYROBALAN'** 1A Blossom
PLUM CULTIVARS
2 **'RIVER'S EARLY PROLIFIC** 3 **'PERSHORE EGG'**
4 **'POND'S SEEDLING'** 5 **'MONARCH'** 6 **'PRUNE D'AGEN'** 6A Dried prune

Plums (3): Dessert cultivars

'VICTORIA' (1) This is the most popular plum in the United Kingdom, both in the commercial orchard and the home garden. It has large red fruit which are good for a number of purposes—jam, canning, bottling—but it is generally agreed that it is only of moderate quality as a dessert plum. The plant is self-fertile and also a good pollinator of other cultivars. It originated as a seedling in Sussex in about 1840.

'COE'S GOLDEN DROP' (2) This is an excellent dessert plum, possessing yellow fruits with reddish-brown spots.

'KIRKE'S BLUE' (3) An attractive black dessert plum.

'LAXTON'S DELICIOUS' (4) This arose as a hybrid between 'Coe's Golden Drop' and 'Pond's Seedling'.

'JEFFERSON' (5) 'Jefferson' was raised in America around 1825. It possesses lemon-green fruits.

As in the case of apples and pears, plums are available as columnar (single-stem) types. Plums to be found on the international market include some described above (e.g. 'Victoria', 'Czar', and 'D'Agen'), but also many others (e.g. 'Santa Rosa', 'Ruby Nel', and 'President').

TWO-THIRDS LIFE SIZE

PLUM CULTIVARS
1 'VICTORIA' 1A Blossom 2 'COE'S GOLDEN DROP'
3 'KIRKE'S BLUE' 4 'LAXTON'S DELICIOUS' 5 'JEFFERSON'

Peaches and apricots

PEACH (1–2) *Prunus persica* syn. *Persica vulgaris.* Peaches arose in the mountainous areas of Tibet and western China. The fruit was cultivated in China as early as 2000 BC and reached Greece at about 300 BC from Persia. Cultivation by the Romans commenced in the first century AD. Large-scale cultivation of the plant now takes place in Italy, the United States, Greece, Spain, France, Mexico, Turkey, Japan, Argentina, and Chile. It was introduced into the United Kingdom in the middle of the sixteenth century. On a worldwide basis there are about 2000 peach cultivars.

The fruit is consumed fresh, canned, frozen, dried, and processed into jelly, jam, juice, and wine. Fruits are sometimes classified as 'freestone' (an easily detachable stone and usually soft flesh) and 'clingstone' (a well-embedded stone and firm flesh). The fruit contains about 8 per cent total sugar, over half of which is sucrose. There are high percentages of potassium (160 mg/100 g) and vitamin C (31 mg/100 g). Malic and citric acids exist in roughly equal quantities in the fruit.

Peach is a small tree, 6–7 m in height, and in the warmer, southern parts of the United Kingdom it can be grown in the open as a tree or bush to produce fruit. Elsewhere in the United Kingdom it requires protection. It may be grown on peach or plum seedling rootstock.

NECTARINE (3) var. *nectarina.* This is a variety of peach and may even appear as a bud sport on peach trees. Its fruit has a smooth skin (not downy as in peach) and also, compared to the peach, is of a richer colour. Its nutritional composition is very much like that of peach.

APRICOT (4) *Prunus armeniaca* syn. *Armeniaca vulgaris.* This has an origin very similar to that of peach, namely in China. Apart from the common apricot, there are a number of other apricot species in China. It is now cultivated in the Mediterranean region (e.g. Spain, France, Turkey, and Italy), in parts of the former USSR (e.g. Armenia, central Asian republics, and North Caucasus), the United States, Canada, South Africa, New Zealand, and Australia. In China there are over 2000 cultivars. Its fruit is utilized in the same way as peach and also contains about the same total sugar concentration (7 per cent). However, it contains a higher percentage of potassium (270 mg/100 g) and several times the carotene content, but its vitamin C content (3–6 mg/100 g) is much lower than that of peach. The fruit has about three times as much malic as citric acid.

Apricot is a tree up to 10 m in height.

TWO-THIRDS LIFE SIZE

1 **PEACH 'PEREGRINE'** 1A Blossom 2 **PEACH 'ROCHESTER'**
3 **NECTARINE 'LORD NAPIER'**
4 **APRICOT 'MOORPARK'** 4A Blossom

Strawberries

STRAWBERRY (1) *Fragaria* spp. The juicy edible part of the strawberry is an enlarged receptacle on the surface of which the yellow pips or 'seeds' (botanically speaking, fruits known as achenes) are embedded. The strawberry plant is a perennial herb, with a leafy crown from which radiate prostrate stems or runners bearing small leaf clusters which take root and will grow into new plants.

ALPINE STRAWBERRY (2) *F. vesca* var. *semperflorens*. This is a variety of one of the wild strawberries (*F. vesca*) which is found in Europe, Asia, and North America as far north as the subarctic region. The fruit of *F. vesca* is smaller than the cultivated types, blunt conical in shape, has a pleasant flavour, and ripens in early July. 'Baron Solemacher' is the best-known cultivar of the alpine strawberry. It has rather elongate fruit, high in pectin and therefore useful for making jam; it has no runners. 'Hautbois' (*F. moschata*), a native of central Europe, is another wild strawberry with more vigorous growth than *F. vesca* and not so many runners. Its fruit is roundish or blunt conical, very dark red, with the achenes concentrated towards the tip; it has a 'musky' flavour. Another wild strawberry is *F. viridis*. In Europe, attempts were made to domesticate local strawberries, starting in the fourteenth and fifteenth centuries, but there was little improvement in productivity until the eighteenth century when the present cultivated strawberry appeared. Wild strawberry 'seeds' have been found in neolithic, Bronze Age, Iron Age, and Roman sites in central and northern Europe.

CULTIVATED STRAWBERRIES (3–5) *Fragaria* × *ananassa*. The designation *ananassa* is said to refer to the pineapple-like flavour, odour, and shape. The cultivated strawberry arose as a hybrid in Europe, from two imported American species. *Fragaria virginiana*, a woodland species from the eastern United States, was introduced into Europe soon after 1600; one of its cultivars—'Little Scarlet'—has been grown for a long time and is useful for canning and jam. More than a century later, the West Coast Pine strawberry (*F. chiloensis*), from some countries along the Pacific seaboard of North and South America, was also introduced. This species has male and female plants which, of course, require pollinators to produce the dark-red fruit with a white, pineapple-flavoured flesh. Because of geographical isolation, no hybridization took place in the Americas, but a chance cross between the two species in France about 1750 produced the cultivated strawberry. However, it was the English amateur horticulturalist, Thomas Knight, who carried out the first controlled hybridization, giving rise to the famous cultivars 'Townton' (1821) and 'Elton' (1828). In the United Kingdom a very well–known cultivar, 'Royal Sovereign', was raised by the Laxton Brothers in 1892.

Today, there are hundreds of cultivars and commercial plantings take place in every continent from the subarctic (Finland) to the tropics (Equador) because the plant can be adapted to a wide range of climatic and soil conditions, although most production takes place in the northern hemisphere. Regions with a Mediterranean-type climate (e.g. Spain, Italy, France, and California) are particularly suitable for strawberry production. Strawberries have sometimes been classified into three groups:

(1) main crop, fruit ripening June/July;
(2) perpetual (remoutant), fruit ripening August until autumn;
(3) alpine, fruit ripening June to first frost.

Because of the numerous cultivars available, the many countries involved in production, and present-day ease of transport, strawberries are available virtually all the year round.

The fruit is utilized for dessert, canning, freezing, jams, jellies, ice-cream, syrups, juices, and bakery products. It has a very high content of water (virtually 90 per cent); the sugar content is mainly glucose and fructose; there is a high percentage of vitamin C (77 mg/100 g), and there is more citric than malic acid.

Strawberries are particularly susceptible to virus infection. Many methods of cultivation have been devised but the traditional way is on the flat with straw underneath the fruit to keep it clean. A net covering might also be used for protection against birds—the wild strawberries are not particularly affected by birds.

PLANT × ¼ FRUITS AND FLOWERS LIFE SIZE FLOWER SECTION × 2

1 **STRAWBERRY PLANT** 1A Flowers 1B Flower section 1C Section of fruit 2 **ALPINE STRAWBERRY 'BARON SOLEMACHER'**
STRAWBERRY CULTIVARS
3 **'ROYAL SOVEREIGN'** 4 **'CAMBRIDGE FAVOURITE'** 5 **'CAMBRIDGE VIGOUR'**

Raspberries

Raspberries, blackberries (see p. 84), and other related berries, are all species of the genus Rubus. *The fruit is an aggregate of small drupes or drupelets attached to a conical receptacle.*

RASPBERRY (1–3) *Rubus idaeus*. This red raspberry grows wild in Europe and western Asia. It was introduced into cultivation about 500 years ago in Europe, but seeds of this and other *Rubus* species have been found in archaeological sites dating back to the neolithic period, fruits no doubt being collected from the wild. In North America, the wild red raspberry is sometimes interpreted as a separate species (*R. strigosus*) and was used by the early settlers; it was eventually domesticated. Also in North America is found the black raspberry, *R. occidentalis*. Hybrids have been made between red and black raspberries, giving plants with purple fruit and stem (cane) colour. Several other *Rubus* species have been used by plant breeders to improve the normally cultivated ones, for example *R. illecebrosus* (strawberry raspberry)and *R. kuntzeanus* (Chinese raspberry). There are three major commercial producing regions: (1) the former USSR; (2) Europe (e.g. Poland, Hungary, the former Yugoslavia, Germany, and Scotland); and (3) the Pacific Coast region of North America. The black and purple raspberries are grown only in restricted areas of North America. Yellow-fruited *R. idaeus* is cultivated on a limited scale. A number of crosses have been made between material from the United Kingdom and North America.

There are two fruiting patterns in commercial raspberries: (1) 'summer bearing', fruits being found in the summer on second-year canes; (2) 'autumn or fall bearing', fruits being found in the autumn on first-year canes. Raspberries are particularly susceptible to virus infection.

The fruit contains 5–6 per cent sugar (mainly glucose and fructose, with a little sucrose); mainly citric with some malic acid; 13–38 mg vitamin C/100 g. Raspberries are used for dessert, canning, freezing, purées, preserves, juice, jam (they are rich in pectin), jelly, bakery products, and a limited amount of wine.

WINEBERRY (4) *R. phoenicolasius*. This comes from northern China and Japan. It is a clump-forming plant, with long, arching canes, clothed in red hairs, without prickles. The calyx lobes envelop the developing fruits and keep them covered until almost ripe, when they open to reveal the golden or orange berries which are good for dessert, jam, and jellies.

LIFE SIZE

1 **RASPBERRY 'MALLING PROMISE'** 1A Flower section 2 **RASPBERRY 'SEPTEMBER'** 2A Flowers
3 **RASPBERRY 'GOLDEN EVEREST'** 3A Flowers 4 **WINEBERRY** 4A Flowers

*Brambles and related berries (*Rubus *spp.)*

Many species and their hybrids are found growing wild in the northern hemisphere. Some of these have been brought into cultivation, but the commercial importance of the blackberry is relatively low. The fruit is an aggregate of drupelets like raspberry but, when harvested, comes away with the dry receptacle or plug. On the other hand, the raspberry plug is left on the plant (see p. 82). This is one reason for the success of raspberry over blackberry.

BLACKBERRY (1–2) *Rubus ulmifolius,* This is one of the best wild blackberries worth picking. It is a prickly climbing plant which consists of radiating canes, erect at first but then curving downwards. When the cane reaches the ground it may root to form a new plant. Older shoots bear the fruit, after 2 or 3 years' cropping the shoots die.

Although, no doubt, gardeners have for a long time transplanted wild blackberry plants into their gardens, real cultivation began in the early nineteenth century, mainly in North America. American blackberries, derived from *R. alleghanensis,* are grown in Europe but do not always flourish in the United Kingdom as they may be killed by a hard frost. The usual ovate to oblong leaves of the blackberry are divided in the cultivar 'Norwood' of *R. laciniatus.* There is a thornless cultivar, 'Oregon', of this species. Blackberry can be hybridized with raspberry—the cultivar 'Bedford Giant' is important in Europe.

World commercial production is centred on a few areas, for example Washington, Oregon, and Arkansas in the United States, and New Zealand. Blackberries are consumed fresh, also canned, frozen, made into jam, and there is some wine manufacture in Europe. Seed-free blackberry material is used in conserve production.

Blackberries contain about 5 per cent sugar, almost entirely glucose and fructose. The vitamin C content is 15 mg/100 g and there is somewhat more citric than malic acid.

LOGANBERRY (3) *Rubus loganobaccus.* This was discovered wild in California in 1881 and originally thought to be a natural hybrid between dewberry or blackberry and raspberry. It was later described as a species, *R. vitifolius,* but the matter is not clear and it may in fact be a hybrid. The plant was introduced into the United Kingdom in about 1899. Loganberry is very similar to blackberry in its growth habit but the leaves are larger and softer. The dull red fruit is more acid than blackberry with twice as much citric acid, and also twice as much vitamin C. It is rather too acid as a fresh dessert but can be used for the other purposes associated with blackberries. The United States is the world's largest producer.

BOYSENBERRY Said to be a loganberry–blackberry–raspberry hybrid. It was raised in California about 1930. Other hybrids are 'Tayberry', 'Sunberry', and 'Veitchberry'.

DEWBERRY (4) *Rubus caesius.* This is very similar to blackberry but the stems are more or less prostrate and are more slender. The fruits are smaller with a slight whitish bloom. It fruits before blackberry, an advantage. Dewberries are grown in America.

CLOUDBERRY (5) *Rubus chamaemorus.* This is a small, low, perennial herb with golden or orange fruit. It is rarely cultivated but grows wild in Scandinavia, Arctic Russia, Siberia, northern Britain, and Canada, and the fruit can be stewed or used in jam.

LIFE SIZE

1 **WILD BLACKBERRY** 1A Sections of flower and fruit
2 **BLACKBERRY 'NORWOOD'** 2A Flowers 3 **LOGANBERRY**
4 **DEWBERRY** 5 **CLOUDBERRY**

Currants and gooseberries

These belong to the genus Ribes *of the family Grossulariaceae. There are about 150* Ribes *species, found mainly in the northern temperate regions of Europe, America, and Asia. The main commercially grown species are black currant (*R. nigrum*), red and white currants (*R. sativum, R. petraeum, *and* R. rubrum*), and gooseberry (*R. grossularia *syn.* R. uva-crispa*). The major production of these fruits takes place in Europe, with Germany and Poland as the main producers; they are of relatively little commercial significance in North America. Domestication of these plants has taken place in the past 500 years. Their fruit is a round berry with a thin, often translucent, skin enclosing a number of seeds in a juicy flesh. They are shrubs, with three to five foliate leaves, bearing their small greenish flowers from buds on shoots of the previous season's growth. European currants were taken to North America by English immigrants in the seventeenth century.*

BLACK CURRANT (1) *R. nigrum.* Black-currant fruits are utilized for culinary and dessert purposes, also the production of juices, syrups, jams, jellies, flavouring for yoghurt and other dairy products, wine, and liqueurs. They have a high content of citric acid (4 per cent). Sugars present are fructose (2.4–3.7 per cent) and glucose (2.6–3.5 per cent), there is very little sucrose. Black currants are a rich and important source of vitamin C (72–191 mg/100 g, depending on season and cultivar). Less than 50 g fruit meets the recommended daily requirement of vitamin C. The seeds provide γ-linolenic acid—a health preparation. Modern cultivars include 'Ben Lomond' and 'Ben Sarek'. The black currant succeeds best if it has several stems and the opportunity to renew its shoots from basal buds.

RED CURRANT (2) *R. sativum, R. petraeum, R. rubrum.* Cultivated red currants evolved from three species: *R. sativum*, the only one grown in 1542, *R. petraeum* shortly afterwards, and very much later *R. rubrum*. The fruit is utilized in very much the same way as black currant. Its sugar content is very similar to black currant but its citric acid content is lower (2.5 per cent). Also its vitamin C content is lower (40 mg/100 g). Newer cultivars include 'Redstart' and 'Rovada'. Unlike the many-stemmed black currant, the red currant has a stout main stem and may therefore be grown in standard or cordon form, or fan-trained against a wall.

WHITE CURRANT (3) This lacks the red pigment (anthocyanin) of red currant. 'White Versailles' is a well–known cultivar.

GOOSEBERRY (4–6) *R. grossularia* syn. *R. uva-crispa.* Gooseberry grows wild throughout most of northern and central Europe. The shrub is variable in size and growth habit, widely spreading to almost upright, with fruit, according to cultivar, of various colours—yellow, green, or red. The plant was cultivated in the United Kingdom at least as early as the sixteenth century. In the nineteenth century it was a popular fruit in home gardens in Europe, especially the United Kingdom, and this led to the development of many cultivars. In the United Kingdom 'gooseberry clubs' were organized with competitions for the heaviest berries.

The fruit is utilized for the same purposes as the currants. Dessert raw gooseberries contain equal amounts (3.0 per cent) of glucose and fructose, there is very little sucrose. They contain equal amounts (0.7 per cent) of malic and citric acids. Vitamin C content is between 14 and 26 mg/100 g.

In 1905 American gooseberry mildew (*Sphaerotheca mors-uvae*) drastically affected European plants. American species are not affected. However, new mildew-resistant cultivars, bred in the United Kingdom, are now available, for example 'Invicta'.

AMERICAN GOOSEBERRIES (7) The 'Worcesterberry' is an American species, *R. divaricatum,* with small, purplish-black fruit. It is immune to mildew. *Ribes hirtellum,* the 'currant gooseberry', is another American species.

Because of its resistance to mildew, *R. divaricatum* has been used for hybridization; for example, *R. nigrum* × *R. divaricatum* × *R. grossularia* produced *Ribes* × *nidigrolaria,* which has a fruit quality between that of black currant and gooseberry. 'Pixwell' is a hybrid American gooseberry.

LIFE SIZE FLOWER SECTIONS × 3

1 **BLACK CURRANT 'MENDIP'** 1A Flowers 1B Flower section 2 **RED CURRENT 'LAXTON'S NO. 1'**
3 **WHITE CURRANT** 4 **GOOSEBERRY 'CARELESS'** 4A Flowers 4B Flower section
5 **GOOSEBERRY 'WHINHAM'S INDUSTRY'** 6 **GOOSEBERRY 'EARLY SULPHUR'** 7 **GOOSEBERRY 'PIXWELL'**

Fruiting species of the Ericaceae

This family contains some 13 genera with species bearing fleshy berries which are consumed locally in many parts of the world. Relatively few species are of commercial importance. The fruit is utilized in the fresh and dried condition and processed into jam, preserves, pastry products, juice, and wine.

BILBERRY (1) *Vaccinium myrtillus.* Bilberry grows in Europe, Asia, and North America. It is known as bilberry, blaeberry, whortleberry, or winberry. Because of its short stature, fruit collection is tedious and difficult; it is not cultivated commercially but is often harvested locally. The fruit sugar content is 7–14 per cent, with mainly glucose and fructose in roughly equal amounts. Vitamin C is present at 17 mg/100 g.

Bilberry is a low shrub, 20–60 cm tall, growing on heaths, moors, and in acid, open woodlands. It has green, angled twigs, bearing ovate and finely toothed leaves, 1–3 cm long. Flowers, greenish-pink, are solitary or in pairs. Fruits, about 6 mm in diameter, are pale red, ripening to blue-black in colour.

HIGHBUSH BLUEBERRY (2) *Vaccinium corymbosum.* This is by far the most important of the *Vaccinium* crops. It is cultivated in seventeen states of the USA, three Canadian provinces, Europe, Australia, New Zealand, and is being established in Chile. The sugar content (about 15 per cent) is predominantly glucose and fructose; vitamin C is present at 22 mg/100 g. Its main acid (1–2 per cent) is citric, but reasonable amounts of ellagic acid, possibly a cancer-reducing agent, are also present. Highbush blueberries, meant for processing, are harvested mechanically. Cultivars of this plant are available in the United Kingdom.

Highbush blueberry is similar to bilberry but the plants (1 m or higher) and fruits (8–12 mm in diameter) are larger. Cultivars of this blueberry often incorporate genetic material from two American species—*V. ashei* (rabbiteye blueberry) and *V. australe.* Commercial production of *V. ashei* is largely confined to the south-eastern United States.

LOWBUSH BLUEBERRY *V. angustifolium.* Commercial production of this blueberry is largely confined to Maine (USA), Quebec, and the maritime provinces of Canada. It is a low shrub, about 30 cm high, with finely toothed, lanceolate leaves, up to 2.5 cm in length. The flower is greenish white, sometimes with reddish streaks. The bluish-black fruits are up to 1.25 cm across.

CRANBERRY (3) *V. oxycoccus.* Cranberry is a native of northern Europe, northern Asia, and North America. Its fruit is used to make cranberry jelly or sauce, a traditional addition to venison and turkey. The berry contains about 3.5 per cent sugars, with more glucose than fructose; the vitamin C content is 13 mg/100 g. It has high levels of several acids—citric (1.1 per cent), malic (0.9 per cent), quinic (1.3 per cent), and benzoic (0.6 per cent). The ingestion of cranberries (and lingonberries) leads to an increased urine acidity which may relieve urinary tract infections and reduce some types of kidney stones.

Cranberry is a small, prostrate plant with thin stems and oblong or elliptical leaves, 4–10 mm long. The flowers are found singly or in pairs with dark pink petals, 5–6 mm long, and red rounded or oval fruit, 6–8 mm across.

LARGE OR AMERICAN CRANBERRY (4) *V. macrocarpon.* This is a larger version of the previous species, with fruits 12–20 mm across. It is produced commercially in some parts of the United States and Canada, and also in the United Kingdom, where it sometimes may be found as an escape. Its uses are as for *V. oxycoccus.*

COWBERRY, MOUNTAIN CRANBERRY, OR LINGONBERRY *V. vitis-idaea.* This grows wild in Europe, Asia, and North America, especially in mountainous areas. The fruits, used in jams and jellies, are normally collected from the wild, but recently there has been some domestication. The berries have high levels of benzoic acid. Cowberry has creeping stems giving rise to erect shoots, up to 30 cm in height. The bell-shaped flowers are white or flushed with pink, giving rise to red berries, 6–8 mm in diameter.

STRAWBERRY TREE (5) *Arbutus unedo.* This small tree or shrub (up to 10 m in height) has a natural distribution around the Mediterranean, and is also found in Ireland. The fruits are made into jams and liqueurs (e.g. the Portuguese *madrongho*). Its greenish-white flowers are found in clusters and appear from October to December. The globose red fruit (1–2 cm in diameter) has a warty skin and matures a year later.

LIFE SIZE   FLOWER SECTIONS × 3

1 **BILBERRY** 1A Immature fruits 2 **BLUEBERRY** 2A Flowers 2B Section of flower
3 **CRANBERRY** 3A Flowers 3B Section of flower 4 **LARGE CRANBERRY**
5 **ARBUTUS** 5A Flowers 5B Section of flower

Citrus fruits (1)

Certain Citrus *species, belonging to the family Rutaceae, are a most important group of tropical and subtropical fruits. The only other genus in the family of food importance is* Fortunella *(kumquat).* Citrus *species are evergreen shrubs or trees up to 10 m or more in height. They may bear spines or thorns and the leaf stalks may be winged or flattened. Most commercial production takes place in subtropical regions with a Mediterranean-type climate (latitudes 45°N–35°S). They do not grow well in the humid tropics. Most* Citrus *cultivars will only stand light frost for short periods. As regards commercial production, sweet orange (*Citrus sinensis*) is produced in greatest amount, then tangerines (mandarins) (*C. reticulata*), then lemons (*C. limon*) and limes (*C. aurantifolia*), and finally grapefruits (*C. paradisi*) and pummelos (*C. grandis*).*

The fruit is known botanically as the 'hesperidium'. The outer rind or peel consists of the coloured 'flavedo' and the inner white spongy 'albedo'. In the immature state, the green pigment is chlorophyll; as the fruit matures this gives rise to the yellow or orange carotenes. Citrus *fruits in the tropics often remain green, even in the mature state, but this can be changed with suitable gas treatment. Within the peel is the juicy pulp containing the seeds, although some cultivars are seedless because fruit development takes place without fertilization ('parthenocarpy'). Individual* Citrus *seeds often contain more than one embryo ('polyembryony').*

As with other fruit types, Citrus *fruits contain a very large amount of water (almost 90 per cent); the usual three sugars (sucrose, glucose, and fructose); a good deal of potassium but little sodium; some carotenes, B vitamins and vitamin E, but rich in vitamin C. The main acid is citric. Some dietary fibre is provided by these fruits.*

Citrus *fruits are utilized for dessert purposes, may be canned, and also provide juice (sometimes more important commercially than the fresh fruit). Essential oils (not to be confused with the triglyceride oils described earlier in the book) are extracted from the peel (also the flowers and leaves) and used as food flavourings and in cosmetics. Pectin is also extracted from peel.* Citrus *fruit is utilized in marmalade production. After juice extraction, the fruit residue is marketed as an animal feed. Citric acid is included in certain manufactured foods and was once extracted from citrus fruits, but is now produced by fermentation from maize (corn).*

The citrus group probably evolved in China, South-East Asia, and possibly India. The earliest records of cultivation are from China about 2200 BC. *Many leading Asian cultivars were not imported into the western hemisphere until the nineteenth or twentieth centuries.*

SWEET ORANGE (1–2) *Citrus sinensis*. This is the most important of the *Citrus* fruits and is a tree growing to some 15 m in height. It originated in southern China and was taken to Europe in the fifteenth century. Columbus carried orange seeds on his second voyage to America in 1493. Sweet orange is grown throughout the subtropics and tropics, but Brazil and the United States of America produce the greatest quantities of this fruit. In these two countries, the bulk of production is used to manufacture orange juice. Spain is the world's largest exporter of fresh oranges. The species may be classified into:

1. Common orange—a well-known cultivar is 'Valencia'.
2. Blood orange, the flesh having a blood-red tint. The pigment responsible for this tint is anthocyanin. A well-known cultivar is 'Maltese'.
3. Navel orange, known because of a navel-like mark at its apex. A well-known cultivar is 'Washington'.

Sweet oranges contain 6–9 per cent total sugars and 44–79 mg vitamin C/100 g. In addition to the fresh fruit, orange juice is also a good source of sugars, vitamin C, and potassium.

SEVILLE, BITTER, OR SOUR ORANGE (3) *Citrus aurantium*. The plant bears fruits with a bitter taste (the bitter compound is 'neohesperidin') which makes them unattractive for eating raw. The species originated in South-East Asia and was introduced into Europe in the eleventh century, well before the sweet orange. Spain is the major producer of this fruit, although it has spread to many tropical and subtropical countries. The fruit is used to make the conserve, marmalade, and as a flavouring, and in liqueurs (Curaçao). Essential oils are prepared from the fruits (also the leaves and flowers); bergamot oil comes from *C. bergamia*, sometimes regarded as a subspecies of Seville orange. Seville orange has been used as a rootstock for lemon, sweet orange, and grapefruit, but it is susceptible to the virus disease tristeza.

LEMON (4) *Citrus limon*. This is a small tree 3–6 m in height, with fruit that is too acid to be consumed as a dessert but is of great importance in providing juice for culinary and confectionery purposes, also drinks (e.g. lemonade). The candied peel is a food constituent. The plant originated in South-East Asia but the exact region is not known. Today, the United States and Italy produce most lemons, but other important producers are Argentina, Brazil, Spain, and Greece. The fruit has only about 3 per cent total sugars but a high content, nearly 5 per cent, of citric acid; there is 58 mg/100 g of vitamin C. Well-known cultivars are 'Eureka', 'Lisbon', and 'Villafranca'. Rough lemon (a hybrid with *C. medica*) is used as a rootstock for other *Citrus* species.

TWO-THIRDS LIFE SIZE   SECTION LIFE SIZE

1 **SWEET ORANGE** 1A Flowers 1B Section of flower
2 **BLOOD ORANGE** 3 **SEVILLE ORANGE**
4 **LEMON** 4A Flowers

Citrus fruits (2)

GRAPEFRUIT (1) *Citrus paradisi.* This is a tree up to 15 m in height. Its origin is not known with certainty but it probably arose in the West Indies as a hybrid between pummelo (*Citrus maxima*) and the sweet orange, or as a mutation from pummelo. Grapefruit was described in Barbados in 1750. Today, the major producing country is the United States of America, but it is also produced in other countries such as Israel, the West Indies, Cuba, Argentina, and South Africa. Grapefruit is generally a breakfast fruit, segments are canned and juice is a commercial item. It is unique among the citrus fruits in that it stores well on the tree and may be harvested all the year round. Cultivars are white, pink, or red-fleshed, also with seeds or seedless. Some cultivars are 'Marsh' (seedless and white flesh), 'Duncan' (seeded and white flesh), 'Thompson' (seedless and pink flesh), and 'Ruby' (seedless, deep flesh colour, reddish peel). The flesh pigment is a carotene—'lycopene'. The fruit has a total sugar content of about 7 per cent and a citric acid content of about 1 per cent (similar to orange) but there is a bitter compound present—'naringin'. Vitamin C is present at 36 mg/100 g—rather low in relation to other citrus fruit.

LIME (2) *Citrus aurantifolia.* This is a small, much-branched tree up to 5 m in height. It may have originated in north-eastern India, but has now spread throughout the tropics, where it takes the place of lemon. The plant is the most frost sensitive of the commercial *Citrus* species. Its fruit is small, up to 6 cm in diameter, and is greenish yellow when ripe. Some countries involved in commercial production of the fruit are Brazil, Egypt, Mexico, and the West Indies. Like lemon, it is not a dessert fruit but it is used in marmalade, as a food flavouring (e.g. chutney, pickles, and sauces), and a source of juice for drinks; oil is prepared from the peel. Its total sugar content is only 0.8 per cent and the fruit is somewhat more acid than lemon. In earlier times, limes were carried on sailing ships to be consumed as a preventative measure against scurvy. Its vitamin C content is 46 mg/100 g.

MANDARIN OR TANGERINE (3) *Citrus reticulata.* A small tree, up to 8 m in height, this is the hardiest of the *Citrus* species. It possibly originated in Indo-China. Classification of forms within the species is difficult but one view is that the species includes 'mandarins', 'tangerines', 'satsumas', 'King mandarins', and 'willow-leaf mandarins'. One possible distinction between the common mandarin and tangerine is that the former is yellow-fruited, the latter deep orange. In southern Africa, the Afrikaans name *haartje* is often used. The species is cultivated in many countries but major producing areas include Japan, Brazil, the United States, and Mediterranean countries (it was introduced into Europe at the beginning of the nineteenth century). The fruit is consumed as a dessert and the segments are canned. Its peel or rind is easy to remove from the flesh. Total sugar content is 8 per cent and vitamin C is 30 mg/100 g.

'Calamondin' (× *Citrofortunella microcarpa*) is a hybrid, originating in China, of mandarin and kumquat (*Fortunella*). Its fruit has a number of food uses and is cultivated commercially in the Philippines. 'Rangpur lime', although an acid fruit, is a form of *C. reticulata.*

TWO-THIRDS LIFE SIZE

1 **GRAPEFRUIT** 2 **LIME**
3 **TANGERINE**

Citrus fruits (3)

CITRON (1) *Citrus medica.* This is a small tree or shrub up to 3 m in height. It probably originated in the sub-Himalayan region of north-eastern India and Upper Burma, spreading through Persia to the western world and also eastwards to China. It was the first of the *Citrus* species to reach Europe, about 300 BC. Although it has been cultivated in many tropical countries, commercial planting is restricted to certain Mediterranean islands of Italy, Greece, and France, and in the mountainous coffee regions of Puerto Rico. The fruit, 10–20 cm long, is elongated with a lumpy surface, thick peel, and mildly acid or acid flesh. Its most important use today is the production of candied peel for confectionery and cakes. For this purpose the green immature fruit is sliced into halves, fermented in brine, and soaked in strong sugar solution. The 'Etrog' cultivar is used in the Jewish Feast of Tabernacles; another cultivar is the 'Fingered' citron, cultivated in China and other eastern countries, with a fruit split into a number of finger-like sections and used as a perfume source and medicine.

KUMQUAT (2) *Fortunella* spp. The two most important species are *F. margarita* and *F. japonica.* They are not true *Citrus* species but do belong to the same family, Rutaceae. The species probably originated in China but are now cultivated not only in that country, but also in Japan, Taiwan, and in a number of other countries, for example Argentina, Brazil, Cyprus, and the United States. Kumquat is a shrub or small tree (2–4 m tall) with small (1–4.5 cm in diameter) ovoid or round orange, or golden-yellow fruit. The fruit can be eaten fresh (the skin is edible), cooked, or made into chutneys, marmalades, jellies, or preserved in syrup or candied. Kumquats have quite a high total sugar (about 9 per cent) and the vitamin C content is 39 mg/100 g.

CLEMENTINE (3) This is often regarded as a cultivar of tangerine, or possibly a hybrid between tangerine and sweet orange. Its peel is easily removed. It may have originated in North Africa but is also cultivated in other Mediterranean countries and South Africa. The total sugar is about 9 per cent with a predominance of sucrose. Its vitamin C content is quite high—54 mg/100 g.

UGLI OR HOOGLY (4) This rather misshapen *Citrus* fruit is a cross between tangerine and grapefruit. Its flesh is sweeter than most grapefruits. It is grown in and exported from Jamaica.

OTHER CITRUS FRUITS The ortanique is a cultivar of a hybrid between tangerine and sweet orange. It resembles the sweet orange in size and juice content but, like the tangerine, has a thin, easily peeled skin. It is exported from Jamaica. The fruit has a high percentage of total sugar (almost 12 per cent) and a high content of vitamin C (50 mg/100 g).

The pummelo, pomelo, or shaddock (*Citrus maxima* syn. *C. grandis*) is quite a well-known *Citrus* crop of South-East Asia. Its yellowish fruit is the largest (10–30 cm in diameter) of the *Citrus* group. It was introduced into Barbados in the seventeenth century by a Captain Shaddock and was the ancestor of the grapefruit (see p. 92). The fruit is exported from Israel.

Papeda or wild lime (*Citrus hystrix*) is grown in Malaysia, Sri Lanka, and Burma. It is a small tree (3–5 m tall) bearing yellow, very wrinkled and rough fruit. The fruit is not eaten fresh but used as a food flavouring.

Many hybrids have been created between various *Citrus* species, also kumquat: tangelos (*C. reticulata* × *C. paradisi*), for example ugli; tangors (*C. reticulata* × *C. sinensis*), for example ortanique; and limequats, orangequats, and citrangequats (hybrids between kumquat and various *Citrus* species).

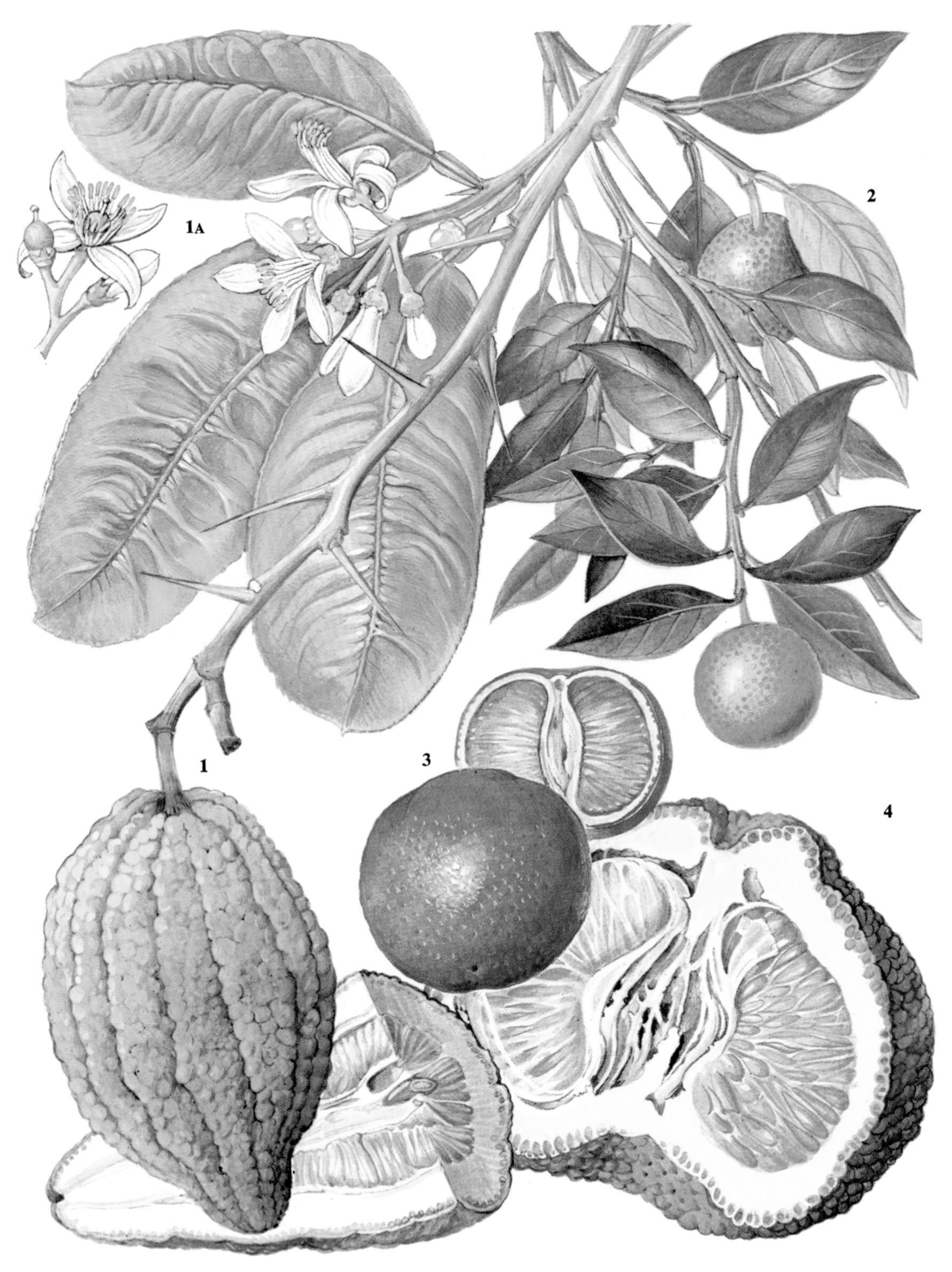

TWO-THIRDS LIFE SIZE

1 **CITRON** 1A Flowers 2 **KUMQUAT**
3 **CLEMENTINE**
4 **UGLI**

Grapes (1)

The genus Vitis *(family Vitaceae) contains up to 60 species which are native to the temperate zones of the northern hemisphere, with a few species reaching the tropics. In North America there are about 25 species; a single species,* Vitis vinifera, *in Eurasia, and a number of species in Asia.* Vitis vinifera *is the species of outstanding importance. Some 68 per cent of grape production is involved in wine production, 1 per cent in non-alcoholic juice, 20 per cent for table or dessert purposes, and 11 per cent for dried grapes (raisins, currants, and sultanas).*

VITIS VINIFERA *Vitis vinifera* (**1**) is a vigorous climber, growing to a height of 16–20 m if left unpruned, but usually restricted by pruning. It climbs by means of forked tendrils, produced intermittently at two out of three vegetative nodes. Its leaves are 9–28 cm wide, long-stalked, palmately lobed, and coarsely toothed. The petals of the small, greenish flowers are joined at the tips. The fruits are berries, with seeds or seedless.

Like other fruit crops, no doubt in the first instance grapes were collected for food in the wild—pips or seeds from the wild *vinifera* (it extends from the Atlantic coast to the western Himalayas) have been discovered at a late neolithic site (4500 BC) in Cyprus. As regards the domestication and cultivation of *vinifera* (viticulture), there is no doubt that it is one of the oldest fruit crops of the Old World—seeds have been found at early Bronze Age sites at Jericho (about 3200 BC) and other places in the Levant. Viticulture, including wine production, was practised in Egypt certainly as early as 2400 BC, as recorded in the hieroglyphics of the time. There is some doubt about the exact area of domestication, present-day Armenia has often been quoted but, on archaeobotanical evidence, it has also been claimed that viticulture started in areas close to the eastern shores of the Mediterranean and was taken to other Mediterranean countries and the Black Sea area, by Phoenician and Greek colonists. The Romans took the crop to temperate European countries, including Britain. At a later date, grape cultivation was often associated with the monasteries, but in Britain there was a decline in activity starting as early as the thirteenth century because of wine imports from France and a general decline in agricultural prosperity resulting from a shortage of labour—the Black Death (1348–49) was responsible for great mortality. In eighteenth- and nineteenth-century Britain there was an emphasis on the production of table grapes, rather than wine, and this took place in the heated greenhouses and conservatories of the great houses of the time. Since the end of the Second World War there has been renewed interest in wine production in the United Kingdom.

Vitis vinifera was taken by Columbus in 1492 to the New World and later by Spanish and Portuguese explorers to North and South America. At a still later date the vine was transported to the Atlantic seaboard of North America by British, French, and Dutch settlers. *Vitis rotundifolia* (the muscadine grape) and *Vitis labrusca* (the slip skin grape) are indigenous North American species, having fruit with a peculiar (musky) flavour, astringency, and lack of sweetness. Some hybridization no doubt took place between these species and the imported *V. vinifera.*

Grapes contain a large amount of sugar (15–25 per cent) which is composed of roughly equal amounts of glucose and fructose, there is only a trace of sucrose. The vitamin C content is low—3 mg/100 g. For fruits, an unusual feature is the presence of tartaric acid (0.5 g/100 g), there is also the same amount of malic acid. The red and black grape pigments are anthocyanins.

Grapes can be classified in various ways. Colour as black (including red and purple) or white (including yellow and green) is one criterion. Then there is the use:

(1) table or dessert grapes: firm flesh and low acidity;
(2) wine grapes: soft flesh and high acidity; and
(3) dried grapes: firm flesh, high sugar, and moderate to low acidity.

Cultivars of the crop are extremely numerous—one estimate is as great as 10 000. Among the leading countries producing table or dessert grapes are Italy, the United States, Chile, and Spain. Some cultivars of dessert grapes are 'Black Hamburgh' (**2**) at one time the most widely grown dessert cultivar in the United Kingdom. It is of German/Italian origin, introduced into England about 1720. 'Perle de Czaba' (**3**), suitable for outdoor cultivation if in a warm site. 'Chasselas Doré' or 'Chasselas d'Or' (**4**) or great antiquity, a famous French dessert grape of the sixteenth century.

TWO-THIRDS LIFE SIZE

1 **GRAPE VINE**, flowers, and leaves
2 **'BLACK HAMBURGH'** 3 **'PERLE DE CZABA'** 4 **'CHASSELAS DORÉ'**

Grapes (2)

Wine is produced by yeast fermentation of sugars in grape juice to produce alcohol (ethanol) and carbon dioxide. The carbon dioxide is retained in champagne and other sparkling wines to give the characteristic bubbles. Yeast occurs naturally on grape skins but yeast cultures may also be added. Wine is distilled to give spirits such as brandy and cognac. The fortified wines of Spain (sherry) and Portugal (port) have brandy added to wine, which increases the alcohol content. A number of conditions, like climate and soil, affect wine flavour. Wine production takes place in many countries. Among the leading producers are France, Germany, Spain, and Italy. Wine has always had strong connections with a number of the world's great religions, such as monastery production, communion wine, etc.

Some of the many cultivars available are:

'CHARDONNAY' AND 'PINEAU (PINOT) BLANC' (1) These bear the rather small, roundish, yellowish-green fruits from which white burgundy and Chablis are made. The same grape, grown in chalky soil in the more northerly district of Champagne, produces the famous sparkling wine. Although cultivars originated in a particular area, they may now be grown in many countries.

'PINEAU (PINOT) NOIR' (2) This is a purplish-black grape, used traditionally for making red burgundy and partly for champagne.

'CABERNET' (3) A bluish-black grape, used for making Bordeaux red wines, sometimes known as clarets.

'RIESLING' (4) A collective term for a group of white (green) cultivars yielding the hocks of the Rhine wine districts. Riesling cultivars have been planted in a number of countries.

In 1863 the aphid-like insect, *Phylloxera*, was noted in England, likewise in France in 1868. It lives in harmony on American wild vines but could have destroyed *V. vinifera* roots in a few years. Consequently, roots of American wild vines (e.g. *V. riparia*, *V. rupestris*, and *V. berlandieri*) were imported into Europe and used as rootstocks for European cultivars. A disaster was consequently avoided.

DRIED GRAPES (RAISINS, SULTANAS, AND CURRANTS) These are produced from about 11 per cent of the world's grape harvest. Turkey, the United States, Greece, and Australia are important producers. Drying may take place in the sun or by mechanical means. Currants are derived from a small-fruited, black, seedless cultivar grown in Greece for more than 2000 years. Raisins and sultanas (seedless or with seeds) are larger and more succulent than currants.

GRAPESEED OIL This is extracted from seeds (6–20 per cent oil) which are a by-product of various forms of grape utilization. It is a polyunsaturated oil. Although not one of the most important vegetable oils, it is available commercially. The seed residue is used as an animal feed.

VINE LEAVES These are used in Turkish, Greek, and Middle Eastern cooking, for example in *dolmades* (vine leaves stuffed with rice and minced meat).

SEA GRAPE *Coccoloba uvifera*. This is completely unrelated to the grape vine, its family is the Polygonaceae. It is found on many Caribbean beaches and is reputed to have been the first Caribbean plant seen by Columbus. Sea grape is too sour to be used as a dessert but has large amounts of pectin and makes a good jelly and jam.

TWO-THIRDS LIFE SIZE

1 '**CHARDONNAY**' 2 '**PINEAU NOIR**'
3 '**CABERNET**' 4 '**RIESLING**'

Fig, mulberry, and pomegranate

FIG (1) *Ficus carica*. This is a deciduous, spreading large shrub or tree, 2–5 m tall (maybe up to 10 m), with large (30 cm long and 25 cm broad), palmately three- or five-lobed leaves. The genus *Ficus* contains over 1000 species which are spread throughout the tropical, subtropical, and warm temperate areas of the world. Some provide fruit, others are ornamentals, for example *F. benghalensis* (Indian banyan or weeping fig) and *F. elastica* (Indian rubber plant), but *F. carica* is of greatest economic importance as a fruit crop. It is cultivated in subtropical and warm temperate regions, the major commercial production taking place in Spain, Italy, Turkey, Greece, Portugal, and the United States. Fig is one of the oldest fruit crops (together with grape and olive) recorded in the Mediterranean basin. Its fruit pips have been found in several eastern Mediterranean early neolithic sites (7800–6600 BC), but these could have been from wild trees. There is evidence that fig cultivation took place in Mesopotamia and Egypt about 2750 BC. Wild fig has a wide distribution in the Mediterranean basin, but the evolution of the cultivated from the wild form probably took place in the eastern part of the region. The fig was taken to England, also the New World, in the sixteenth century, and to California in the eighteenth century.

Its numerous purplish-red flowers are found inside a flask-shaped inflorescence opening to the outside by a minute opening or 'ostiole'. In a large number of cultivars of the common fig this inflorescence develops into the green, brown, or purple fruit (a 'syconium') but without pollination and fertilization (parthenocarpy). The pips are actually drupelets. A well-known group of cultivars that constitute the Smyrna fig do require pollination to produce fruit. This process, known as 'caprification', is carried out by the minute fig wasp (*Blastophaga psenes*) for which purpose baskets of wild caprifig fruits, which have a role in an essential stage in the wasp's life history, are placed near the developing Smyrna figs.

Figs may be consumed fresh or canned, dried, or used in various bakery products. Roasted figs have been added to coffee—the so-called Viennese coffee. A fresh fig contains about 10 per cent total sugar (mainly glucose and fructose) but when dried this rises to about 50 per cent. Fresh figs do not always travel well, so the dried product provides a good dietary substitute. Similarly, the potassium content rises from 200 mg/100 g in the fresh fig to 970 mg/100 g in the dried product. The vitamin C content is low: 2 mg/100 g in the fresh state, 1 mg/100 g when dried. Figs have a mild laxative action.

Figs belong to the family Moraceae and contain a milky latex which, in fresh figs, causes mouth skin eruptions in some individuals.

MULBERRY (2) *Morus nigra*. The common or black mulberry also belongs to the Moraceae. It probably originated in Iran and was known to the Greeks and Romans before the Christian era; it was also cultivated in ancient Egypt. The Romans have been credited with bringing the plant to England. It enjoyed popularity in Europe in medieval times.

The fruit can be eaten as a dessert, used for tarts, pies, jams, and processed into wine, but it is not widely available. Mulberries contain about 8 per cent total sugars (roughly equal amounts of glucose and fructose), quite a large amount of potassium (260 mg/100 g) and 19 mg/100 g of vitamin C. The main acid is citric with a smaller amount of malic.

The common mulberry is a small spreading tree up to 10 m in height with toothed leaves, up to 20 cm long. The flowers are unisexual, found in drooping catkins on the same tree. Its multiple fruits, purple when ripe, are a dense cluster of fleshy flower bases and perianth parts—superficially resembling a raspberry.

WHITE MULBERRY *Morus alba*. This native of Asia is grown primarily for its light-green leaves which are the food of the common silkworm. It is grown in the tropics to a limited extent for its edible fruit; in Iran the fruit is dried and sold as a food product. *Morus alba* is a low-branched tree up to 9 m in height.

POMEGRANATE (3) *Punica granatum*. A common view is that pomegranate is a native of Iran and was grown in the hanging gardens of Babylon, and in ancient Egypt. It spread around the Mediterranean and eastwards to India and China. The movement into the Mediterranean area must have been at a very early time because pomegranate remains have been found at early Bronze Age sites in Israel (wild pomegranate plants do not occur in this region, suggesting early cultivation). Pomegranate is now cultivated throughout the tropics and subtropics.

The fruit, 6–12 cm in diameter, has a leathery skin and is dark yellow to crimson. Internally the juicy pink or crimson pulp contains numerous seeds. The pulp is the edible part of the fruit and may be eaten fresh when scooped or sucked out. Also, the juice can be used for drinks, wine, and syrup (e.g. 'grenadine'). In the Middle East, the fruit is combined with walnuts to make a sauce. Pomegranates contain about 12 per cent total sugars (essentially glucose and fructose) but are quite acid (1 per cent citric acid). The vitamin C content is 13 mg/100 g.

The plant is a large shrub, 2–4 m in height, deciduous in Europe but evergreen in some tropical countries. Its opposite leaves (4–8 cm long) are oblong-lanceolate. Its orange-red flowers have five to seven sepals (which persist at the apex of the fruit), five to seven petals, and numerous stamens. The fruit, botanically speaking, is a berry and is divided internally into a number of compartments.

The pomegranate can be grown out of doors in the United Kingdom, but its fruit rarely ripens. There are a number of cultivars, with white or red flowers. Some countries involved in the export of pomegranates are Egypt, India, Iran, Israel, Pakistan, Peru, and Spain.

TWO-THIRDS LIFE SIZE DETAIL × 2

1 **FIG**
2 **MULBERRY** 2A Female flower-head 2B Detail of flower
3 **POMEGRANATE** 3A Flowers

Tropical fruits of the Americas (1)

PINEAPPLE (1) *Ananas comosus.* Pineapple belongs to the family Bromeliaceae and was domesticated in tropical South America in pre-Columbian times. It was first seen by Europeans on the island of Guadelope during the second voyage of Columbus (1493–96). Pineapple is now widely grown in tropical and subtropical countries, the major producing regions being the United States, Mexico, Formosa, Thailand, the Philippines, Malaysia, Ivory Coast, South Africa, and Australia. It cannot tolerate frost or prolonged cold and is therefore best cultivated in coastal or near-coastal areas of low or moderate elevation.

The plant is a perennial or biennial, up to 1.5 m in height, with tough, spiny leaves which can cause difficulties during cultivation. The cultivar 'Smooth Cayenne' has smooth leaves, for this and other reasons it is the most popular of the pineapple cultivars. Pineapple's inflorescence consists of up to 200 reddish-purple flowers, each subtended by a pointed bract. The fruit (about 20 cm long and 14 cm in diameter) normally develops by parthenocarpy (no pollination), although pollination sometimes takes place between species other than *A. comosus* and *A. comosus* cultivars by humming birds to give small, hard seeds. It is a compound fruit (a sorosis) where the axis thickens and the small berry-like fruits fuse. The rind on the outside forms from the bracts and sepals. On the top of the fruit is a 'crown' of leaves which, together with the 'slips' (shoots below the fruit) and 'suckers' (shoots below in the leaf axils), can be used for propagation. Pineapple has a unique feature in that flowering (and fruit set) can be controlled by externally applied growth hormones so that harvest can be organized to fall on any day of the year. Miniature pineapples (about 7 cm in length) now appear in markets.

Pineapples are consumed fresh, canned, and processed to give juice. The juice may be fermented to form vinegar or further distilled to give an alcoholic spirit. Cannery waste is included in animal feed. Pineapple flesh contains about 10 per cent total sugars, half of which is sucrose, the rest being composed of glucose and fructose. In dried pineapple, total sugars rise to almost 70 per cent. The vitamin C content of fresh pineapple is 12 mg/100 g but there is only a trace in the dried product. Citric is the main acid present.

Both the fruit and stem of the plant contain a protein-digesting enzyme,'bromelain', which has been used in the pharmaceutical industry, and to prevent a proteinaceous haze in 'chill-proof' beer when refrigerated. It may possibly moderate tumour growth and blood coagulation. Because of bromelain's protein-digesting ability, fresh pineapple juice added to jelly will prevent setting.

CERIMAN (2) *Monstera deliciosa.* Said to be a native of Mexico, this is now distributed widely throughout the tropics. It is a tall climber on trees and has hanging roots. Its leaves are perforated, an unusual feature. The cone-like fruit (a spadix) is 12–25 cm long and sometimes eaten, but it must be consumed in the fully ripe state, unripe fruits contain crystals which irritate the mouth. *Monstera* belongs to the family Araceae and is a popular house plant.

ANNONACEOUS FRUITS (3–4) A number of species of the genus *Annona* of the family Annonaceae are utilized as fruits. They originated in subtropical and tropical America but are now spread throughout the tropics. They are small trees (up to 7 m tall). The fruit forms by the fusion of a large number of carpels and the receptacle. It contains a custard-like, whitish, edible flesh with seeds. The flesh is eaten as a dessert or utilized in fruit salads, sherbets, ice-cream, milk shakes, and yoghurts. Unfortunately the fruit does not travel well and therefore is of limited international importance. The name 'custard apple' has been applied to a number of species. The main commercial species are cherimoya (*Annona cherimola*), sugar apple or sweet sop (*A. squamosa*), sour sop (*A. muricata*), and atemoya (a hybrid of *A. cherimola* and *A. squamosa*). These fruits contain about 15 per cent total sugars (more or less equally divided between glucose, fructose, and sucrose) and up to 40 mg/100 g vitamin C.

SOUR SOP (3) *Annona muricata.* This has a dark-green fruit up to 35 cm in length and covered in soft spines. A certain amount is exported from Indonesia, Jamaica, Malaysia, and Thailand.

CHERIMOYA (4) *Annona cherimola.* This is a native of the subtropical highlands of Peru and Ecuador but is now grown commercially in Chile, Spain, California, and New Zealand.

SWEET SOP OR SUGAR APPLE *Annona squamosa.* This is cultivated commercially in Thailand, the Philippines, and Malaysia.

ATEMOYA This is a hybrid of *A. cherimola* and *A. squamosa*, grown in Florida, Australia, South Africa, and Israel.

BULLOCK'S HEART *Annona reticulata.* The reddish-brown fruit is heart-shaped (hence the common name). Its flavour is not considered as good as that of other *Annona* species.

ILAMA *Annona diversifolia.* It resembles the cherimoya but can be grown in the lowlands. It is particularly appreciated in Mexico and Central America.

SONCOYA *Annona purpurea.* This tree, grown almost exclusively in Mexico and Central America, has particularly large fruits.

PLANTS AND BRANCHES × 1/8 FRUITS × 1/2 FLOWER DETAILS × 1

1 **PINEAPPLE** flowering and fruiting plants 1A Fruit 1B Flower detail
2 **CERIMAN** fruiting branch
3 **SOUR SOP** fruiting branch 4 **CHERIMOYA** fruiting branch 4A Fruit 4B Flower detail

Tropical fruits of the Americas(2)

PASSION FRUIT OR PURPLE GRANADILLA (1) *Passiflora edulis*. The genus *Passiflora* contains about 450 species but only a few are exploited commercially. The best known is *Passiflora edulis*. It originated in Brazil but is now cultivated in many other countries, including Peru, Sri Lanka, Australia, New Zealand, and South Africa. The plant is a vigorous climber up to 15 m long with fascinating flowers. The early Spanish explorers of South America named it 'passion' plant because the flower parts seemed to represent the passion of Christ. There are two forms of the species: *P. edulis*, purple passion fruit (the more usual form), and *P. flavicarpa*, the yellow passion fruit. The purple fruit is a globose or ovoid berry (4–6 cm long) containing many seeds surrounded by yellowish juicy 'arils'. The yellow passion fruit is somewhat larger and better suited to the tropical lowlands. When ripe, the skin of the purple passion fruit assumes a wrinkled appearance. The fruit contains reasonable amounts of carotenes and niacin, also 23 mg/100 g vitamin C. The acid present is mainly citric. Some part of the fruit production is utilized for dessert purposes but the major part is processed into juice which may be canned or frozen and has a number of uses (flavouring cocktails, ice-cream, other fruit juices).

GIANT GRANADILLA (2) *Passiflora quadrangularis*. This is not so well known as *P. edulis* but its fruit is used for essentially the same purposes. In addition, the unripe fruits may be boiled and eaten as a vegetable. Its yellowish-green fruit is large (20–30 cm in length). The plant originated in tropical South America.

Some other *Passiflora* species are utilized in a minor way, either in cultivation or in the wild state. These include water lemon (*P. laurifolia*), sweet granadilla (*P. ligularis*), and wild watermelon (*P. foetida*).

SAPODILLA (3) *Manilkara zapota* syn. *Achras zapota*. This is an evergreen forest tree (up to 20 m in height) which originated in Central America, Mexico, and the West Indies. It is now cultivated throughout the tropics. The fruit (3–8 cm in length) is brown with black, shining seeds embedded in the pulp, and should only be consumed when ripe because the immature fruit produces a latex which is unpleasant to the palate. Prior to consumption, the seeds should be removed because it is said that they are liable to lodge in the throat. Sapodilla is essentially a dessert fruit. There is some international trade in the product. The fruit contains about 15 per cent total sugars, roughly equal amounts of glucose and fructose, and 10 mg vitamin C/100 g. The main acid is malic.

Wild and cultivated trees in America can be tapped for their milky latex which coagulates into 'chicle' which at one time was most important in the manufacture of chewing gum.

GUAVA (4) *Psidium guajava*. This originated in Central America but is now found throughout the subtropics and tropics. It is a shrub or small tree (6–10 m in height); its younger branches have a characteristic reddish-brown bark which peels off in thin flakes. The fruit is a round, oval, or pear-shaped berry (2–8 cm in diameter) with a yellow skin and white, pink, yellow, salmon-coloured, or carmine flesh containing numerous seeds. Guava is a popular tropical fruit. India is the leading producer but other producing countries include the United States, Mexico, Pakistan, Colombia, and Egypt. The fruit is eaten fresh, canned, and made into preserves, jam, jelly, paste, juice, and nectar. Guava jelly is well known. The fruit contains about 5 per cent total sugars with roughly equal amounts of glucose and fructose. There are considerable amounts of carotenes and niacin, and the fruit also may contain a large amount of vitamin C but there is variation (23–486 mg/100 g) according to cultivar and environmental conditions.

Other *Psidium* species occasionally cultivated include the strawberry guava (*P. littorale*) and *P. guineense*.

TWO-THIRDS LIFE SIZE VINES AND Fig. 2 × ¼

1 **PASSION FRUIT** 1A Flowering vine 2 **GIANT GRANADILLA** 2A Flowering vine
3 **SAPODILLA** 3A Flowers 4 **GUAVA** 4A Flower 4B Immature fruit

Tropical fruits of India and Malaysia

MANGO (1) *Mangifera indica.* This is one of the most important of the tropical fruits. It originated in the foothills of the Himalayas of India and Burma and has been in domestication for some 4000 years. Mango now grows in most tropical countries and some subtropical ones. Early on it was taken to Malaysia and other East Asian countries, also to East and West Africa, and finally the New World. India is a very important producing country, but other countries involved in commercial production include Indonesia, Mexico, Brazil, and Thailand. There is increasing export of the fruit to western countries. The plant plays an important part in Hindu culture and religion.

In the tropics, mangoes grow at altitudes up to an elevation of 1200 m. The tree, up to 40 m or more in height, bears rosettes of evergreen leaves (red or yellow when immature), and reddish or yellowish flower sprays. The fruit (a drupe up to 30 cm in length) varies in shape (round, oval, egg-shaped, kidney-shaped), and colour (green, yellow, red, purple) with a dotted skin. Its flesh, surrounding the stone containing the seed, is yellow or orange with variable amounts of fibres. There are many mango cultivars. Mangoes can be propagated by seed, although the fruit on the resulting plants may have a turpentine-like flavour and be fibrous.

The fruit is utilized in many ways, as dessert, canned, dried, the source of juice, in jams, jellies, and preserves. Immature fruits are used in pickles and chutneys. In India, in times of scarcity, the seed within the stone has been used as human food. Mango flesh contains about 14 per cent total sugars, the main constituent being sucrose; there is a considerable amount of vitamin C (37 mg/100 g) and, likewise, a considerable quantity of carotenes. The acid present is mainly citric. Mangoes (family Anacardiaceae) occasionally cause human dermatitis.

RAMBUTAN (2) *Nephelium lapaceum.* This is a tree, 4–7 m tall, with its largest commercial plantings in Malaysia, Indonesia, Thailand, and the Philippines. Its place of origin is not known. The characteristic fruits (3–8 cm long) are covered with red or yellow spines (spinterns). Within the fruit wall is the edible off-white or rose-tinted flesh (aril or sarcotesta). This encloses the seed. The fruit flavour relates to the cultivar—sweet fruits are consumed fresh, the more acid ones are stewed. There is some international trade in the fresh fruit, and canning of the product takes place in Thailand and Malaysia. Sweet rambutan flesh can have total sugars of 16 per cent, with sucrose being the major sugar. Vitamin C content is high at 78 mg/100 g.

A related species, *N. mutabile,* the pulasan, is planted on a smaller scale in the same region, especially in western Java. The fruit in this case is covered by short blunt red or yellow tubercles. It is not as important as rambutan. Both species belong to the family Sapindaceae.

MANGOSTEEN (3) *Garcinia mangostana.* This is a slow-growing evergreen tree, 6–25 m tall, which likes a hot and humid climate. It is possibly a native of Malaysia and is cultivated mainly in South-East Asia. The rounded dark-purple fruit (4–7 cm across) has a persistent calyx and stigma. Within the shell are the edible whitish segments (arils) containing seeds. The edible portion has quite a high content of sugars (about 16 per cent), but the vitamin C content is low (3 mg/100 g). The fruit has a very acceptable flavour but commercial exploitation is difficult because seed germination and vegetative propagation are not always successful.

FRUITS × ⅔ BRANCHES × ⅛ FLOWER DETAILS × 3

1 **MANGO** fruit 1A Fruiting and flowering branches 1B Flower detail
2 **RAMBUTAN** fruit 2A Fruiting branch 2B Flower detail
3 **MANGOSTEEN** fruit 3A Fruiting branch 3B Young fruit

Some other tropical fruits

CARAMBOLA OR STAR FRUIT (1) *Averrhoa carambola.* This originated in South-East Asia and is said to occur wild in Indonesia. It is now found throughout the humid tropics and subtropics. The plant is a small tree (up to 15 m in height) bearing rose-coloured flowers which give rise to yellow fruits, 8–12 cm in length and star-shaped in cross-section. The juicy (over 90 per cent water) fruit exists in sweet and acid forms and is used in fruit salads, tarts, preserves, and drinks. Total sugars are about 7 per cent (roughly half glucose and fructose), and the vitamin C content is 31 mg/100 g. The acid present is mainly citric. Carambola fruits are exported to western countries.

Bilimbi (*A. bilimbi*) is a related species but of less importance. Its very acid fruits are used in pickles and curries.

DURIAN (2) *Durio zibethinus.* Durian is a tree, up to 40 m in height, with a famous and characteristic green-brownish fruit which is about 25 cm long and 20 cm in diameter, and covered with numerous sharp spines. It contains seeds, up to 4 cm long, which are covered with sweet, cream-coloured arils—the edible portion of the fruit. The mature fruit gives out an abominable smell (due in part to sulphides) which has been described in a number of ways, including reference to bad drains; it has been banned on some airlines. Those who can overcome the odour find the arils most acceptable. The arils (20–35 per cent of the fruit weight) are consumed fresh or preserved by drying, fermentation, salting, or deep-freezing. Durian flavour is most acceptable in ice-cream and cookies. Wild durian trees are found in Borneo and Sumatra, but cultivation of the plant takes place throughout South-East Asia, with commercial production in Thailand, Indonesia, Malaysia, and the Philippines. The product is exported to the United States and Europe. The aril contains a large amount of total sugars (about 23 per cent) with 41 mg/100 g of vitamin C. Durian belongs to the family Bombacaceae which includes well-known plants such as kapok (*Ceiba*) and baobab (*Adansonia*).

AKEE (3) *Blighia sapida.* An evergreen tree, 7–25 m in height, which occurs wild and is grown in West Africa. It is also cultivated in the West Indies, particularly Jamaica, where it was introduced in the late eighteenth century. The fruits, about 6 cm in length, are red or yellow and, when ripe, split to expose three shining black seeds surrounded by fleshy arils. The edible arils are usually cooked but great care must be taken in the preparation because the pink tissue connecting the aril to the seed is highly poisonous (the toxic constituent is a peptide), and therefore must be removed. As the unripe arils are also poisonous, those from unripe, damaged, or fallen fruits must not be eaten. A popular dish in the West Indies is akee and saltfish. The Latin name of the plant refers to Captain Bligh of HMS *Bounty*; he could have introduced it to the West Indies.

TWO-THIRDS LIFE SIZE

1 **CARAMBOLA** 1A Flowers
2 **DURIAN**
3 **AKEE** 3A Flowers

Chinese and Japanese fruits

LYCHEE, LITCHI, OR LITCHEE (1) *Litchi chinensis.* It is a member of the family Sapindaceae and related to the rambutan. Lychee originated in southern China where it still grows wild. It was possibly brought into cultivation as early as 1500 BC. The main producing areas are now Taiwan, India, China, Madagascar, Thailand, South Africa, Australia, Mauritius, and Réunion. The plant can be grown in the tropics and subtropics but is very sensitive to its environment; it produces most fruit when the winters are short, dry, cool, and frost-free, and the summers are long, hot, with high rainfall and high humidity.

Lychee is an evergreen tree up to 30 m in height with small yellowish-white flowers. The rounded fruits (about 3 cm in diameter) are red, borne in clusters, and have a warty rind. Within the rind is the flesh—a white edible aril (possibly comprising 70–80 per cent of the fruit) enclosing the brown seed. The aril is consumed fresh or it may be canned. In China, the aril is dried to form 'litchi nuts'. The aril contains about 14 per cent total sugars (glucose and fructose in equal quantities) and 45 mg vitamin C/100 g. There is a small amount of malic acid. Fresh and canned lychees are articles of international commerce.

JAPANESE PERSIMMON, ORIENTAL PERSIMMON, OR DATE PLUM (2) *Diospyros kaki.* This is thought to be of subtropical origin but is well adapted to warm temperate regions. It is of ancient origin in China and Japan, those countries remaining the main areas of cultivation, but it is also cultivated in Italy, Israel, Brazil, California, and South-East Asia. Israel exports the product to Europe as 'Sharon fruit'. The plant is a deciduous tree growing to 15 m in height. Its fruit is a yellowish-red berry, externally rather like a tomato, with a persistent calyx and about 6 cm or more in diameter. The fruit can be eaten fresh, cooked, candied, or processed into ice-cream, jelly, and jam. There are many cultivars of the crop. The fruit of some is astringent because of tannin ('kaki-tannin'). This is not always a desirable feature for the consumer so the non-astringent cultivars are usually preferred. The fruit contains about 19 per cent total sugars (equal amounts of glucose and fructose) and 19 mg vitamin C/100 g. It also contains an appreciable amount of carotenes.

Other *Diospyros* species include *D. blancoi*, *D. digyne*, *D. virginiana*, and *D.lotus.*

LOQUAT OR JAPANESE MEDLAR (3) *Eriobotrya japonica.* This belongs to the family Rosaceae (apples, pears, plums, and other species already described), but is one of the few subtropical representatives. It probably originated in south-eastern China and has been cultivated in China and Japan since ancient times. Loquat is now also cultivated in the Mediterranean region, Australia, South Africa, South America, California, and India. The plant is an evergreen shrub or small tree (5–10 m in height) with narrow leaves, dark green on the upper surface, and a lighter woolly undersurface. Its white flowers give rise to pale-yellow or deep-orange pear-shaped fruit (pomes) some 3–8 cm in length. They can be eaten fresh or processed into jam or jelly. Loquats contain about 6 per cent total sugars (glucose and fructose), a considerable amount of carotenes, but a very small amount of vitamin C (3 mg/100 g). Malic acid is present.

FRUITS AND FLOWERS LIFE SIZE FRUITING BRANCHES × ¼

1 **LITCHI** fruits 1A Flowers 1B Fruiting branch
2 **PERSIMMON** fruit 2A Flowers 2B Fruiting branch
3 **LOQUAT** fruits 3A Flowers 3B Fruiting branch

Date and Palmyra palms

DATE PALM (1) *Phoenix dactylifera.* Date palm is one of the most ancient of food plants in that it was probably domesticated about 5000–6000 years ago in the Middle East. Wild trees still exist in the area. Today it is a major crop in Iraq, Saudi Arabia, Egypt, Iran, Algeria and, on a smaller scale, in other countries of the Arabian peninsula, North Africa, also India, Spain, Mexico, and the United States. The tree flourishes in dry, hot climates, the conditions necessary for successful growth being a moderate winter, a hot summer, little rainfall, and a low atmospheric humidity. For the best fruit production some irrigation is usually required. There are many cultivars available—over 2000 in the Arabian peninsula. Date palm is essentially a plant of dry subtropical areas.

The tree grows to a height of 25 m and its stem is completely covered by the persistent leaf bases of long-dead leaves. It is surmounted by a crown of large, feather-shaped leaves. Suckers or offsets are produced at the base of the young stem. These can be removed and used for propagation, thus ensuring the same genetic constitution of the tree; normally all the suckers are removed because cultivated date palms are preferred with one stem. Date palms are male or female and it is only necessary to plant one male tree to 50–100 females. Fruit set can be improved by artificial pollination by cutting off clusters of male flowers and fixing them among the branches of the female flower-bunch. Full fruit crops are not usually obtained until the tree is 5–8 years old and an adult tree can produce 20–100 kg of fruit per annum. Fruiting may continue until the tree is 80 years old.

The fleshy fruit varies in colour (yellow, brown, red), shape (spherical, oval, elongated, egg-shaped), and size (up to 7.5 cm in length). It contains one hard seed. Because ripening times vary, fresh dates are available for about 8 months of the year. With reference to their moisture content, dates may be classified as soft, semi-dry, or dry. This may relate to a particular cultivar or to the time of harvesting. For Arab populations, dates have always constituted a most important food source. The total sugar concentration is remarkably high but varies, according to the soft or dry date type, from about 30 per cent (soft) to 80 per cent (dry). Also, the constituent sugars vary according to cultivar, some contain half glucose and fructose and half sucrose, others mainly glucose and fructose. As in other fruits, carotenes, the vitamin B complex (apart from B_{12}), and vitamin E are present in small quantities. Vitamin C is present at about 14 mg/100 g but drying reduces it to a trace, although the drying process does not seem to affect the other vitamins.

Although dates are utilized in the western world as appetizers, desserts, in cakes, snack food, and biscuits, a much greater range of uses can be attributed to Arab populations which, in addition to the usage already described, produce date juice as a substitute for sugar in tea or coffee, various syrups, and vinegar. Date seeds have been crushed and included in animal feed and the tree wood used for various building purposes. As with other palms (see p. 18), the tree may be tapped for its sap which can be fermented to give 'toddy' or palm wine; this might be distilled to give a spirit. There is interest in other potential forms of utilization, such as the extraction of various sugars and the inclusion of date flesh in bread. Dates can be utilized as a famine food.

PALMYRA OR BORASSUS PALM (2) *Borassus flabellifer.* This is about the same height as the date palm and, in dry areas where it normally grows, it takes the place of the coconut, which requires a higher rainfall. The palm has characteristic fan-shaped leaves and the trees are male or female. Apart from sugar and toddy production already described (see p. 18), other food products of the palm are: nuts, which contain a sap used as a refreshing drink; the soft kernel of the fruit; and germinated nuts, which have an enlarged fleshy embryo used as a vegetable.

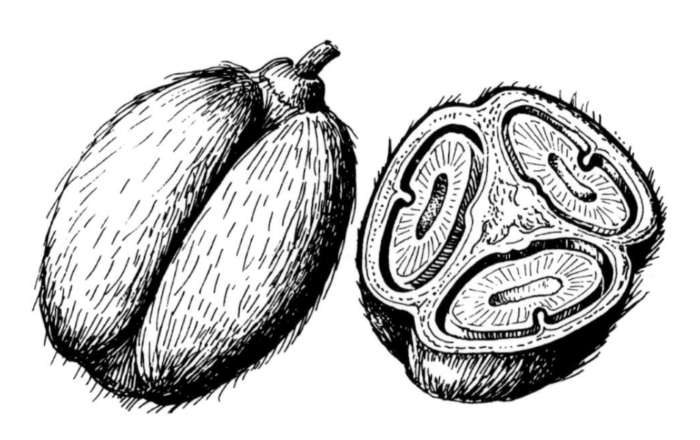

PALMYRA PALM fruit (× ⅓, approximately).

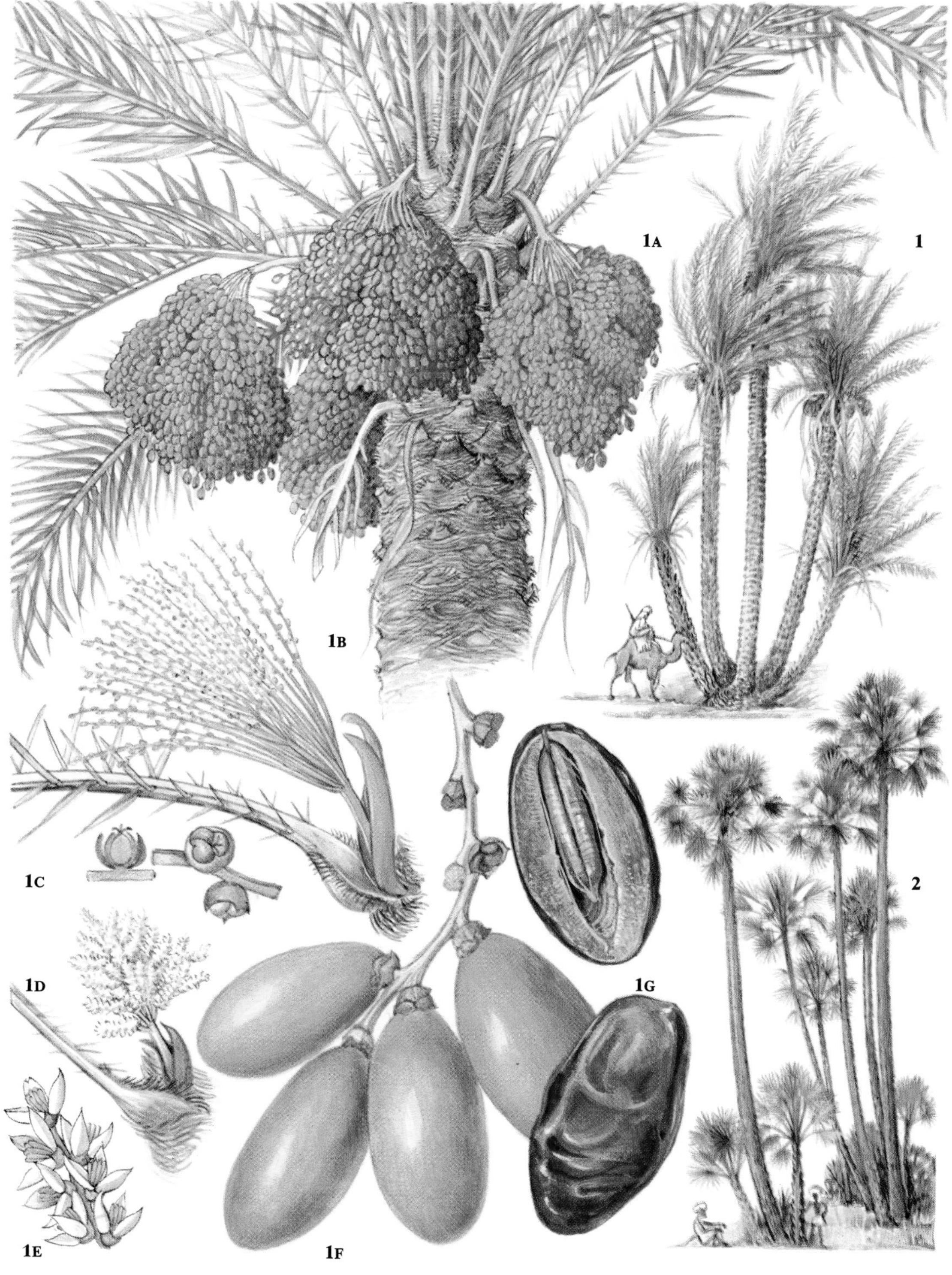

FRUITING SPADICES × 1/20 FLOWERING SPADICES × ⅛ FLOWER AND FRUIT DETAILS × 1

1 **DATE PALM** (small scale) 1A Fruiting spadices 1B Female spadix in flower 1C Detail of female flowers 1D Male spadix in flower 1E Detail of male flowers 1F Ripe fruit 1G Dried fruit
2 **PALMYRA PALM** (small scale)

Banana

The bananas belong to the genus *Musa* but as the edible cultivars are sterile (seedless) hybrid forms, they cannot be given exact species names. They probably evolved in South-East Asia (India to New Guinea) in prehistoric times from wild species (*Musa acuminata* and *Musa balbisiana* are often quoted as the precursors of most banana cultivars), but have now spread throughout the tropics and subtropics. Banana plants are essentially of tropical humid lowlands. The plant is a tree-like herb (2–9 m tall). Its aerial 'stem' is actually composed of the overlapping bases of the leaves above and is known as a 'pseudostem'. The very large leaves often become torn by the wind. Its true stem is an underground 'corm' (about 30 cm long and wide) from which the leaves develop. Suckers also develop from the corm. They are used to propagate the plant. Alternatively, corm pieces bearing buds can be used.

Within a year after the sucker has been planted, the flowering stem will emerge at the apex of the plant and will gradually bend over to hang downwards. At its end this stem carries the sterile male flowers protected by large red bracts which are a conspicuous feature until they wither and fall away. Higher up the stem are borne the groups of female flowers which give rise to the seedless fruit without fertilization (parthenocarpy). Occasionally dark, hard seeds are formed in the fruit if pollen is received from nearby wild banana species. Banana fruits vary in length (6–35 cm) and colour (green, yellow, or red). Although seeds are not actually formed, brown specks which represent the remains of ovules (ovules are structures which, after pollination and fertilization, give rise to seeds) may be seen. Terms employed are: (1) 'stem'—the complete fruit bunch; (2) 'hand'—a cluster of fruits within a bunch; and (3) 'finger'—an individual fruit. A bunch may contain 5 five hands, each with 5–20 fingers. After fruiting the pseudostem is cut down but the plant is continued by suckers. In commercial cultivation, fields of bananas are most usually kept in existence for 5–20 years before replanting is undertaken, but many small farmers in the tropics maintain patches of bananas for 50–60 years.

Bananas are of enormous importance in terms of production, consumption, and trade. Roughly half the production is consumed fresh as dessert, the other half is cooked (fried, boiled, roasted, or baked). Sometimes green bananas presented for cooking are referred to as 'plantains' but it is not always possible to maintain the distinction. Dessert bananas only contain about 2 per cent starch but about 20 per cent total sugars (roughly half sucrose, the rest consisting of glucose and fructose). Vitamin C is present at 11 mg/100 g. In bananas for cooking the situation is reversed as regards the starch/sugar content, there being far more starch than sugar. Banana fruits have many other uses, such as being canned, made into purées, and included in bakery items and ice-creams. Thin slices of unripe fruit are made into chips (a snack food) and ripe fruits are sometimes dried (banana figs). Banana chips contain about 38 per cent starch and 22 per cent total sugars (mainly sucrose), but there is little vitamin C. In dried bananas there is about 50 per cent total sugars but the vitamin C content is much reduced. Dessert bananas are often considered a suitable and easily digestible food for babies and the elderly, also for those with intestinal complaints. In East Africa, beer is made from bananas. The male buds are eaten as a boiled vegetable in parts of South-East Asia (they are sold in specialized shops in the West). The chopped pseudostems and other plant parts are fed to cattle in East Africa.

In East Africa (Africa produces about 50 per cent of the world's bananas), cooking bananas are a staple food—consumption can be as much as 400 kg/person/year. The fruit is now an important item of international commerce, although this international trade only began somewhat over 100 years ago. Bananas for export to North America, Europe, and Japan are produced in Central and South America, the Caribbean, the Philippines, and Africa. Samoa and Fiji export to New Zealand. Production in Australia and South Africa is consumed locally. There are hundreds of banana cultivars. Well-known ones include 'Gros Michel' and 'Dwarf Cavendish'. Short fruit (apple bananas or 'Lady's Fingers') and red bananas are now available to the consumer in western countries. Bananas for export are carried in the green condition in refrigerated ships. In the importing country they are ripened to the yellow state, sometimes with the help of ethylene gas.

A crop closely related to banana is 'ensete' (*Ensete ventricosa*), grown as a staple in some parts of southern and south-eastern Ethiopia at altitudes of 1500–3000 m. The pseudostems and corms are pulped. Either the pulp is cooked fresh or fermented and then made into bread.

PLANTS SMALL SCALE DETAILS × ¼

1 **BANANA** plant 1A Young plant 1B Inflorescence 1B Leaf detail 1D Ripe fruits
2 **'LADY'S FINGERS'** 3 **RED BANANAS**

Kiwifruit and prickly pear

KIWIFRUIT, KIWI, OR CHINESE GOOSEBERRY (1) *Actinidia deliciosa* syn. *A. chinensis.* This is a native of south-western China. The genus contains 50–60 species, the fruits of which were traditionally collected in the wild. In the late nineteenth and early twentieth centuries, plant collectors took some species to other countries, but they were grown mainly for ornamental purposes. Seed of *Actinidia deliciosa* from China was taken to New Zealand in 1904 and that country was primarily responsible for developing the fruit as an article of international importance—hence the name 'kiwifruit'. Although introduced in the early part of the century, it was not until the 1970s that New Zealand produced fruit in significant quantities for the international market. The crop is now also grown in Italy, France, the United States, Chile, Australia, and some other countries. New Zealand restricts itself essentially to one cultivar of kiwifruit—'Hayward'. One advantage of kiwifruit on a worldwide scale is the availability of fruit all through the year because of its excellent keeping qualities, also its production in both northern and southern hemispheres.

Kiwifruit is utilized as a dessert item; it can also be included in salads. As with other fruits, kiwifruit contains very little protein and fat, but there is about 10 per cent total sugars, mainly glucose and fructose. There is a range of minerals and B vitamins, also an appreciable amount of carotenes and a good supply of vitamin C (59 mg/100 g). Another advantage of kiwifruit is that, even after 6 months' storage, 90 per cent of the vitamin C remains. The main organic acid of the fruit is citric.

The plant (1) is a woody climber, up to 10 m in length, with heart-shaped leaves. Plants are male or female. The flowers (**1A**) are white to cream in colour. The fruit is an oval berry (**1B**), 55–70 mm in length, with a light-brown skin and containing up to 1400 small, black seeds. The flesh is of a characteristic green colour. A warm temperate climate suits the plant best but it can tolerate a range of conditions. Sometimes pollen is transferred to female plants by spray application, because the fruit size is affected by the number of fertilized seeds. Kiwifruit belongs to the family Actinidiaceae.

PRICKLY PEAR (2) *Opuntia ficus-indica.* A native of the New World, this was probably introduced to Europe at the beginning of the sixteenth century. It is now naturalized in parts of the Mediterranean region, Africa, Asia, and Australia, where it sometimes becomes a pest. The edible fruit is collected in the wild and also from cultivated plants.

The fruit are covered with minute spines which can cause intense irritation, but these are often removed before sale by rubbing on a rough cloth. As a food item, the fruits may be eaten raw, cooked in oil, or included in a stew. The fruit contains very little protein and fat but about 12 per cent total sugars (glucose and fructose). There is some carotene, a range of B vitamins and 22 mg/100 g of vitamin C. The immature, soft stem segments (minus spines) have been cooked as vegetables, known as *nopitos.*

Prickly pear (2) is a leafless, succulent spiny cactus growing to a height, possibly, of 5 m. The stem consists of joints or pads, covered with spines. Its large flowers (**2A**) are yellow and the fruits (**2B**), 5–15 cm in length, are green, yellow, red, or purple, with numerous seeds.

Prickly pear fruits are now found in western supermarkets.

PITAHAYA *Celenicereus* sp. This is another cactus, the fruit of which is sometimes seen in western supermarkets. It can be used in fruit salads.

HALF LIFE SIZE

1 **KIWIFRUIT** plant 1A Flowers 1B Fruits
2 **PRICKLY PEAR** plant 2A Flowers 2B Fruits

Coffee

Coffee is one of the world's most important crops. The genus Coffea *(family, Rubiaceae) contains a number of species but only three are of economic importance.* Coffea arabica *(arabica coffee) accounts for about 90 per cent of the world's coffee production,* Coffea canephora *(robusta coffee) about 9 per cent, and* Coffea liberica *(liberica coffee) the remainder. The coffee seeds (beans) are roasted, ground, and infused or brewed in hot water to provide a stimulating non-alcoholic drink, the stimulant being the alkaloid, caffeine. Some other plant materials have been used as substitutes for, or added to, coffee.*

ARABICA COFFEE (1) *Coffea arabica.* The plant originated in the south-western highlands of Ethiopia and still occurs naturally there. It was taken to Arabia at an early date, then the Middle East, and into Europe in the seventeenth century. During the eighteenth century, the species was transported to the Caribbean, Central and South America. There was also early transportation of the crop to India, Ceylon (Sri Lanka), the Dutch East Indies, the Philippines, and East Africa. *Coffea arabica* is an evergreen shrub or small tree up to 5 m in height when unpruned. It is an upland species which bears white, fragrant flowers (**1A**) giving rise eventually to crimson fruits (**1**), often described as berries but, in strict botanical terms, actually drupes. Within the fruit skin (**1B**) is the pulp (mesocarp), then the horny parchment (endocarp) which contains, usually, two seeds (green coffee). After harvest, the seeds (beans) are removed from the berries by one of two processes:

1. The wet process, which involves pulping, fermentation in water, and drying. This leaves the beans still enclosed in the parchment which is ultimately removed by mechanical hulling. In commercial terms, these beans are considered the best and are described as 'mild'.
2. The dry process, which involves drying and the removal of the pulp and parchment by mechanical hulling to give 'naked' seeds which are described as 'hard'.

'Instant' coffee is the dried soluble (powder or granules) portion of roasted coffee—the residue from the process is used as an animal feed. Arabica coffee is grown in several regions, including Central and South America (particularly Brazil), India, Kenya, and Tanzania.

Brewed coffee is consumed primarily for sensory pleasure and its stimulatory effects (see p. 207) due to caffeine (1–1.5 per cent). Nevertheless it has some nutritive components, for example in one cup of coffee there are 1–3 mg of niacin and 80–160 mg of potassium. For those who find caffeine too stimulating, there is decaffeinated coffee, available since about 1905. Coffee is also added to desserts, confectionery, ice-cream, and drinks (e.g. Tia Maria).

ROBUSTA COFFEE (2) *Coffea canephora.* This grows wild in African equatorial forests. It is cultivated mainly in West Africa, also in Uganda and Indonesia. The tree grows to a height of 10 m (larger than arabica coffee). Also, compared to arabica coffee, it is cultivated at lower altitudes, it is more tolerant of adverse conditions, and its beans, usually prepared by the dry process, are smaller. The flavour quality of the beans, after roasting and brewing, is considered inferior to arabica coffee but it is less expensive and has been used for some instant and brewed (including expresso) coffees in various countries, including France, Italy, and Spain. Its caffeine content is 2–2.5 per cent.

LIBERICA COFFEE *Coffea liberica.* This is of little economic importance but has been used as a filler with other coffees.

Some plant materials (various seeds and roots) have been, and are, used as substitutes for, or added to, coffee. They relate to coffee because drinks made by infusing them with hot water have coffee-like characteristics, but there is no caffeine. This lack of caffeine appeals to some consumers and the substitute materials are generally cheaper than coffee. Shortage of coffee supplies, for example during wars, has led to the acceptance of substitutes.

CHICORY (3) *Cichorium intybus* Chicory has probably been used as a beverage for hundreds of years but, together with other coffee substitutes, only became important in the eighteenth century. The part used is the root (**3B**) which is chopped, roasted, and ground; the powder may be added to ground coffee or the chicory may be extracted and the extract added to liquid coffee extract. The chicory plant grows wild but since about 300 BC has also been cultivated as a food plant. Chicory brews have a very bitter taste. Countries producing chicory roots include France, Germany, Belgium, Holland, and Poland. Cultivars grown for their roots include 'Magdeburg', 'Soncino', and 'Chiavari'; other cultivars are salad or vegetable plants (see p. 160).

DANDELION (4) *Taraxacum officinale* agg. The root (**4B**) of this plant has been used for making coffee—dandelion coffee was known in 1855. The brew smells like coffee but with a chicory flavour. On the continent of Europe, the leaves (often blanched) are used in salads; in France in the nineteenth century improved forms (giant, curled, and thick-leaved) were developed. The flower-heads (**4A**) are used for making wine.

Dandelion is a perennial herb belonging to the Compositae (Asteraceae) family. It is said to be an 'aggregate' (agg.) species because it has been divided into many species based on the great variation of leaf form and other characters.

In addition to chicory and dandelion, many other different roots and seeds have been used as substitutes. As early as 1886, 42 substitutes were listed. Common substitutes include barley and rye (p. 4), fig (Viennese coffee, p. 100), beetroot (p. 181), lupin (p. 50), and oak acorns.

MATÉ This is a popular drink of South America. It is prepared in a similar manner to tea from the green leaves of *Ilex paraguariensis.* The leaves contain about 2 per cent caffeine. *Ilex paraguariensis* is a small tree, about 6 m in height in the wild state. Little maté is exported.

TWO-THIRDS LIFE SIZE PLANTS × ⅛

1 **COFFEE** fruits 1A Flowers 1B **'ARABICA' COFFEE** berry (in section) and beans
2 **'ROBUSTA' COFFEE** beans 3 **CHICORY** plant 3A Flower-head 3B Root
4 **DANDELION** plant 4A Flower-head 4B Root

Cocoa and tea

COCOA (1) *Theobroma cacao*. Cocoa was cultivated from ancient times by the Indians of tropical Central America, although the crop possibly evolved from wild cocoa in the Amazon forests. Beans (seeds) of the plant were taken by Columbus to Europe. Cocoa beverage became popular in Europe in the seventeenth century, although it was not until the beginning of the nineteenth century that the two modern products of the crop became established, namely slab chocolate and cocoa powder. Today, the plant (family, Sterculiaceae) is cultivated in Central and South America, the West Indies, West Africa, and South-East Asia.

Cocoa is a small, evergreen tree growing to a height of 6–8 m. The white to reddish flowers (**1B**) (and subsequently the pods) are strange in that they grow directly out of the main stem and branches. Pods (10–32 cm long) (**1A**) are green, yellow, red, or purple and contain 20–60 seeds (beans) per pod. The seeds are enclosed in a mucilaginous pulp. In the country of origin, the ripe pods are harvested and split open for the extraction of the wet beans. These are placed in covered heaps, baskets, or wooden boxes ('sweat boxes'). Bacteria and yeasts in the pulp bring about fermentation of the beans (**1C**) ('curing'). This process goes on for some 5–7 days, during which the degraded pulp runs away as 'sweatings'. Fermentation causes chemical changes in the beans, these changes are essential for the development of the characteristic cocoa flavour. Following fermentation, the beans are dried. Although cocoa is a tropical crop, the dried beans are exported to temperate countries for processing in factories. The beans are roasted and their shells (seed coats) are removed to give the cocoa 'nibs' (embryos) which are ground to produce the 'cocoa mass'. At this point there is diversification. To produce cocoa powder, some of the nib fat or 'cocoa butter' (which comprises about 55 per cent of the nib) is removed from the cocoa mass, to leave about 25 per cent fat. The resulting material is cocoa powder. To produce chocolate: cocoa butter, sugar, and milk powder are added to the cocoa mass. The waste bean shells are included in animal feed (there is a legal limit) or used as a fertilizer and mulch. Cocoa-butter substitutes (e.g. palm kernel fat) are utilized.

Cocoa powder is made into a drink, and can be added to milk, cakes, and ice-cream. The beans contain the stimulant alkaloid (see p. 207) theobromine (about 2.5 per cent) and about 0.8 per cent caffeine, but these quantities are reduced after processing. Cocoa differs from the other common beverages (coffee and tea) in that it has a marked nutritional composition, for example cocoa powder contains about 25 per cent fat (saturated), 16 per cent protein, and 12 per cent carbohydrates (about half are sugars).

TEA (2) *Camellia sinensis*. This is one of the world's great beverages and is manufactured from the leaves of a plant which originated near the source of the Irawaddy River and then spread eastwards into south-east China and westwards into Upper Burma and Assam, resulting in the China and Assam tea types, the former being smaller plants with narrower leaves than the latter. Tea has been used as a beverage in China for 2000–3000 years. It was introduced into Japan about 600 AD and into Europe in the seventeenth century. Tea is grown mainly in the subtropics and the mountainous areas of the tropics in countries that include China, India, Sri Lanka, Russia, East Africa, Japan, and Indonesia. Under cultivation, the plant is pruned down to 0.5–1.5 m in height for easier leaf picking but otherwise it can become an evergreen tree up to 15 m in height. The crop appreciates a high rainfall and an acid soil. It is normally propagated by seed, but cuttings are also used.

Harvesting involves plucking the terminal bud and the two or three leaves immediately below it, a shoot with a bud and four leaves gives a poorer-quality product (**2B**). Plucking is usually carried out by hand but there is some mechanized plucking. The process takes place every 7–14 days. A tea bush is normally harvested for 40–50 years, sometimes up to 70–100 years.

About 75 per cent of world production is 'black' tea. The plucked leaves are subjected to:

(1) withering—loss of water;
(2) rolling—maceration of the leaves;
(3) fermentation—enzyme modification of polyphenols, no microorganisms are involved;
(4) firing—a drying process which halts fermentation; and
(5) grading and sorting.

The tea is packed for export in chests lined with aluminium foil. The value of tea as a beverage depends on a number of factors. There is caffeine present (3–4 per cent in the fresh leaf, although this amount is reduced in the final brew). The fermentation process changes the leaf polyphenols to coloured astringent compounds often known as tannins. Various volatile compounds are also important for tea aroma. The production of 'green' tea, mainly in China and Japan, does not involve a fermentation process. *Oolong* is a partially fermented tea. Teas may be flavoured with various essential oils (e.g. lemon, bergamot) or blended with flower petals, spices, dried leaves (e.g. jasmine, camomile, or peppermint). Sometimes *Camellia sinensis* is omitted so that the product is caffeine-free. Instant teas are available.

CAROB (ST. JOHN'S BREAD) *Ceratonia siliqua*. This is a tree up to 15 m in height and a member of the family Leguminosae. It is grown in a number of countries, for example Greece, Italy, Spain, Cyprus, and Israel. The pod is lined with a soft brownish pulp, high in sugar (up to 50 per cent), and is converted into carob flour used in bread, cakes, and 'chocolate'. The seed is a source of a food gum.

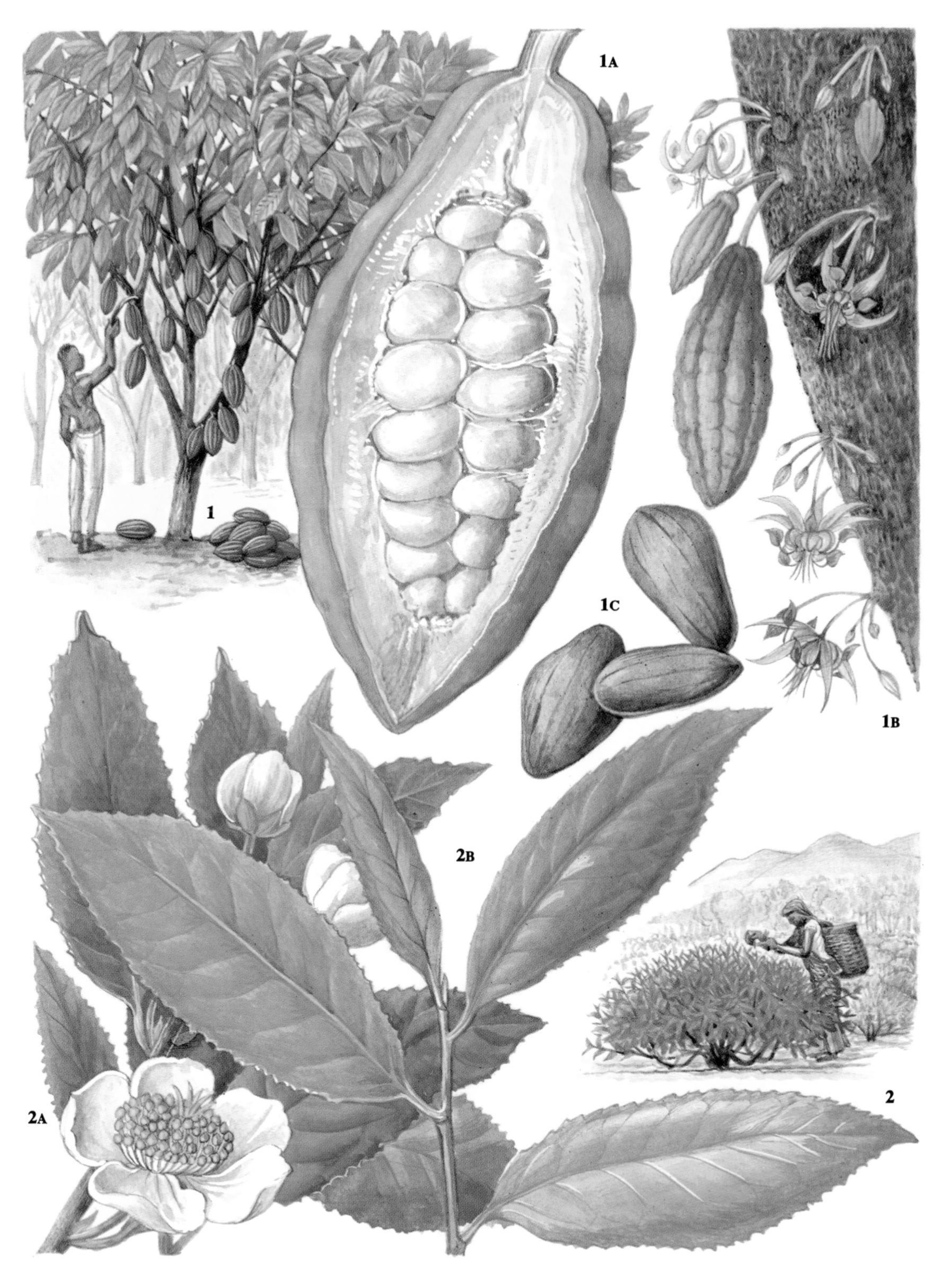

TREE AND SHRUB SMALL SCALE DETAILS LIFE SIZE

1 **COCOA TREE** 1A Unripe fruit 1B Flowers and immature fruits 1C Fermented cocoa beans
2 **TEA SHRUB** 2A Flowering shoot 2B Tea leaves

Tropical vegetable fruits

PAPAYA OR PAWPAW (1) *Carica papaya.* Papaya probably originated in southern Mexico and Costa Rica but is now widely distributed throughout the tropics. Spanish explorers brought the plant to the Caribbean and South-East Asia in the sixteenth century. Among the major producers of the fruit are Hawaii, Brazil, Mexico, Indonesia, and Zaire; it is now commonly available in western countries. Papaya is a fast-growing, tree-like herb, 2–10 m in height (**1A**), with a crown of deeply lobed leaves. Plants are normally male (**1C**) or female (**1B**) with creamy white or yellow flowers, but some hermaphrodite forms do occur. One male is sufficient to pollinate 15–20 females. The fruit (**1**) varies in shape (ovoid, nearly spherical, pear-shaped, or cylindrical) and is 7–30 cm in length. In ripe fruit, the skin is yellowish or orange with an internal flesh of very much the same colour containing numerous seeds enclosed in mucilaginous envelopes.

The ripe fruits are utilized as dessert, in fruit salads, in soft drinks, jam, ice-cream, as crystallized and dried fruit, and are also canned. The plant bears fruit throughout the year and therefore is an important nutritive source in the tropics. The total sugar content is about 9 per cent (equally divided between glucose, fructose, and sucrose). Compared with most fruits, there is a high content of carotenes, and the vitamin C content is 60 mg/100 g. The acid content is low (0.2 per cent—equal quantities of citric and malic acids) and there is therefore a lack of tartness. Unripe fruits are sometimes boiled as a vegetable.

The enzyme 'papain', which is used as a meat tenderizer and beer clarifier, is extracted from the immature fruits by lancing their surfaces and collecting the white latex that exudes, then drying it down to a powder. On a domestic scale the leaves may be used to tenderize meat.

MOUNTAIN PAPAW *Carica candamarcensis.* A native of the Andes, this has smaller fruits that need to be cooked before eating, or it may be made into jam; it can be grown at higher altitudes in the tropics than *C. papaya.*

BREADFRUIT OR BREADNUT (2) *Artocarpus altilis* syn. *A. communis.* This is now found throughout the tropics but is of greatest importance in the Pacific islands where it is a staple or subsistence crop. The fruits (**2**) are unusual in that they contain about 20 per cent starch and may be roasted, boiled, or fried before consumption. The plant is a tree, up to 20 m in height, with deeply lobed leaves and bearing multiple fruits (formed from whole inflorescences) arranged in groups of two or three. The fruit (10–30 cm in diameter) is rounded to cylindrical with a yellow to green rind. Most fruits are seedless—those with seeds, which are edible, are known as breadnuts. The breadfruit is associated with the famous historical incident, the mutiny on the *Bounty* in 1789, which occurred during Captain Bligh's voyage from Tahiti to the West Indies. He had been commissioned to carry breadfruit plants which might be developed as a food source in the West Indies. In 1792–93, Bligh was successful in transporting breadfruit to St. Vincent and Jamaica. Today, the fruit is a popular food item in the Caribbean and is exported to West Indian communities abroad.

JACKFRUIT OR JAKFRUIT *Artocarpus heterophyllus* syn. *A. integrifolia.* This is related to breadfruit and is the largest of the cultivated fruits (30–100 cm in length × 25–50 cm in diameter). It is of ancient cultivation in India but is now distributed throughout the tropics. The fruit can be eaten raw, salted as a pickle, or cooked.

AVOCADO PEAR (3) *Persea americana.* In recent times this has become an important food crop. It originated in Central America but is now grown in nearly all subtropical/tropical regions, including California, Florida, Brazil, South Africa, Israel, Australia, and South-East Asia. The plant is an evergreen tree which may grow to 20 m in height with small green-yellow flowers (**3B**). Its fruit (**3**), 7–20 cm in length, can vary in shape (pear-shaped or rounded). The fruit skin may be green, yellowish, or crimson, within which is the green-yellow flesh, or pulp, containing a single seed. The flesh is highly nutritious with a composition very different to that of most fruits. It contains 15–25 per cent of a monounsaturated fat, 1–2 per cent protein, but little sugar. Vitamin C is present at 6 mg/100 g and useful quantities of the vitamin B complex and vitamin E are reported. There are some carotenes. Avocado is often served as a half-fruit with the addition of various constituents, for example lemon juice or vinegar, or it may be included in salads. The flesh has been used as a sandwich filling, dip, or spread (*guacamole*), and in ice-cream and milk shakes. Because of the value of avocado as a savoury fruit, probably very little fat or oil is extracted on a commercial scale but, because of its mild flavour, it would mix well with other foods. The oil has also been used in cosmetics and toiletries.

The cocktail avocado (imported from Chile, Israel, and South Africa) is occasionally available. It is bullet-shaped, 5–6 cm in length, and lacking the seed.

FRUITS AND FLOWER DETAILS × ½ PLANT AND BRANCHES × 1/18

1 **PAWPAW** fruit 1A Plant 1B Female flower detail 1C Male inflorescence and flower detail
2 **BREAD FRUIT** 2A Male and female flowering branches 2B Detail of female inflorescence
3 **AVOCADO PEAR** 3A Fruiting branch 3B Detail of flowers

Cucumbers and gherkins

The Cucurbitaceae is an important food plant family and includes cucumbers, gherkins, melons, gourds, marrows, squashes, and pumpkins. A number were domesticated in the Americas and were important in pre-Columbian diets but others originated in Africa and Asia. According to species, the fruit (which botanically is normally a 'pepo') is utilized as a vegetable or dessert, but other parts of the plant (young leaves, shoots, seeds, and root) may also be consumed as food. The fruit contains a large amount of water, reasonable amounts of vitamin C, sometimes carotenes, but small quantities of protein, fat, carbohydrate, and the vitamin B complex. The cucurbits are usually climbing plants. Bitter compounds, sometimes present, are 'cucurbitacins'.

CUCUMBER *Cucumis sativus.* Cucumber is thought to have originated in the foothills of the Himalayas, possibly from the wild *Cucumis hardwickii.* In India, the cucumber was cultivated some 3000 years ago and was known in ancient Egypt, Greece, and Rome; it was cultivated in China in the sixth century AD. Cucumbers are now found worldwide.

It is a hairy, trailing, or climbing plant (**1**), climbing by means of unbranched tendrils borne in the axils of the alternate, triangular–ovate, 3–5 angled, irregularly toothed leaves, 7–20 cm long. The yellow flowers (3–4 cm in diameter) are unisexual, the males (**1A**) are in clusters, while the females (**1B**) are solitary or in pairs. The pendulous fruit varies in size (may be up to 90 cm in length), shape (usually elongated but some cultivars, e.g. 'Crystal Apple', have globular fruits), skin colour (in Europe and North America usually green, but elsewhere maybe yellow or rusty brown), and skin texture (some cultivars have warts or soft spines). There are many cucumber cultivars (e.g. **2**, **3**, **4**). Cucumbers like a warm climate and, in temperate countries, the best yields are obtained under glass but, in these countries, some cultivars are grown outside in fields. These are known as 'ridge' cucumbers (**5**) although it is not necessary to grow them on ridges—they may be cultivated as ordinary field crops. Cucumbers grown under glass set fruit by parthenocarpy (without pollination or fertilization). Pollination is avoided by either removing the male flowers or using bee-proof greenhouses because bees and large flies are often the pollinating agents. Fertilized indoor cucumbers become swollen at the end and often taste bitter (cucurbitacins).

Cucumbers are normally harvested before they are mature. They are eaten raw, initially being peeled, sliced, and included in salads. Cucumber can be added to yoghurt to make *raita*, or *mast ó khiao*, useful for tempering the hot effects of curry. They may be pickled. In the East, the large yellow types are boiled and included in stews. Also, the young shoots can be eaten raw or steamed. The seed kernels may be consumed as a snack food.

Cucumbers contain about 96 per cent water with 2 per cent sugars (glucose and fructose). There is little protein, fat, or vitamin B complex, but 2–8 mg/100 g vitamin C, and unpeeled cucumbers contain reasonable amounts of carotenes.

GHERKIN (7) *Cucumis anguria.* This is the West Indian gherkin, evolved from the African *C. longipes*, and probably introduced to the New World through the slave trade. Its fruits, 4–5 cm long, are mostly used in pickles. The plant has deeply lobed leaves.

Gherkins, as known in Europe, are normally small-fruited ridge cucumbers (**6**).

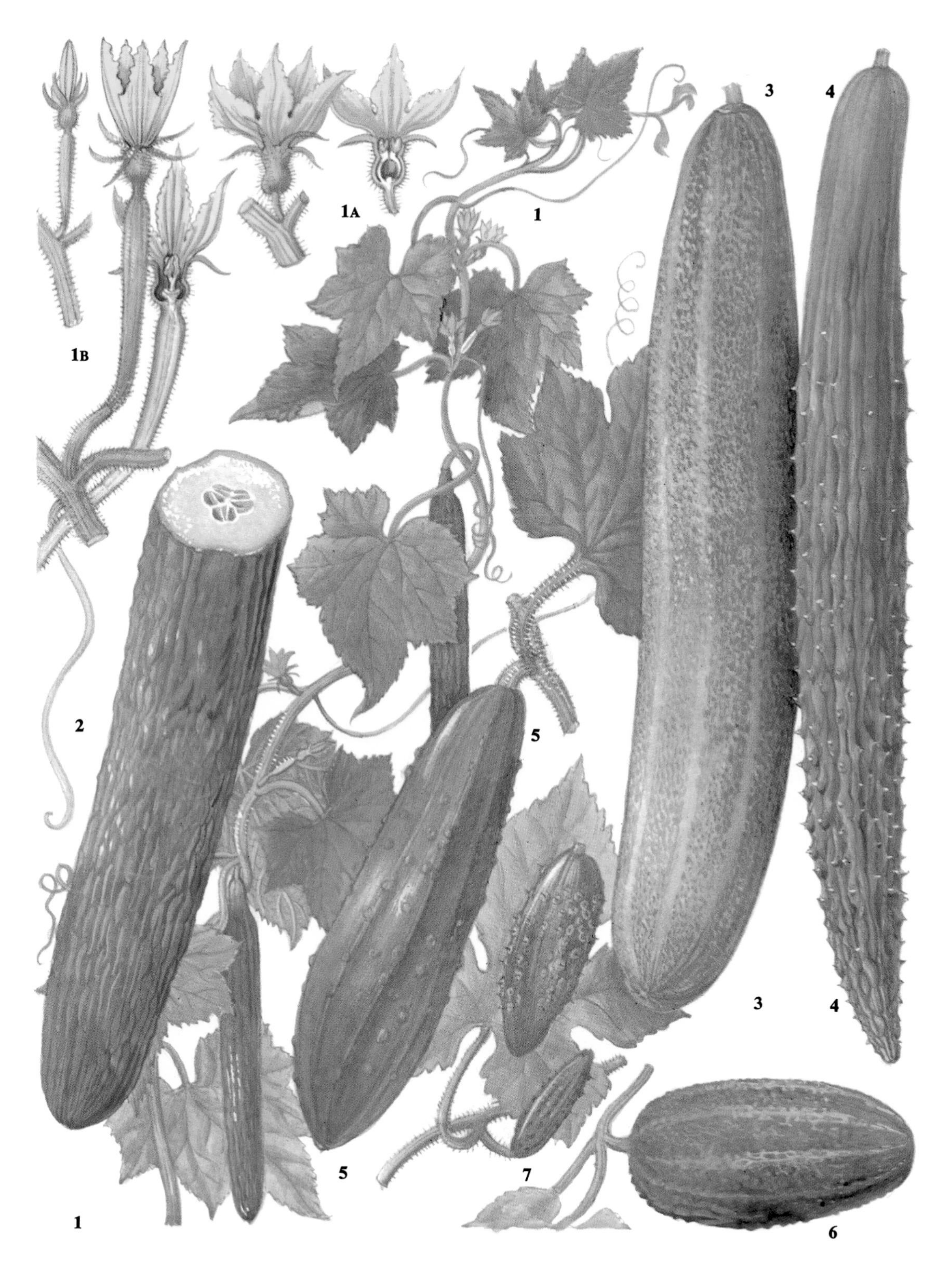

PLANT × ⅛ FRUITS AND FLOWERS × ½

1 **CUCUMBER** plant 1A Male flowers 1B Female flowers
2 '**IMPROVED TELEGRAPH**' 3 '**KAGA**' 4 '**SUYO**'
5 **RIDGE CUCUMBER** 6 '**VENLO PICKLING**' 7 **GHERKIN**

Melons

MELON (1) *Cucumis melo.* Melon probably originated in Africa. However, it does not appear to have been known to the ancient Egyptians and Greeks but came to Europe towards the decline of the Roman Empire. It was then introduced to Asia and is now worldwide. Major producing countries include China, Turkey, India, Spain, the United States, Japan, Italy, and France. Melon is a climbing or trailing, softly hairy annual with five to seven lobed leaves (8–15 cm in diameter). The yellow flowers are male, in clusters (1), or female (**1A**), usually solitary, but hermaphrodite forms do exist. Melons are grown from the tropics to temperate countries. They appreciate much sunshine and heat. In Europe, melons are cultivated in all Mediterranean countries and in France as far north as the Loire on a field scale. Further north (including the United Kingdom) they are grown under glass. They can be planted in pits, on the flat, on hills, or on ridges at a distance up to 1.5 m apart. Fruits ripen about 3–4 months after planting.

The classification of melons is based on fruit characters (surface and flesh) but, because of hybridization, it is not always easy to distinguish exactly between the groups. Nevertheless the following system is often used:

'WINTER MELONS' (2) These are either smooth or shallowly corrugated but not netted. Their green flesh is not strongly scented. They ripen late, are hard-skinned, and stored for a month or more. Consequently, they are popular with growers in warm countries who export to other countries. Included here are 'Honey Dew' melons with a smooth, white skin and pale-green flesh, also 'Spanish' melons with a dark green, shallowly corrugated skin.

'MUSK-MELONS', 'NETTED MELONS', OR 'NUTMEG-MELONS' (3) These are usually distinctly netted, with a raised network on the surface of the skin generally lighter than the overall colour of the fruit, which may be yellowish or green. The surface of the fruit may be smooth (apart from the network) or segmented into broad ribs and grooves. The aromatic flesh is green to salmon-orange. 'Galia' melons belong here.

'CANTALOUPE MELONS' (4) These have a warty or scaly skin but are not netted. They are often deeply grooved and usually have orange-coloured, rarely green, aromatic flesh. Cantaloupes are commercially grown in Europe. The French 'Charantais' belongs here.

'OGEN MELON' (5) This was developed in Israel from the cantaloupe type. The fruit is relatively small, bright orange-yellow ribbed with green, and with sweet, green, aromatic flesh.

In India, China, and Japan, elongated melon fruits (somewhat like cucumbers) are used as vegetables.

The juicy and sweet-tasting flesh of the melon is consumed as a dessert, sometimes with added powdered ginger or lemon juice. As with other cucurbit fruits, melon flesh contains over 90 per cent water but small quantities of protein and fat. However, there is 6–15 per cent total sugars (glucose, fructose, and sucrose) and 6–26 mg vitamin C /100 g. Melons with a pink, orange, or salmon-coloured flesh have a high percentage of carotenes—maybe 5.4 mg/100 g.

HALF LIFE SIZE

1 **MELON** male flowers 1A Female flower 2 '**WINTER MELONS**'
3 '**MUSK-MELONS**' 4 '**CANTALOUPE MELON**' 5 '**OGEN MELONS**'

Water-melon and gourds

WATER-MELON (1) *Citrullus lanatus* syn. *C. vulgaris*. Water-melon is a native of the drier, open areas of tropical and subtropical Africa. It was cultivated in the Mediterranean region starting some 3000 years ago, also in India; China in the tenth century, South-East Asia in the fifteenth century, Japan in the sixteenth century, and the Americas in post-Columbian times. Water melon is grown in all tropical and subtropical countries, and in temperate countries with a continental climate. The plant is an annual climber, with more or less hairy leaves, which are often deeply three- or five-lobed, with the lobes themselves pinnately divided. The yellow flowers (2.5–3 cm in diameter) (**1A**) are unisexual. The fruit is globose or oblong and large (up to 70 cm in length) with a skin varying in colour from golden-yellow to light or dark green (uniform, mottled, or striped). Its flesh is usually red or yellow, but sometimes pink, orange, or white. The seeds vary in colour (black, brown, red, yellow, sometimes white). Water-melon flesh contains over 90 per cent water, a useful source of moisture in arid regions, but little protein and fat. Total sugars are about 7 per cent; there are some carotenes and the vitamin C content is 8 mg/100 g. In China and other Asian countries, the seeds (containing some 40 per cent unsaturated oil, 40 per cent protein) are eaten dry or roasted. Water-melons are found in international trade.

BALSAM PEAR OR BITTER GOURD (2) *Momordica charantia*. This is a tropical climber with orange-yellow fruits, ribbed, and with tubercles, 5–25 cm in length. It is grown in India and other parts of South-East Asia. The fruits are cooked as a vegetable, if mature, usually after treatment with salt-water to remove the bitterness. The fruit may also be included in pickles and curries. Compared with other Cucurbitaceae the fruit is richer in minerals and vitamins, for example vitamin C, 38–70 mg/100 g. The tender shoots and leaves can be cooked as a kind of spinach.

SNAKE-GOURD (3) *Trichosanthes cucumerina*. This is a tropical climbing gourd which grows wild in Asia and Australia but is mostly cultivated in India and the Far East. The slender fruit is greenish white when immature, dark red when mature. It can grow to a length of 180 cm—a weight is often attached to the end to prevent twisting. The immature fruits are boiled as a vegetable or used in curries.

OTHER TROPICAL CUCURBITACEAE The bottle gourd, *Lagenaria siceraria*, is grown primarily for the dry, hard shells of the fruits which are used as containers, but the young fruits may be boiled as vegetables. *Luffa cylindrica* provides loofahs but, again, the young fruit may be utilized as a vegetable. *Benincasa hispida*, the wax- or ash-gourd, is grown chiefly in Asia, where its fruits are cooked as vegetables.

Cucumeropsis edulis and *C. manii* are the 'egusi' melons of West Africa, cultivated mainly for their oily seeds which are cooked and eaten (see text figure).

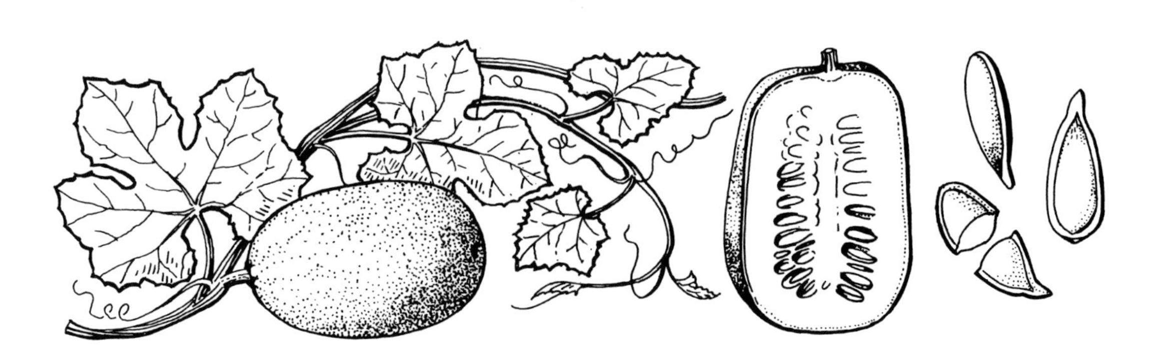

EGUSI MELON fruits and vine (× ¼); seeds (life size).

HALF LIFE SIZE

1 **WATER-MELON** 1A Female flower
2 **BALSAM PEAR** 3 **SNAKE-GOURD**

Marrows, squashes, and pumpkins (1)

Marrows, squashes, and pumpkins belong to the genus Cucurbita *which contains about 25 species but only a few are well known as cultivated plants (*Cucurbita pepo, C. maxima, C. moschata, C. mixta, *and* C. ficifolia*).* Cucurbita *plants have long, trailing stems or can be bushy; the leaves and stems are bristly to the touch; the leaves have a heart-shaped base and are sometimes deeply lobed; and the yellow flowers are unisexual. The classification of the* Cucurbita *species is not easy to understand and the vernacular names (marrow, squash, pumpkin) have often been applied somewhat indiscriminately. The term 'summer squash' is usually applied to an immature fruit with a soft skin used as a table vegetable; the term 'winter squash' to mature fruits used for a variety of culinary purposes.* Cucurbita *fruits are utilized as vegetables, in jams, for pies, are canned, and are included in animal feed. The leaves and flowers can be cooked for consumption. The seeds are eaten and sometimes processed for oil.* Cucurbita *fruits contain a large amount of water (from just under to well over 90 per cent), small amounts of starch, sugars, protein, fat, and vitamin B complex, some carotenes, and 11–21 mg vitamin C/100 g.*

Cucurbita pepo *contains the vegetable marrows, courgettes, some squashes, and some pumpkins. The oldest archaeological remains are from Mexico, dated 7000–5500* BC *and the species formed an integral part of the pre-Columbian maize–bean–squash diet complex. Its cultivars are now widely distributed in the world and are hardier than other* Cucurbita *species. Consequently they are cultivated in cooler countries, for example the United Kingdom and northern Europe.*

VEGETABLE MARROWS (1–2) These have trailing stems or are bushy. They have large cylindrical or round fruits with skins of varying colours (green, cream, or yellow) and greenish-white, rarely yellow, flesh. Vegetable marrow can be eaten as a boiled vegetable on its own or stuffed with ingredients such as meat, onions, and tomatoes.

COURGETTE OR ZUCCHINI (3) These are names applied to 'baby' squashes or marrows, the fruit being cut when young (12–25 cm in length). If allowed to grow on the plant they become vegetable marrows (1–2).

CUSTARD-MARROW, SCALLOPED SUMMER SQUASH, OR PATTY PAN (4) This has white or yellow fruits with scalloped edges (about 15 cm across). It has been grown in the United Kingdom and the continent of Europe for about 400 years but has never achieved much popularity.

SUMMER SQUASHES (5) A well-known example is 'Summer Crookneck' with curved yellow or orange fruits (20–35 cm in length).

PUMPKINS (6) Pumpkins can be large (8–12 kg in weight) or small (2.5–3 kg in weight). The skin is usually orange with cream, yellow, or orange flesh. The small pumpkins are best for pies. These fruits are associated with Hallowe'en.

'Little Gem' is a well-known cultivar of *C. pepo* and is often seen in western supermarkets (see p. 133, **6**). 'Vegetable Spaghetti', another cultivar, is interesting in that the fruit flesh forms loose strands (see p. 133, **2**).

Cucurbita pepo seed kernels contain about 45 per cent of an unsaturated fat, 25 per cent protein, and useful amounts of minerals and the vitamin B complex. They can be consumed raw, roasted, or fried (*pepitos*) and are now sold in western countries. In some parts of Central America they are included in a sweet confection known as *pepitorio*. In the past there has been some industrial extraction of the oil.

Cucurbita maxima includes the winter squashes and pumpkins. Seeds, dated to AD 1200, have been found in Peru but no remains have been found in Mexico and Central America. It was probably domesticated in South America but is now found worldwide. The species is less rough to the touch than *C. pepo* and the leaves are kidney-shaped and not deeply lobed.

WINTER SQUASHES (7) These include 'Hubbard', 'Turban' (the ovaries protrude above the fruit apex, see p. 113, **4**), 'Acorn' (see p. 113, **3**), and 'Banana' squashes (see p. 113, **5**). The 'Mammoth' squashes are very large (16–50 kg).

CHAYOTE OR CHRISTOPHINE (8) *Sechium edule*. This probably originated in Central America and was cultivated as a vegetable by the Aztecs in pre-Columbian times. It has now spread throughout the subtropics and tropics. Chayote has some commercial importance in countries such as Brazil, Mexico, and Costa Rica and is an export article. It is popular in the West Indies.

Chayote is a vigorous, herbaceous, perennial climber (up to 10–15 m in height) with heart-shaped, angled, or lobed leaves, 7–25 cm across. The greenish or cream-coloured flowers are axillary; the male in small clusters, the female solitary. The whitish or green fruit (10–20 cm in length) is usually pear-shaped with longitudinal furrows. Its whitish flesh contains a single, flat, white seed (3–5 cm in length). The fruit structure is different to that of melons, gourds, squashes, and pumpkins. Its root is large and tuberous.

The fruit can be boiled, baked, or fried as a vegetable and included in sauces, puddings, tarts, and salads. As a food source, the flesh composition is very similar to that of the other cucurbits already described. Other parts of the plant (root (with about 20 per cent carbohydrate), young leaves, and tender shoots) are also eaten.

FRUITS AND LEAVES × 1/8 FLOWER DETAILS × 1/2

1 **MARROW** male flower 1A Detail 1B Detail of female flower
2 **VEGETABLE MARROW** 2A Part of plant
3 **COURGETTE** 4 **CUSTARD-MARROW** 5 **SQUASH 'SUMMER CROOKNECK'**
6 **PUMPKIN** 7 **SQUASH 'HUBBARD TRUE'** 8 **CHAYOTE**

Marrows, squashes, and pumpkins (2)

CUCURBITA MOSCHATA Includes pumpkins and winter squashes. This is probably the earliest *Cucurbita* species to have been domesticated, remains have been found in Mexico (Tehuacán) dated to about 5000 BC and in Peru to about 3000 BC. It will tolerate hotter conditions than other cultivated *Cucurbita* species and is the most widely cultivated of these species in the tropics today. The plant has a trailing stem which is softly hairy, not prickly. Cultivars include 'Crookneck', 'Large Cheese', and 'Butternut' (**1**).

CUCURBITA MIXTA Includes pumpkins and winter squashes. It is very similar to *C. moschata* but its cultivation is not so ancient. Archaeological material (dated AD 100–760) has been found in Mexico. It seems to have been widely distributed in Mexico and the south-western United States in pre-Columbian times. The cultivars include 'Cushaw Pumpkin' and 'Tennessee Sweet Potato'.

CUCURBITA FICIFOLIA Known specifically as fig-leaf gourd or Malabar gourd or generally as pumpkin or squash. Remains, dated to 4000–3000 BC, have been found in Peru and it has been cultivated since ancient times in the highlands of Mexico, Central, and South America. It is a vigorous perennial climber with fig-like leaves (18–25 cm in diameter), solitary yellow-orange flowers (up to 7.5 cm in diameter), and a globular or cylindrical fruit (15–50 cm in length) with a white to green skin, with white stripes, containing a coarse flesh and black seeds. The young fruits are consumed like summer squash or cucumbers; the flesh of older fruits is candied or may be fermented to give an alcoholic drink. As with some other cucurbits, the young leaves, shoots, flowers, and seeds are consumed as food. Its nutrient composition is similar to other members of the Cucurbitaceae already described.

FRUITS × ½

1 '**BUTTERNUT**' 2 '**VEGETABLE SPAGHETTI**'
3 '**ACORN**' 4 '**TURBAN**' 5 '**BANANA**' 6 '**LITTLE GEM**'

Tomatoes

*The family Solanaceae contains a number of very important food plants, including tomato, potato (see p. 186), aubergine, and sweet pepper (*Capsicum*). There are also some drug plants: tobacco (*Nicotiana tabacum*), belladonna (*Atropa belladonna*), thorn-apple (*Datura stramonium*), and henbane (*Hyoscyamus niger*).*

TOMATO (1–2) *Lycopersicon esculentum.* This is thought to have evolved from the cherry tomato (*Lycopersicon esculentum* var. *cerasiforme*) (3) which occurs wild in Peru, Ecuador, and other parts of tropical America. The actual area of tomato domestication is considered to be Mexico. Early in the sixteenth century the Spaniards took the plant to Europe and it then spread throughout the Old World. It was known as the 'golden apple', 'love apple', or 'Peruvian apple'. In the centuries that followed the tomato was taken to many other parts of the world, for example towards the end of the eighteenth century it was introduced into the United States of America from Europe.

On a worldwide basis, tomato is one of the most important vegetable or salad plants. Its fruits are consumed raw, or cooked and processed into a great variety of products (juice, soup, sauce, ketchup, purée, paste, and powder). Tomatoes may be canned (both fruit and juice) or dried. Green tomatoes are utilized in pickles and preserves. In the production of a number of tomato products, a considerable amount of waste is produced. This can be used as animal feed or possibly the source of seed oil, which is unsaturated.

There are many tomato cultivars and the plant shows a wide climatic tolerance, being grown in tropical and temperate regions in the field, under plastic shelter, or in greenhouses. Although cultivated worldwide, countries in which large-scale production takes place include Russia, China, the United States, Egypt, and Italy. The tomato is a herbaceous plant capable of perennial growth but normally cultivated as an annual. Under natural conditions it forms a straggling bush; under cultivation some cultivars are grown as bushes but it is best if the fruits do not come into contact with the soil, consequently a mulch of straw may be used. Other cultivars are grown as single stems, usually supported by string or canes. Side shoots are normally removed and, when a limited number of fruit tresses are produced, the tip of the main stem is pinched off to prevent further (indeterminate) growth, although some cultivars have been produced with determinate growth. Tomatoes are harvested by hand or by mechanized harvesting.

The plant is 0.7–2 m in height with pinnate or bipinnate leaves. All green parts of the plant bear glands or hairs and give off a characteristic strong odour. The flowers, borne in racemes of 3–11 or more, have a six-lobed green calyx and six yellow petals (1A) (some cultivars have five sepals and five petals). In a number of cultivars the yellow anthers enclose the stigma, thus ensuring self-pollination; in others, particularly in tropical regions, the stigma protrudes beyond the anthers, thus allowing some cross-pollination. Cross-pollination is purposely carried out to produce F_1 hybrid seed.

The fruit is a fleshy berry and shows variation in colour—red (1) or yellow (2). There is variation in size (1.5–10 cm in diameter); cherry tomatoes (3) are small, while 'beef' tomatoes are large. The fruit surface is smooth or furrowed. Tomatoes are usually globose in shape but 'plum' and 'pear' (4) types do exist. The fruit contains a large amount of water (over 90 per cent); very little protein and fat; about 3 per cent carbohydrate (glucose and fructose); a range of minerals (particularly potassium—250 mg/100 g); a considerable amount of carotenes; small quantities of the vitamin B complex and vitamin E; vitamin C content is 17 mg/100 g. The carotenes include lycopene and β-carotene, but it has been reported that lycopene has no pro-vitamin A activity. As with other fruits, the sensory quality of the tomato is related to the interaction of sugars, acids (malic and citric), and a large number of volatiles. Tomatoes contain the alkaloid 'tomatine' which decreases in quantity as the fruit ripens. Tomatine, in contrast to some other alkaloids in the Solanaceae, is not toxic. The tomato is an important dietary constituent, particularly in communities where large amounts of the fruit are consumed. An interesting application of genetic engineering has been to reduce the amounts of cell wall softening enzymes. Such treatment leads to a longer shelf-life (see p. xv).

TREE TOMATO OR TAMARILLO (5) *Cyphomandra betacea.* This is a native of the Andes but is now distributed throughout the subtropics; it is popular in New Zealand. The plant is a shrub or small tree up to 4 m in height. Its egg-shaped fruits (about 10 cm in length) have a yellow, orange, or deep-red skin. The flesh, containing numerous seeds, is yellowish-orange, deep red, or purple. It is utilized fresh as an ordinary tomato, particularly if the rather bitter skin is removed, and is processed to give juice, jam, chutneys, sauces, and flavouring for ice-cream. The fruit contains about 85 per cent water; small amounts of protein and fat; about 5 per cent total sugars (half of which is sucrose, the rest glucose and fructose); a range of minerals, with 300 mg potassium/100 g; a large quantity of carotenes, the usual range of B vitamins, a significant amount of vitamin E (1.86 mg/100 g) and vitamin C (23 mg/100 g). The fruit is an article of international commerce.

FRUITS AND FLOWERS × ⅔   FRUITING BRANCHES × ¼

1 **TOMATO** 1A Flowers 1B Fruiting branch
2 **'GOLDEN TOMATO'** 3 **'CHERRY TOMATO'** 4 **'PEAR TOMATO'**
5 **TREE TOMATO** 5A Flowers 5B Fruiting branch

Some plants of the potato family with edible fruits

GARDEN HUCKLEBERRY OR SUNBERRY (1) *Solanum intrusum* syn. *Solanum nigrum* var. *guineense.* This is said by some to have originated in Africa but it could have evolved in America as a hybrid. The plant is an annual which can grow from seed to a height of 1 m in a season. The purple-black fruits, about 2 cm across, can be included in pies, jellies, and jams but are bitter unless sugar is added. Garden huckleberry is related to the black nightshade (*Solanum nigrum*); however, it is a taller plant, the leaves are larger, ovate, and entire (**1B**), it has brownish-yellow anthers (**1A**), and larger fruits. Apparently its fruits are harmless, yet those of black nightshade should be avoided because of solanine alkaloids (see p. 207). The plant is not of commercial importance.

AUBERGINE, EGG PLANT, OR BRINJAL (2) *Solanum melongena.* Aubergine is a native of tropical Asia. It was first brought into cultivation in India, where wild forms occur, at an early date. The Arabs introduced it to Spain in the eighth century AD; the Persians to Africa. It is now cultivated throughout tropical, subtropical, and warm temperate countries as a field crop. In cool temperate regions it is grown under glass. The fruit is cooked as a vegetable, boiled, fried, or stuffed, and included in curries or dishes such as 'moussaka'. The fresh fruit contains over 90 per cent water, very little protein and fat, about 2 per cent of total sugars (mainly glucose and fructose), a range of minerals with a large amount of potassium, some carotenes, small quantities of the E and B vitamins, and a little vitamin C (4 mg/100 g).

Aubergine is a perennial, usually grown as an annual with erect or spreading, tough, herbaceous, branching stems (**2B**), 0.5–1.5 m in height. All parts of the plant are covered with a woolly, greyish felt of hairs; there are sometimes spines. The leaves are large (7–15 cm in length), ovate, wavy-edged or lobed. The flowers (3–5 cm in diameter), borne singly or in groups, have a deeply lobed, toothed calyx and purple or violet petals. Its fruit (5–15 cm in length) may be egg-shaped (hence the name egg-plant), oblong, or sausage shaped with a white, yellow, or purple skin and containing many seeds.

Quite a number of other *Solanum* species are grown in the tropics for their fruits and leaves, which are used as pot-herbs. These include: naranjilla (*S. quitoense*), a very popular fruit in Colombia and Equador; pepino (*S. muricatum*), a common fruit in Colombia, Equador, Peru, Bolivia, and Chile but also now introduced to countries such as Morocco, Spain, Israel, Kenya, and New Zealand; and *S. aethiopicum* and *S. macrocarpon* in Africa.

GROUND CHERRY, HUSK TOMATO, STRAWBERRY TOMATO, OR DWARF CAPE GOOSEBERRY (3) *Physalis pruinosa.* This is a native of eastern and central North America. It is not of commercial importance. The plant is an annual with spreading branches, up to about 1 m in height. The leaves (**3B**) are heart-shaped, shallowly toothed, greyish-green when young, softly hairy, with one basal lobe larger than the other. Its yellow flowers (**3A**) (about 1 cm in diameter) have five brownish patches in the throat. The fruit (about 2 cm across) is a roundish, yellow berry (3) enclosed in a lantern-like, light-brown calyx or husk. It is sweet, slightly acid, and it utilized in jams, jellies, and tarts.

CAPE GOOSEBERRY OR GOLDENBERRY (4) *Physalis peruviana.* This originated in South America but is now cultivated in Hawaii, California, South Africa (hence the name 'Cape'), East Africa, India, New Zealand, Australia, and the United Kingdom. It is found in international commerce. The plant is similar to ground cherry (*Physalis pruinosa*) but with leaves (**4A**) which are equal at the base, and slightly longer flowers of a bright yellow. The fruiting calyx or husk is considerably thicker and larger (4). Cape gooseberry fruits are eaten fresh, used in jams and jellies, canned, and included in *petits fours.* They contain about 11 per cent total sugars; the usual range of minerals, with a high content of potassium (320 mg/100 g); similarly a high content of carotenes; some vitamin B complex and vitamin E, also a significant amount of vitamin C (49 mg/100 g).

The bladder cherry or Chinese lantern plant (*Physalis alkekengi*) is grown mainly for the ornamental value of its large, red, fruiting calyx. The red berry is edible but the calyx should not be eaten.

TOMATILLO OR JAMBERRY (5) *Physalis ixocarpa.* Tomatillo is a native of Mexico and was cultivated there before the tomato. It is a perennial, but often grown as an annual, less hairy than Cape gooseberry, and bearing generally smaller leaves which are either toothed or not toothed (**5B**). The flowers (**5A**) are yellow (about 2 cm across) with purple-brown blotches. The fruit is large, yellow or purple in colour, and fills the yellowish husk completely. It may be used in sauces and preserves. A number of hybrids are available, including the cultivar 'New Sugar Giant' (5).

FRUITS AND FLOWER DETAILS LIFE SIZE BRANCHES × ¼

1 **GARDEN HUCKLEBERRY** 1A Flowers 1B Branch 2 **AUBERGINE** 2A Flower 2B Branch
3 **GROUND CHERRY** 3A Flower 3B Branch 4 **CAPE GOOSEBERRY** 4A Branch
5 **JAMBERRY 'NEW SUGAR GIANT'** 5A Flower 5B Branch

Chillies and peppers

The word 'pepper' is used in English to describe a group of spices which are usually derived from two quite different kinds of plants and from two different families. Chillies and other spice types come from the bushy Capsicum *species (family, Solanaceae) and 'white' or 'black' pepper come from climbing vine,* Piper nigrum *(family, Piperaceae).*

In pre-Columbian times, Capsicum *was widely used in Central and South America, the Caribbean area, and Mexico. Archaeological evidence suggests that the Indians ate chillies as early as 7000* BC, *although these were probably wild plants. It has been stated that* Capsicum *was brought into cultivation between 5000 and 3000* BC. *On the return of his first voyage (1492), Columbus took the plant to Europe and it quickly spread to Africa and Asia. As a spice, it gradually became more important than* Piper nigrum *in the East, which constituted a marked change in food culture. Because of the very long period of its cultivation and human selection, many* Capsicum *forms exist and this makes classification difficult. However, the two important species normally recognized are* Capsicum annuum *and* C. frutescens.

SWEET PEPPER, PAPRIKA, AND CHILLI (1) *Capsicum annuum.* This is an annual plant grown from seed. It can be cultivated in tropical countries up to an altitude of about 2000 m and, although sensitive to frost, can also be cultivated in warm temperate countries and in cooler climates under protection. The plant (1) grows to a height of 1–5 m with leaves variable in size (1.5–12 cm in length) and white or greenish flowers (10–15 mm across) borne singly. The fruit (**1A**) is a hollow, many-seeded berry which is very variable in size (0.8–30 cm in length), shape (elongated, top-shaped, almost spherical), colour (yellow, red, brownish-purple, and often picked when still green), and pungency. The pungent principles are 'capsaicinoids'. Sweet or bell peppers (**1,1A**) are large, mild, and eaten raw or cooked. 'Paprikas' are mild or slighly pungent cultivars with a thinner flesh than sweet pepper and varying in shape. The dried fruits are ground to a powder which is added as a flavouring to foods such as eggs, cheese, and potatoes. Spanish paprika (sometimes known as 'pimiento'—not to be confused with allspice, see p. 142) and Hungarian paprika (included in the famous dish 'goulash') are well known but the type is grown in other countries. 'Chillies' are hot or strong-tasting cultivars of *C. annuum,* cultivated in many tropical countries, included in curry powder, and added to enhance the flavour of bland dishes. *Capsicum* fruits contain small amounts of protein, fat, and sugars, but a large quantity of carotenes (maybe as much as 4770 µg/100 g) and are an excellent vitamin C source (up to 340 mg/100 g in fresh material). *Capsicum* 'oleoresins' are fruit extracts used to flavour and colour various foods and beverages. Extracts are also included in aerosols to deter 'muggers'.

CHILLI OR BIRD CHILLI (2–3) *Capsicum frutescens.* This is a perennial usually grown as an annual with its flowers borne in groups rather than singly. The fruits are small and conical (2–3 cm in length) and extremely pungent. They are cultivated in many tropical countries and utilized as *C. annuum* chillies. 'Tabasco' sauce is made by pickling the fruit pulp in brine or vinegar and 'cayenne' papper is made from the powdered dried fruits. 'Red' pepper is not so pungent as cayenne.

WHITE OR BLACK PEPPER (4) *Piper nigrum. Piper nigrum* originated in the Western Ghats of India where it still grows wild. It may have been in cultivation for some 3000 years. Pepper was known to the ancient Greeks and Romans. The plant was probably taken to Java by Hindu colonists between 100 BC and AD 600. In the Middle Ages it was well known in Europe. Today it is one of the most important spices, being used as a condiment and with many culinary uses such as seasoning dishes and being included in sausages and similar products. As explained above, the introduction of *Capsicum* early on into the East reduced the usage of white or black pepper in that area. Major cultivation is carried out in India, Sarawak, Indonesia, and Brazil.

It is a climbing vine which, under cultivation, grows to a height of about 4 m. The small fruits (4–6 mm in diameter) are borne in long, hanging spikes (**4A**). The fruits (peppercorns), from which both white and black pepper are made, turn red when ripe. To make black pepper, unripe peppercorns are sun-dried leading to black and wrinkled structures (**4B**); to make white pepper, the outer coverings are removed from the ripe fruit (**4C**). Both products are finally ground to a powder.

The pungency of pepper is related to alkaloids (see p. 207) of which piperine is the most important. These, together with a volatile oil, constitute the sensory qualities of pepper. Pepper extracts (oleoresins) are used in food.

Other pepper-like plants are *Schinus molle* (pepper tree, Andes); *Xylopia aethiopica* (Guinea pepper, tropical Africa); *Zanthoxylum* spp. (Szechuan pepper, Far East); and *Aframomum melegueta* (Melegueta pepper, West Africa).

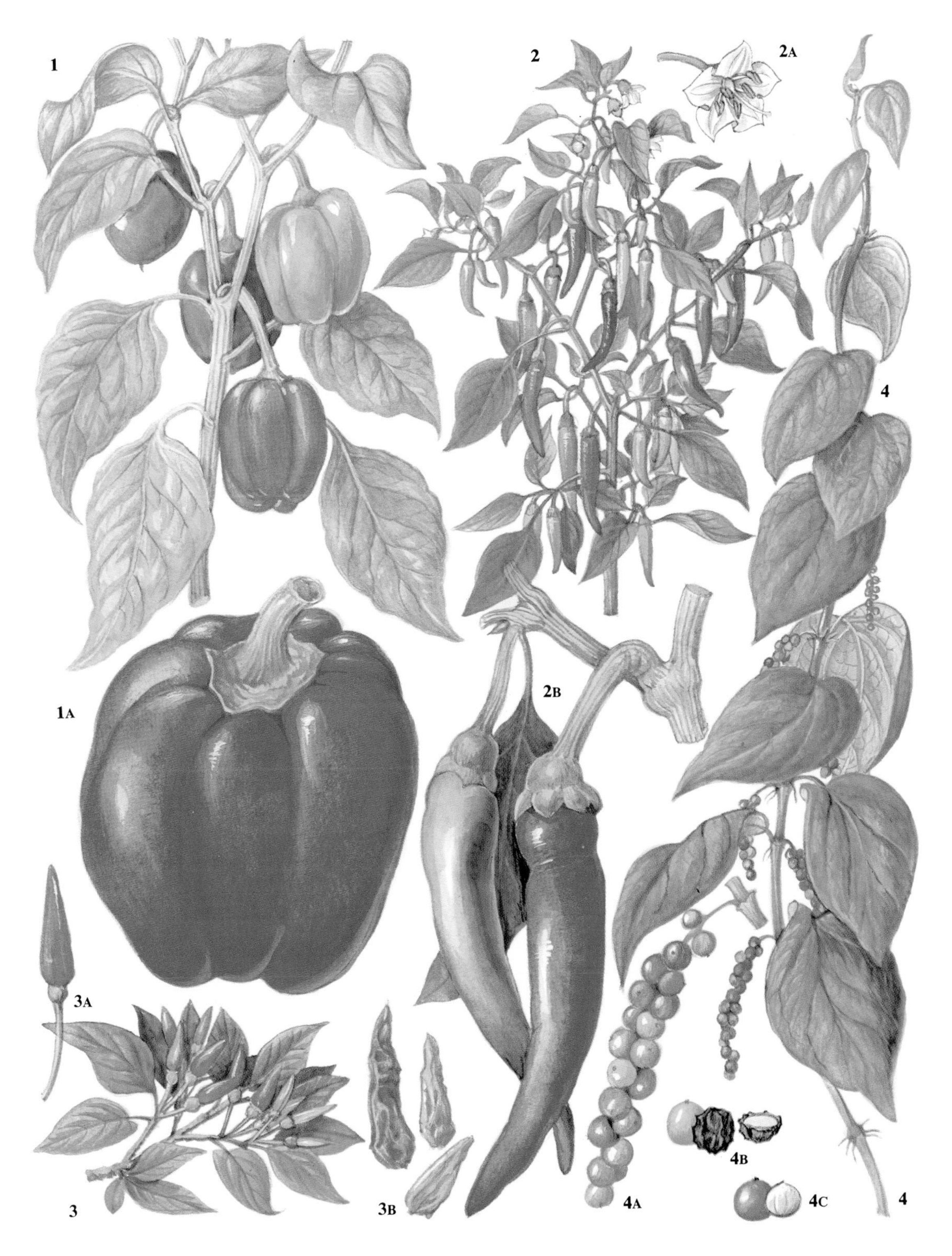

PLANTS × ¼ FLOWER AND FRUITS × ⅔ PEPPERCORNS × 1

1 **SWEET PEPPER** plant 1A Fruit 2 **RED PEPPER** plant 2A Flower 2B Fruits
3 **CHILLI** plant 3A Fruit 3B Dried chillies
4 **PEPPER** plant 4A Panicle of fruits 4B Black peppercorn 4C White peppercorn

Spices and flavourings (1)

(For general information on spices, see p. xix)

VANILLA (1) *Vanilla fragrans* syn. *V. planifolia.* Vanilla is a fleshy orchid climbing by means of aerial roots on various trees or other supports to a height of 10–15 m in natural conditions, but under cultivation to a height which will allow hand-pollination and harvesting. The plant is native to Mexico and Central America where the fruits or pods, commonly known as 'beans', were utilized by the Aztecs. The Spaniards took the beans to Spain in the sixteenth century. It is now found throughout the tropics but grows best in climates with frequent rain. Major producers include Madagascar (Malagasy Republic), the Comoro Islands, Mexico, Réunion, and Indonesia. Extracts (in alcohol or oleoresins) of the cured bean are used to flavour ice-cream, chocolate, and other confectionery, also in perfumery and medicine.

In Mexico and Central America the greenish-yellow flowers are bee-pollinated to some extent but in all areas hand-pollination is necessary for full pod production. This is carried out by transferring the pollen masses to the stigmatic surface with a small, pointed stick (often of bamboo). The green pods (10–25 cm in length) are usually harvested before they are fully ripe. To develop their aroma and flavour, pods are subjected to a 'curing' process which involves 'killing' or wilting (by sun or oven treatment), 'sweating' (raising the temperature), drying, and storage (3 months or longer). The cured pods (**1A**) are dark brown or black and may show vanillin crystals on their surface. They are packed into tin boxes lined with waxed paper for export.

The aroma and flavour is related to many constituents of the pod but the most important is vanillin, a volatile ester. Synthetic vanillin (made from a number of sources, including wood pulp) supplies 95 per cent of the world demand for vanilla flavour but it does not have the full sensory qualities of the natural product.

NUTMEG AND MACE (2) *Myristica fragrans.* This is an evergreen tree which grows to a height of 4–10 m (sometimes 20 m) and is unique among spice plants in that it produces two different products. It originated in the Moluccas but is probably never found wild. Both spices were well known in Europe by the end of the twelfth century. The crop has been introduced into almost every tropical country but most commercial supplies of nutmeg and mace come from Indonesia and Grenada (West Indies).

Trees are usually male or female with creamy-yellow flowers (up to 1 cm in length) (**2**). The yellow, pear-shaped fruit (6–9 cm in length) (**2A**) splits open when ripe to expose the purplish-brown seed enclosed by a scarlet network known as the 'aril'. The seed, after its coat or shell is removed (**2B**), is the nutmeg spice; the aril is the mace spice which, after drying, turns brown (**2C**). Nutmeg (often powdered) is used to flavour milk dishes, cakes, and punches; mace in savoury dishes, pickles, and ketchups.

Damaged nutmegs have sometimes been processed to make nutmeg butter (triglyceride oil) and nutmeg oil (an essential oil).

Both nutmegs and mace contain myristicin and elemicin, which are poisonous. Consequently the spices should only be used in very small quantities.

CINNAMON (3) *Cinnamomum verum* syn. *C. zeylandicum.* This is said to be one of the oldest spices but the proposed evidence is confusing, although it was certainly known to the ancient Greeks and Romans. The species originated in Sri Lanka and is now a major commercial crop in that country, the Seychelles, and the Malagasy Republic. In the wild state it is a tree growing to a height of 17 m; when cultivated, to a height of 2.5 m. The plant can be propagated from seed or cuttings. The spice is the bark, an unusual feature among spice plants. Every 2 years the young shoots are cut close to the ground and the bark removed in long strips. The outer bark is scraped off and the strips slowly dried to form 'quills' which are pale brown in colour, curl into a semi-tubular shape when dry, and can be folded inside each other for packing (**3A**). This harvesting every 2 years produces a bushy plant. In the Seychelles fragments of bark are produced rather than quills. Ground cinnamon is used in bakery products, sauces, pickles, puddings, curry powder, and confectionery. Essential oils are produced from the leaves and bark.

CASSIA This is produced from *Cinnamomum* species other than *C. verum*; its utilization is the same as cinnamon and it is cultivated in China, Indonesia, and Vietnam.

CARDAMOMS (4) *Elettaria cardamomum.* These are the fruits of a tall (2–5.5 m in height) herbaceous perennial belonging to the ginger family, Zingiberaceae. The species grows wild in the monsoon forests of South India and Sri Lanka. Major cultivation also takes place in these countries, together with Guatemala. Seeds constitute the spice element but they retain their pleasant aroma/characteristic slightly pungent taste if kept within the capsules (fruits) (**4A**) which are harvested before they are fully ripe. Cardamoms are utilized in rice, vegetable, and meat dishes; as a coffee flavouring; in baked goods and curry powder. Cardamom essential oil is used as a food flavouring, in perfumery, and for flavouring liqueurs. Substitutes for the true cardamom are melegueta pepper (*Aframomum melegueta*) and *Amomum* spp.

FLOWERING AND LEAFY SHOOTS × ⅔ FRUITS AND BARK × 1

1 **VANILLA ORCHID** 1A Dried pods
2 **MYRISTICA FRAGRANS** flowers 2A Fruit 2B Seed, **NUTMEG** 2C Aril, **MACE**
3 **CINNAMON** leaves 3A Bark 4 **CARDAMOM** flowers 4A Fruits

Spices and flavourings (2)

BAY LAUREL (1) *Laurus nobilis.* This belongs to the avocado family (Lauraceae) and originated in Asia Minor and the Mediterranean area. The leaves are used in European, especially Mediterranean, cooking. The important volatile oil is cineole. Fresh leaves are often bitter and dried leaves, if kept for too long, lose much of the cineole. In classical times wreaths of laurel leaves were used for crowning the victorious—hence titles such as Nobel or poet laureate.

It is an evergreen shrub or tree, up to 20 m in height, with dark-green leaves (4–10 cm in length) which often have wavy margins. The inconspicuous greenish-yellow flowers are unisexual and give rise to glossy black berries (about 1 cm across). Bay laurel should not be confused with the cherry laurel (*Prunus laurocerasus*, Rosaceae) or Japanese laurel (*Aucuba japonica*, Cornaceae).

SAFFRON (2) *Crocus sativus.* The dried stigmas (**2A**) of this member of the Iridaceae family constitute the world's most expensive spice (about 200 000 dried stigmas from 70 000 flowers give 1 lb (0.45 kg) of pure saffron). The plant has an underground corm, producing leaves and lilac-purple flowers with protruding red stigmas in the autumn. It probably originated in Asia Minor but has been cultivated in the Mediterranean area since ancient times. Today the main areas of cultivation are Spain, Turkey, and India. It has long been considered a very desirable reddish-yellow colouring agent (the pigment is crocin) and flavouring agent (essential oils) for food—cheeses, butter, pastry, confectionery—and is found in the French *bouillabaisse* and the Spanish 'paella'. Because of its high price, saffron has often been adulterated with turmeric (see p. 144) or safflower (*Carthamus tinctorius*). Saffron should not be confused with the autumn crocus (*Colchicum autumnale*, Liliaceae).

CAPERS (3) *Capparis spinosa.* Capers are the unopened flower buds (**3A**) of a straggling, spiny bush which occurs wild in the Mediterranean area but also in North Africa and Asia Minor to the Gobi Desert. It has been known for thousands of years and is cultivated in the Mediterranean countries. The buds are pickled in vinegar and used as a condiment, also included in sauces (e.g. 'tartare') and with raw meat (steak tartare).

It is a small shrub (family Capparaceae) with thick leaves, each with a pair of spines at its base. The white or pinkish flowers are 4–6 cm across.

BLACK MUSTARD (4) *Brassica nigra.* Black mustard seed flour was once included in table mustard, together with white mustard (5), but since the 1950s this has been largely replaced by brown or Indian mustard (*Brassica juncea*), which is an annual herb growing to a height of 30–160 cm and with brown or yellow seeds. In the East particularly, *B. juncea* is also utilized as a salad, vegetable, and oilseed plant.

WHITE MUSTARD (5) *Sinapis alba* syn. *Brassica hirta.* White mustard seed (yellowish) flour, together with that of *B. juncea*, is found in table mustard, of which there are many kinds. Both seed types produce essential oils (*B. juncea*, allyl isothiocyanate; *S. alba*, parahydroxybenzyl isothiocyanate) which are responsible for the 'hotness' of mustard; however, the latter is milder. White mustard seedlings may be part of 'mustard and cress' (see p. 162).

Mustards belong to the family Cruciferae (Brassicaceae) characterized by pod-like fruits (**4**, **5**, **5A**) and four petals arranged in the form of a cross.

CLOVES (6) *Syzygium aromaticum* syn. *Eugenia caryophyllus.* Cloves are the dried unopened flower buds (**6A**) of an evergreen tree, growing to a height of 15 m, belonging to the family Myrtaceae, which is indigenous to the Moluccas (Indonesia). The spice was introduced into China and India in very early times and was known throughout Europe by the Middle Ages. By the nineteenth century, Zanzibar (Tanzania) and the Malagasy Republic (Madagascar) had become important producers. The fungus disease 'sudden death' has led to significant losses of the plant in Zanzibar for over a century. The spice is used for domestic culinary purposes; in sauces and pickles, also the flavouring for *kretek* cigarettes in Indonesia. Clove oil (the main constituent is eugenol) is distilled from buds, leaves, and stalks. The oil (or eugenol) is used in dentistry and perfumery, and as a flavouring, but an earlier use for vanillin production has decreased.

ALLSPICE OR PIMENTO (7) *Pimenta dioica.* This is a small tropical tree whose unripe dried berries (**7A**) (in which eugenol is the main volatile oil) provide the spice called allspice because it seems to combine the flavour of several spices (cinnamon, cloves, and nutmeg). Exports come mainly from Jamaica where the plant grows in the semi-wild state.

FENUGREEK *Trigonella foenum-graecum.* This is indigenous to western Asia and south-eastern Europe and has long been cultivated in the Mediterranean area, India, and North Africa. Its pungent seeds have been used medicinally since ancient time and are included in curries. The plant is eaten as a vegetable (containing a very large amount of carotenes). The seed extract is the principal flavouring ingredient of imitation maple syrup.

It is an annual herb (family Leguminosae/Fabaceae) which grows to a height of possibly 90 cm. Its whitish flowers give rise to slender, curved, predominantly beaked pods (8–15 cm in length) containing 10–20 characteristically shaped (oblong or rhomboidal) brownish seeds.

FENUGREEK: A flowers; B fruits; C seeds.

HALF LIFE SIZE DETAILS LIFE SIZE

1 **BAY LAUREL** 2 **SAFFRON** 2A Style 3 **CAPER** 3A Pickled bud
4 **BLACK MUSTARD** fruits and seeds 5 **WHITE MUSTARD** 5A Fruit and seed
6 **CLOVE BRANCH** 6A Dried buds 7 **ALLSPICE** 7A Fruits

Tuberous flavouring plants

GINGER (1) *Zingiber officinale.* Ginger belongs to the family Zingiberaceae and is a slender perennial herb (30–100 cm in height) usually grown as an annual. Flowers are seldom seen. It evolved in South-East Asia but is never found in the wild state. The underground tuberous stem or rhizome (**1A**) constitutes the spice. It has been used as a spice and medicine in India and China since ancient times, was known to the Greeks and Romans, and generally throughout Europe by the tenth century. Today it can be found in most tropical countries; Jamaica, India, Nigeria, Sierra Leone, Australia, and China are well-known sources. It is propagated from rhizome pieces. After 5–7 months the new young rhizomes ('green') are dug up manually or by mechanical means and they are preserved in sugar syrup or used for crystallized ginger. The older rhizomes (harvested after 8–10 months) become dried ginger, utilized in biscuits, puddings, cakes, gingerbread, soups, pickles, curry powder, ginger beer, ginger ale, and ginger wine. Oleoresin and essential oil are produced from ginger.

TURMERIC (2) *Curcuma domestica* syn. *C. longa.* This also belongs to the ginger family, Zingiberaceae, and is a similar plant to ginger, although the leaves are much broader. Again, the underground tuberous stem or rhizome constitutes the spice. Turmeric was domesticated in South-East Asia, possibly from *C. aromatica.* It is an important constituent of curry powder, not only does it provide flavour but it contains strong yellow pigments (the main one is curcumin) that colour the food. Turmeric is an important yellow dye in southern Asia, its use in Europe was superseded by aniline dyes. It is used to colour various food products, for example mustard powder. An oleoresin is produced from turmeric.

LIQUORICE (3) *Glycyrrhiza glabra.* This grows wild throughout southern Europe, and in Russia, the Middle East, and Afghanistan. Today the main cultivation takes place in Spain, Italy, Russia, and Turkey. The rhizomes and roots are used as a flavouring, also in medicine. They contain glycyrhizin, which is 50 times sweeter than sucrose sugar. The dried roots may still be sold as a sweet. Usually the juice is obtained from the root and concentrated by boiling. The solid extract thus obtained has important uses in confectionery, for example liquorice sticks, candy, chewing gum. There are also some medicinal applications. Liquorice was known to the ancient Greeks and Romans. In the sixteenth century Spanish monks living in Pontefract (Yorkshire) had the monopoly of growing liquorice in England.

Liquorice is a perennial herb belonging to the pea family (Leguminosae/Fabaceae), growing to a height just over 1 m. Its pinnate leaves are composed of 9–17 leaflets. The numerous blue flowers (1–1.5 cm in length) are borne in long conical heads and give rise to reddish-brown pods (1.5–2.5 cm in length) which contain three or four seeds. Under cultivation the crop is allowed to grow for 3–5 years before being harvested, by which time it will have formed an extensive system of rhizomes and roots (**3A**) in well-drained soils, reaching a depth of about 1 m and spreading for several metres.

HORSE-RADISH (4) *Armoracia rusticana.* This provides 'horse-radish sauce', eaten as a condiment with meat and fish and made by crushing, mincing, or powdering the root (**4A**), simmering it with vinegar, milk, and seasoning. It is pungent (as in mustards, the pungent compounds are isothiocyanates) with a distinctive flavour, and off-white in colour.

Horse-radish is a member of the Cruciferae (Brassicaceae) family. It is a perennial herb with a long, yellowish-buff tap-root and bearing long-stalked ovate or oblong leaves (30–60 cm in length) with coarsely toothed, wavy margins. The plant probably originated in south-eastern Europe and western Asia but has long been cultivated in many parts of Europe, also America and the hilly regions of India.

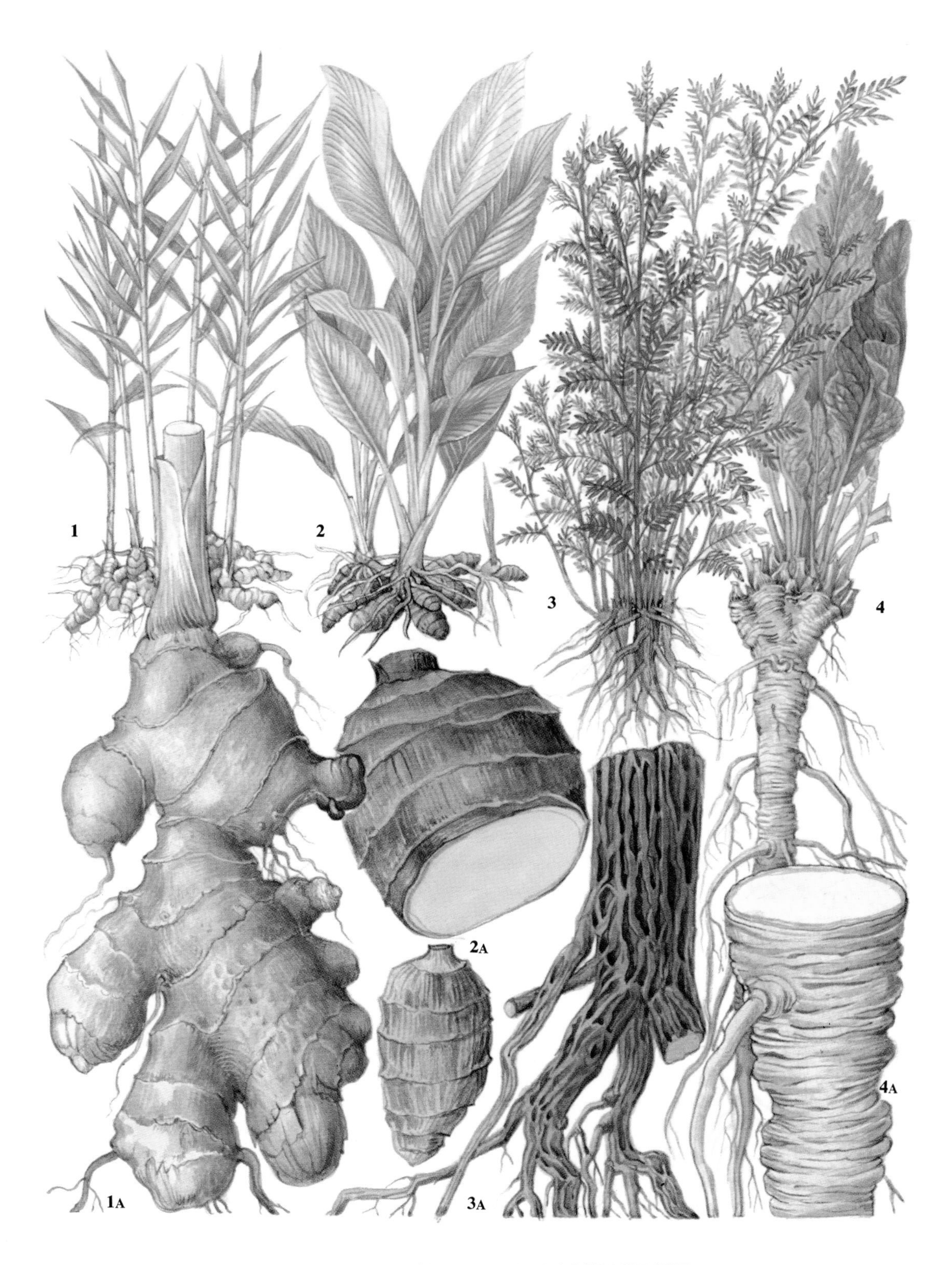

PLANTS × ⅛ TUBERS AND ROOTS LIFE SIZE

1 **GINGER** 1A Tuber 2 **TURMERIC** 2A Tubers
3 **LIQUORICE** 3A Root
4 **HORSE-RADISH** 4A Root

Plants grown for making or flavouring alcoholic drinks

HOP (1) *Humulus lupulus.* Hop is a native of northern Europe, where wild plants are still found, although escapees from cultivation can be confused with these. It is cultivated in a number of areas, for example Europe, America, New Zealand, and Australia. The plant is a perennial vine (family Cannabaceae) which grows to a height of 6 m. Plants are male (**1A**) or female, but the males are usually eliminated because seed set is to be avoided. The part used in brewing (see p. 4) is the female 'cone' (**1**), which consists of a cluster of pale, yellowish-green reduced leaves (bracts and bracteoles) containing the female flowers. On the bases of the bracts and bracteoles are the glistening lupulin glands which contain the essential oils and soft resins that give the aroma and bitter taste to beer, also they promote, through their antiseptic action, the shelf-life of beer. Extracts of these substances may also be used in brewing.

Hops do not seem to have been widely used for brewing beer before the Middle Ages, various bitter herbs were used in previous times. The plant was brought into cultivation in Britain in the sixteenth century. The shoots or 'bines' are trained up a framework of poles, wires, and strings (**1B**); harvesting is now essentially by mechanical means.

The young shoots have been utilized as a vegetable.

JUNIPER (2) *Juniperus communis.* Juniper belongs to the Cypress family (Cupressaceae) and occurs wild throughout the northern hemisphere. It is a shrub or small tree up to a height of 6 m. The evergreen leaves are sharp-pointed or awl-shaped, with a broad white band on the upper surface. Male and female cones are borne on separate plants. The male cones (**2A**) are small and yellow; the female cones, or 'berries', are blue-black (6–10 mm in diameter). Juniper berries are used to flavour gin, liqueurs, cordials, and are also added to various European meat dishes. An essential oil is available by distillation. The berries contain about 25 per cent fermentable sugars which provide alcohol in certain spirits, e.g. *brinjevac* (this is in addition to the juniper flavour). Juniper wood is used for smoking salmon.

ANISE OR ANISEED (3) *Pimpinella anisum.* A native of the Near East, this was utilized by the ancient Egyptians, Greeks, and Romans, and was cultivated in most of Europe by the Middle Ages. It was also introduced into Asia and America. The small, hard, greyish-brown fruits (**3A**) are rich (up to 3.5 per cent) in an essential oil, the main constituent of which is anethole. The fruits and the oil prepared from them have a number of uses, for example in Indian and European cooking, confectionery, and in drinks (French 'pastis', Greek 'ouzo', Middle Eastern 'arrak', 'anisette').

It belongs to the family Umbelliferae (Apiaceae) and is an annual (usually 60–70 cm in height). The lower leaves are heart-shaped but the upper ones are divided. Its small white flowers are in compound umbels.

A substitute for true anise is star anise (*Illicium verum*), with an essential oil similar to true anise but a member of the Magnolia family. It is cultivated in South-East Asia.

WORMWOOD (4) *Artemisia absinthium.* This is found growing wild in temperate regions of Europe and Asia, and is also cultivated to some extent, for example in the central and north-western United States. It is a very bitter herb, the bitterness being due to glycosides. The plant has a long history of use in medicines, such as a vermifuge—hence the vernacular name. It was included in the notorious drink 'absinthe', banned in 1915 in France because of its narcotic effects. The herb is possibly used in the manufacture of vermouth.

Wormwood belongs to the daisy family (Compositae or Asteraceae). It is a perennial herb, 30–100 cm in height, with greyish-green, silky-hairy stems and leaves, the latter pinnately to tripinnately divided. The small, yellowish-green flower-heads are borne numerously on a much-branched terminal inflorescence.

OTHER PLANTS WITH SIMILAR USES A number of other plants are used locally in different parts of the world for making or flavouring intoxicating beverages. 'Kava' (*Piper methysticum*) is cultivated in the Pacific islands for its roots which, after extraction and fermentation, provide the alcoholic drink. The bark of a wild South American tree, *Galipea officinalis*, has been used, together with other spices, to make Angostura bitters (used for 'pink' gin) but, because of adulteration with undesirable materials, is no longer always included. The alcoholic 'tequila' is made from *Agave* species.

LIFE SIZE   DETAIL × 3   BINES AND SHRUB SMALL SCALE

1 **HOP** 1A Male flowers 1B Bines
2 **JUNIPER** 2A Male cones 2B Shrub
3 **ANISE** 3A Detail of seeds 4 **WORMWOOD**

Aromatic umbellifer seeds

The cultivated Umbelliferae (Apiaceae) are essentially plants of temperate regions. They include well-known vegetables, spices, and herbs. The flowers, and consequently the fruits (usually referred to as 'seeds'), are held in 'umbels'—inflorescences with branches radiating like the ribs of an umbrella.

CARAWAY (1) *Carum carvi.* Caraway probably arose in Asia Minor and is found growing wild throughout North and central Europe, also in central Asia. The plant has been cultivated since ancient times. Today it is cultivated in a number of countries, but mainly The Netherlands. The 'seeds' have been used to flavour cakes, bread, cheese, soups, meat dishes, also the liqueur 'kümmel'. The main constituent of the essential oil is carvone. The young leaves have been used in salads, the tap root as a vegetable (like parsnip).

Caraway is a much-branched, hollow-stemmed biennial herb, 30–80 cm in height. Its bipinnate leaves have pinnatifid segments with deep, linear-lanceolate lobes. Its small, white flowers (**1B**) are borne in compound umbels, sometimes with a few bracts. The fruit (**1A**) is 3–6 mm in length, light brown, and each half has five pale ridges.

This species is often confused with cumin.

CORIANDER (2) *Coriandrum sativum.* Coriander probably originated in the eastern Mediterranean area and, from archaeological evidence, was known in neolithic times (Israel). It was used in medicine and utilized by the ancient Greeks and Romans, the latter taking the species to north-western Europe (including Britain). Coriander was also taken to India and China. All parts of the fresh plant, when crushed, give off a foetid odour reminiscent of bed- or shield-bugs. This odour is lost when the 'seed' is dried, which then produces a mild aromatic flavour. It is used in curries, meat dishes, bread, confectionery, and some alcoholic drinks. The leaf, too, is utilized (see p. 156). The plant is cultivated in a number of countries, including India, Morocco, Pakistan, Romania, and the former USSR.

Coriander is a slender, solid-stemmed annual, up to 60 cm in height. The upper leaves are divided into narrow linear segments, the lower with broad leaflets. The small flowers (**2B**) are white or pink; the middle ones in each umbel are infertile, the outer slightly larger and fertile. The globose fruits (**2A**) are about 3 mm in diameter, ridged, and yellowish-brown.

CUMIN (3) *Cuminum cyminum.* Cumin is said to be a native of Egypt, Turkestan, and the eastern Mediterranean region, and is grown today in Iran, India, Morocco, China, Russia, Indonesia, Japan, and Turkey. The fruit resembles caraway in aroma and flavour but is more pungent. Consequently it is used in curries, for pickling, in *sauerkraut*, soups, and stews (widely utilized in Latin America).

Cumin is a small, slender branching, annual herb, about 30 cm in height. The leaves are divided into a few thread-like segments, up to 5 cm in length. The small white or pink flowers are borne in few-flowered umbels, with thread-like bracts. The fruit is 4–8 mm in length and greyish-green to dark grey in colour.

Black cumin is *Nigella sativa*, a member of the buttercup family (Ranunculaceae).

DILL (4) *Anethum graveolens.* This is a native of southern Europe and western Asia and is now cultivated in India, Europe, and the United States. Both fruits and leaves (see p. 156) are utilized. The fruits are used in pickling cucumbers, in dill vinegar, bread, potatoes, vegetables, and dill butter.

Dill is a smooth-stemmed annual or biennial, 50–100 cm in height, with finely divided, fennel-like leaves (their ultimate divisions up to 2 cm in length), and a basal sheath of larger leaves, 1.25–3 cm in length. The fruit (**4A**) is elliptic, flattened, brownish, 4 mm in length, with thin, yellow, dorsal ridges and distinct narrow wings.

FENNEL (5) *Foeniculum vulgare.* This includes a number of forms, the interrelationships of which are in dispute. Here, *F. vulgare* is the plant used for 'seeds' and leaves; *F. vulgare* var. *dulce* for Florence fennel (see p. 158).

Foeniculum vulgare is a native of the Mediterranean area but has become naturalized throughout much of Europe—it is found on the sea cliffs of England, Wales, and Ireland. It is cultivated in a number of countries including India. 'Seeds', or the essential oil (the main constituent is anethole) obtained from them, are used to flavour bread, pastries, confectionery, liqueurs, and fish dishes. The leaves are used to flavour and garnish fish.

Fennel is a perennial herb, up to 2 m in height. The leaf segments (1–5 cm in length) are thread-like; the basal sheath up to 10 cm in length. Its fruit (**5A**) is oblong–ovoid, flattened, greenish or yellowish brown, or greyish, with yellow ridges, about 4 mm long.

LIFE SIZE SEEDS AND FLOWER DETAILS × 3

1 **CARAWAY** 1A Seeds 1B Flower detail 2 **CORIANDER** 2A Seed 2B Flower detail
3 **CUMIN** 3A Seeds
4 **DILL** 4A Seeds 5 **FENNEL** 5A Seeds 5B Flower detail

Aromatic labiate herbs (1)

The mint family (Labiatae or Lamiaceae) contains a number of important herbs and it is concentrated in the Mediterranean region. The plants have a four-angled stem and often a two-lipped corolla.

PEPPERMINT (1) *Mentha* × *piperita.* Peppermint is a hybrid of *M. aquatica* and *M. spicata.* It is cultivated in a number of European countries, North Africa, and the United States of America. Its importance lies in the production of peppermint oil by distillation. The oil is used to flavour sweets, chewing gum, cordials, liqueurs (e.g. Crème de Menthe), toothpastes, and various pharmaceutical products (e.g. indigestion tablets).

The plant has lanceolate, ovate, short-stalked leaves and terminal, oblong spikes of flowers, their stamens more or less concealed within the reddish-lilac corolla.

SPEARMINT (2) *Mentha spicata.* This is well-known for mint sauce or jelly, an accompaniment to lamb. Spearmint oil (rich in carvone) has similar uses to peppermint oil. The species is a native of central Europe but is now cultivated in the United Kingdom (introduced by the Romans), other European countries, China, and North America. The plant (30–90 cm in height) has lanceolate or oblong–lanceolate leaves, which are sessile or very shortly stalked, and cylindrical, usually tapering spikes of flowers—the lower whorls are often more or less distant from each other. Its stamens protrude well beyond the lilac corolla (2A).

Other mints are cultivated, for example horse-mint (*M. longifolia*) and round-leaved mint (*M. rotundifolia*).

SAGE (3) *Salvia officinalis.* This is a native of the northern Mediterranean coast but it is cultivated in many countries. The species shows a great deal of structural variation: narrow-leaved sage produces flowers but broad-leaved sage often does not flower. The fresh or dried herb is used for stuffings (e.g. sage and onion with goose), also in cheese, meat, sausages, meats, and beverages. Its essential oil is produced commercially and used in the food and pharmaceutical industries.

Sage is a low-growing shrub (up to 45 cm in height). It has lanceolate to ovate, long-stalked leaves, sometimes lobed at the base, up to 15 cm in length and either green and rather hairy, or greyish green when densely hairy. The surface is wrinkled and the margin crenulate. The pink or bluish-lilac flowers are borne in distant whorls on erect inflorescences. The corolla is two-lipped, the upper lip hood-like, the lower lobed. There are only two fertile stamens.

A number of forms of sage are grown as ornamentals and a number of species have some economic value, for example pineapple sage (*Salvia rutilans*) and clary sage (*Salvia sclarea*).

OREGANO OR WILD MARJORAM (4) *Origanum vulgare.* This is a native of Europe and is still found wild but is also cultivated in that continent and in North America. The herb was used by the ancient Greeks and Romans and is famous for its use in 'pizza' but is also used in a number of tomato and meat dishes. Its essential oil (high in carvacrol) is used in a number of foods and liqueurs.

The plant is an erect perennial, 30–60 cm in height with stalked, ovate leaves, 1–4.5 cm in length and purplish flowers (4A) in dense, rounded terminal panicles with purplish bracts. This species should not be confused with Mexican oregano (*Lippia graveolens*; family Verbenaceae).

SWEET OR KNOTTED MARJORAM *Marjorana hortensis* syn. *Origanum marjorana.* This is also a native of the Mediterranean region but is less piquant than oregano. It is used in a number of food products, for example soups, stews, stuffings, and pies, and is also a component of a bunch of mixed herbs known as 'bouquet garni'.

The herb is an annual, 30–60 cm in height, with short-stalked, ovate, greyish leaves, up to 2.5 cm in length. The purple or whitish flowers are borne in small axillary clusters or 'knots' along the stem.

POT MARJORAM *Marjorana onites.* This is hardier than sweet marjoram but also comes from the Mediterranean region. It is less fragrant than sweet marjoram. The species (20–35 cm in height) is smaller than the two previously mentioned species and has whorls of mauve to white flowers along the stems.

COMMON OR GARDEN THYME (5) *Thymus vulgaris.* This is a native of southern Europe but is found elsewhere, such as the United Kingdom. It is cultivated in a number of countries, including the United Kingdom. The herb is popular and is added to fish, meat, poultry dishes, or included in 'bouquet garni', or in packeted 'mixed herbs'. The essential oil contains thymol which, being antiseptic, is included in a number of pharmaceutical preparations. Common thyme is a small, bushy sub-shrub, up to 45 cm in height, with very small (4–8 mm in length) greyish or green leaves and white, pink, or violet flowers in rounded or ovoid terminal clusters.

LEMON THYME (6) *Thymus citriodorus.* Lemon thyme is also cultivated. Its leaves have a characteristic lemon-like scent. Other cultivated species are breckland or wild thyme (*T. serpyllum*) and caraway thyme (*T. herba-barona*).

LIFE SIZE DETAILS × 4

1 **PEPPERMINT** 2 **SPEARMINT** 2A Flower detail 3 **SAGE**
4 **MARJORAM** 4A Flower detail
5 **COMMON THYME** 6 **LEMON THYME**

Aromatic labiate herbs (2)

ROSEMARY (1) *Rosmarinus officinalis.* Rosemary is a native of the Mediterranean region and grows well near the sea. At an early date it was introduced into many European countries. As a herb it is used to flavour meat (it is popular with lamb), poultry, savoury dishes, and salads. In herb mixtures it is included in small quantities because of its rather overpowering scent. The essential oil is used in cosmetics and some pharmaceutical preparations.

Rosemary is an erect, bushy shrub, up to 2 m in height. Its evergreen leaves are dark green above, white hairy beneath, 2–3.5 cm in length, and folded inwards along the margins. The violet-blue or whitish flowers (**1A**) are borne in small axillary racemes in April and May, and sporadically at other seasons. The calyx and corolla are two-lipped, the latter about 1.25 cm in length, enclosing two stamens.

BASIL OR SWEET BASIL (2) *Ocimum basilicum.* The origin of basil differs from that of many labiate herbs in that it is thought to have evolved in India. The herb has been cultivated in India and the Middle East since ancient times; it was known to the Greeks and Romans. Basil has some significance in the Hindu religion. The plant has no resistance to frost and therefore is not normally cultivated out of doors in the United Kingdom and northern Europe, but can be grown in pots with protection. It is produced commercially in Italy. It is used with fish and poultry dishes but has a particular affinity for tomato products (e.g. salad, sauces, juice, *provençale*, and *pesto* paste). Basil is an ingredient of the liqueur, chartreuse.

The plant is an erect annual, up to 35 cm in height, bearing ovate, toothed or entire, long-stalked leaves, up to about 8 cm in length. The flowers (**2A**) are white or purple-tinged, about 1 cm in length, borne in whorls in simple, terminal racemes. Some forms have partly red or entirely purple leaves.

BUSH BASIL *Ocimum minimum.* This is a smaller plant than *O. basilicum* and it has an inferior flavour.

SUMMER SAVORY (3) *Satureja hortensis.* Summer savory is a native of the Mediterranean region and was well known to the Romans who mixed it with vinegar and used it as a sauce during feasts. Today it is used as a flavouring for poultry, meats, soups, eggs, and seems to have a special affinity for beans, peas, and lentils. The essential oil may be used in processed foods.

Summer savory is an erect, bushy, rather densely pubescent annual, 10–45 cm in height. The linear–oblong, tapering, indistinctly stalked, opposite leaves are up to about 4 cm in length. The pale-lilac flowers, 4–7 mm in length, are arranged in whorls which are close together near the apex of the spike, more distant below.

WINTER SAVORY (4) *Satureja montana.* A native of the Mediterranean region, as summer savory, this has the same culinary uses but is considered inferior.

The plant differs from summer savory in that it is perennial; there is a small but distinct ridge around the stem between opposite leaf bases; the flowers are larger, up to 14 mm in length.

BALM OR LEMON BALM (5) *Melissa officinalis.* A native of the eastern Mediterranean region, this was distributed through much of Europe in ancient times by the Romans. As many other herbs, it was formerly used for its supposed medical properties but now, because of its lemon-scented leaves, it is included in omelettes, wine cups, wines, and liqueurs (e.g. Benedictine).

Balm is a vigorous, perennial herb, 30–60 cm in height, with rather long-stalked, ovate, toothed, or deeply crenate leaves. The white or pinkish flowers are borne in axillary whorls. The toothed calyx and the campanulate corolla are both two-lipped and the four curved stamens are shorter than the corolla.

LIFE SIZE FLOWER DETAILS × 3

1 **ROSEMARY** 1A Flower detail 2 **BASIL** 2A Flower detail
3 **SUMMER SAVORY** 4 **WINTER SAVORY**
5 **LEMON BALM**

Aromatic composites

The sunflower family or Compositae (Asteraceae) is the largest family of flowering plants, with about 20 000 species. The inflorescence head, or capitulum, is made up of small flowers or florets. In some species all the florets are strap-shaped, in others all are tubular, while a third form exists with tubular florets to the inside and strap-shaped florets to the outside. As regards food, the family contains herbs, spices, oilseed plants, vegetables, and salad plants.

TARRAGON (FRENCH OR GERMAN) (1) *Artemisia dracunculus.* Its leaves are added to white wine vinegar to make tarragon vinegar used for salads, *sauce béarnaise, sauce tartare, sauce hollandaise,* and certain types of mustard. The herb can be added to salads, meats, and stews, and is often a constituent of *fines herbes.* Methyl chavicol is the main constituent of its essential oil. The natural distribution of tarragon is from south-eastern Europe, through north central Asia to North America. It is cultivated in a number of countries, including Russia, the United States (California), The Netherlands, France, Germany, and Italy. Although used in earlier times, it did not become popular until the sixteenth century. Tarragon is a bushy perennial herb, about 60 cm in height, but taller if allowed to grow naturally. The slender, branching stems bear smooth, olive-green, thin, narrow leaves (2–4 cm in length). The small, whitish-green inflorescence heads, 3–5 mm across, are arranged in racemose panicles. They bear florets of two kinds, female and bisexual, the latter being functionally male. Fertile seed is rarely produced and the plants are usually propagated either by cuttings or by division.

RUSSIAN OR FALSE TARRAGON *Artemisia dracunculoides.* This is very similar in appearance to *A. dracunculus* but is a taller plant and has paler, less smooth leaves. It is not so pungent as French or German tarragon and is an inferior substitute.

SOUTHERNWOOD OR LAD'S LOVE (2) *Artemisia abrotanum.* This is used in herbal tea mixtures and as a substitute for wormwood in vermouth and liqueurs. It is a native of western Asia which spread to Europe in medieval times. The plant is a much-branched sub-shrub, up to 2 m in height, with finely divided, greyish-green leaves with an apple- or lemon-like scent. It does not set seed or rarely flowers in the United Kingdom and the north.

TANSY (3) *Tanacetum vulgare.* This is an old-fashioned herb at one time used for culinary and medicinal purposes. The leaves and shoots were included in puddings, omelettes, tansy cakes (made with eggs and eaten at Easter), and tansy tea (a supposed tonic and stimulant drink). Its essential oil contains the irritant substance thujone and therefore the herb must be used with great care. Chopped young leaves have been added sparingly to salads, egg dishes, and stews. Tansy is a common plant of grassland throughout Europe, to Siberia and the Caucasus, also North America, and New Zealand. It is an erect perennial up to 1 m in height with pinnate, feathery, dark-green leaves (15–20 cm in length), and very fragrant. The inflorescence is a compound, flat-topped corymb, with numerous bright-yellow, discoid inflorescences, 6–12 mm across. The greenish-white 'seeds' (achenes) are about 2 mm in length, with the pappus represented by a short, membranous rim.

ALECOST, COSTMARY, BALSAM HERB, MACE, OR BIBLE LEAF (4) *Chrysanthemum balsamita.* This is a native of western Asia. It was cultivated by the ancient Egyptians, Greeks, and Romans, and spread into Europe, although it is not of commercial importance now. Alecost was used to flavour beer (before the widespread use of hops) and has been added to salads and various cooked dishes, and included in herbal tea and liqueurs. It is a perennial herb, with entire, bluntly toothed, oblong or oval leaves (5–20 cm in length), the lower leaves long-stalked, those on the flowering stems sessile and sheathing at the base. The flowering stems may grow to a height just over 1 m and bear rather loose clusters of inflorescences, each 1–1.5 cm across, usually yellowish, with a few white florets. Seed is not set in north-western Europe.

CHAMOMILE *Anthemis nobilis* syn. *Chamaemelum nobile.* Chamomile yields an essential oil used to flavour ice-cream, confectionery, alcoholic and non-alcoholic beverages. The flower-heads are infused to give chamomile tea. Chamomile lawns are sometimes seen. It occurs naturally in western Europe but is local in southern England.

WILD OR GERMAN CHAMOMILE *Matricaria recutica.* This has similar uses to *A. nobilis* but has a far wider distribution, ranging through temperate Europe and well into Asia.

LIFE SIZE

1 **TARRAGON** 2 **SOUTHERNWOOD** 3 **TANSY**
4 **ALECOST**

Herbs: Umbellifers grown for their leaves

PARSLEY (1) *Petroselinum crispum.* This is probably a native of southern Europe but is now cultivated and naturalized in many temperate parts of the world. It was used as a herb by the ancient Greeks and Romans, reaching England in the sixteenth century. There are three forms of cultivated parsley: (1) the usual curled parsley with curled and crisped leaflets, (2) plain-leaved parsley with larger and flat leaflets—more common on the continent of Europe than in the United Kingdom, and (3) Hamburg parsley—a plain-leaved form with a tuberous root, like a small turnip, consumed in some parts of Europe. Parsley is probably the most important herb used in Europe and is included in foods such as sauces, soups, salads, omelettes, and stuffing. It is familiar as a window dressing in butchers' and fishmongers' shops. Fresh parsley has a high content of vitamin C (190 mg/100 g).

The plant is a glabrous biennial, producing in the first year a rosette of ternate to pinnate, long-stalked, bright-green leaves, up to about 28 cm in length, their segments usually curled and crisped. In its second year it produces solid, erect flowering stems, 30–70 cm in height, surmounted by flat-topped, compound umbels of small, yellow flowers (**1A**) which give rise to fruits (2.5 mm in length) characteristic of the Umbelliferae (see p. 148).

DILL (2) *Anethum graveolens.* Dill is described on p. 148. The leaves and 'seeds' are used for the same purposes. In addition to which, dill (essential) oil is extracted from both leaves and fruits; it is used to make dill water, a treatment for flatulence in infants. The percentage of essential oil produced by umbellifer leaves is generally less than that produced by the seed, for example in dill the fresh leaf produces 0.3–0.6 per cent essential oil, the seed 2.6–4 per cent.

CHERVIL (3) *Anthriscus cerefolium.* Chervil is native to Asia Minor, the Caucasus, and southern Russia. The Romans were responsible for spreading the plant throughout most of Europe, where it is now naturalized. The herb is particularly popular in France where it is added to omelettes, salads, and soups, and is a constituent of *fines herbes.*

Chervil is a slightly hairy annual, 40–70 cm in height, with tripinnate leaves (curly in some forms). The small, white flowers (about 2 mm across) are borne on short, hairy stalks in small umbels, 2.5–5 cm in diameter. The fruit is quite large for the family, about 1 cm in length, oblong–ovoid with a slender, ridged beak (**3A**).

SAMPHIRE (4) *Crithmum maritimum.* Samphire is found on sea cliffs, rocks, sand, or shingle on the Channel, Atlantic, Mediterranean, and Black Sea coasts. Although not popular these days, it has been pickled as a savoury or cooked in butter as a vegetable.

It is a fleshy perennial, up to 30 cm in height. The leaves have fleshy, tapering cylindrical segments, up to 5 cm in length. The greenish-yellow flowers (2 mm across) are arranged in terminal, compound umbels (3–6 cm across) with a whorl of lanceolate bracts. Its fruits (6 mm in length) are ovoid.

SWEET CICELY (5) *Myrrhis odorata.* This is probably a native of mountain areas of central and southern Europe but is now generally naturalized in Europe, including the United Kingdom. It is not at present in widespread cultivation but was once used as a herb because of its aniseed/liquorice aroma.

The plant is a perennial growing to a height of 1 m or more with bi- or tripinnate leaves, up to 30 cm in length. The white flowers are of two kinds: short-stalked male and longer-stalked hermaphrodite flowers, in compound umbels. The brown fruits are spindle-shaped, large (25 mm in length), strongly ridged, and with a terminal beak.

LOVAGE (6) *Levisticum officinale.* Lovage is a native of south-western Asia and possibly the Mediterranean region, but spread throughout Europe. It was used as a vegetable like celery (blanched, boiled, or braised) or candied like angelica. The plant is not utilized to a great extent at present. It is a stout perennial, 1 m or more in height with large, bipinnate leaves. The yellowish flowers are borne in compound umbels.

SCOTCH LOVAGE *Ligusticum scoticum.* This is a native plant in northern Britain, occasionally used as a vegetable. It can be distinguished by its ternately pinnate leaves, their segments toothed in the upper half and often lobed. Its flowers are greenish white, sometimes flushed with pink.

CORIANDER (7) *Coriandrum sativum.* Coriander (see p. 148) leaf is now widely available in western countries. Illustrated below.

LIFE SIZE

1 **PARSLEY** 1A Fruits and flowers 2 **DILL** 3 **CHERVIL** 3A Flowers and fruits
4 **SAMPHIRE** 5 **SWEET CICELY** 6 **LOVAGE**

Umbellifers grown for their leaf stalks (petioles)

CELERY (1) *Apium graveolens.* Wild celery is found growing in moist places near the sea in Europe, including the United Kingdom and Asia. Even before the Christian era it was cultivated as a medicinal plant and then later for its leaves (leaf or cutting celery, *A. graveolens* var. *secalinum*) to be used as flavouring. In the sixteenth and seventeenth centuries the milder-tasting forms were selected in France and Italy for the familiar leaf-stalk (petiole) celery (*A. graveolens* var. *dulce*). Celeriac (*A. graveolens* var. *rapaceum*) (see p. 184) was also selected. The leaf celery is still important in South-East Asia.

The best leaf-stalk celery is obtained by 'blanching', either with soil, paper, or black polythene. This produces white leaf bases and reduces bitterness (a glycoside, apiin). Some cultivars can be grown without blanching. Celery can be eaten raw, in salads, or as a cooked vegetable, or in soups. It contains some 95 per cent water and there is little protein, fat, or sugar, but a range of minerals, some carotenes, vitamin E, and the vitamin B complex. The vitamin C content is low (8 mg/100 g). Celery 'seeds' may be used for flavouring and in celery salt.

Celery is a biennial which produces in the first year an upright rosette of leaves (40–60 cm in height) with closely appressed succulent leaf-stalks. In the second year it produces a tall flowering stem with terminal and axillary umbels of small, greenish-white flowers (**1A**) developing into fruits, 1.5 mm in length.

ANGELICA (2) *Angelica archangelica.* Angelica occurs wild in the colder parts of Europe from Iceland, through Scandinavia, to central Russia, also in mountain ranges from the Pyrenees to Syria. It is found as an escape from cultivation in many countries, including the United Kingdom. Pieces of the young stems and leaf-stalks are candied (crystallized with sugar) and used in confectionery, their bright-green colour being attractive. The roots may be used in making gin and the 'seeds' in vermouth and chartreuse. It is also cultivated in a few places, probably Germany and France.

The plant is a biennial, growing to a height of 2 m (when in flower). The green stems bear large (30–70 cm in length) bi- to tripinnate leaves, with oblique, somewhat decurrent, leaf segments. If allowed to grow into the second year, flowering stems carry umbels of greenish-white or green flowers (**2A**), giving rise to fruits (5–7 mm in length).

The cultivated angelica should not be confused with wild angelica (*Angelica sylvestris*), found in the United Kingdom, which usually has purplish stems, white or pink flowers, and is less aromatic.

FLORENCE OR FLORENTINE FENNEL (3) *Foeniculum vulgare* var. *dulce.* This is possibly a native of the Azores but is best known in Italy, from which country it was introduced into England in the early eighteenth century. It is a short, stocky plant (**3A**), about 30 cm in height, with greatly swollen leaf bases forming a kind of false bulb (aniseed flavoured), about the size of a large apple, which is the edible part. This may be eaten raw or cooked, often as an accompaniment to cheese. The 'bulb' contains about 95 per cent water; little protein, fat, or sugar; a large amount of potassium and a range of minerals; carotenes, vitamins E and the B complex, but a small amount of vitamin C (5 mg/100 g). The plant's leaves resemble those of ordinary fennel (see p. 148) and the flowering stems are about 60 cm high, bearing umbels of yellow flowers.

TWO-THIRDS LIFE SIZE PLANTS × 1/8

1 **CELERY** blanched leaf-stalks 1A Flowers 1B **'AMERICAN WINTER GREEN'** plant
2 **ANGELICA** stem 2A Plant
3 **FLORENCE FENNEL** leaf bases 3A Plant

Composite (Asteraceae) salad plants: Lettuce, endive, chicory

LETTUCE (1–3) *Lactuca sativa.* Lettuce is probably the most popular of salad plants and is cultivated in temperate, subtropical, and tropical lands. Sometimes it is cooked as a vegetable. It probably evolved in Asia Minor or the Middle East from the wild *L. serriola* and was known to the ancient Egyptians (4500 BC). The Greeks and Romans cultivated lettuce, the latter introducing the plant to Britain. Plants with firm heads were first recognized in the sixteenth century.

Lettuce leaves are included in salads but in stem lettuce or 'celtuce' (popular in China) the stem is boiled as a vegetable. In the diet, lettuce provides little protein, fat, starch, or sugars, but is useful for fibre, minerals (a large amount of potassium), carotenes (a high content of β-carotene), vitamin E, and vitamin C (5 mg/100 g). Lettuce is available throughout the year, the winter crop being often cultivated under glass. Ideas on the classification of cultivars do vary, but commonly they are divided into 'cabbage' (1 and 2) and 'cos' (3) lettuces. Cabbage lettuces have roundish or somewhat flattened heads with soft or crispy leaves; cos lettuces have longer leaves forming relatively elongated, upright heads. Some cos cultivars may have to be ringed with raffia or other material to produce good-quality hearts. A number of cultivars (e.g. *lollo rosso*) have a marked red pigmentation (anthocyanin).

The lettuce is an annual or biennial belonging to the Compositae (Asteraceae) family. If allowed to run to seed ('bolts') (1A), it produces an erect, branched flowering stem (30–100 cm in height) bearing small, pale-yellow flower heads. Its white, greyish or brown 'seeds' (3–4 mm in length) bear a 'pappus', or parachute, of soft, white hairs for wind dispersal.

The flowering stems contain a white latex which has soporific effects.

ENDIVE (4) *Cichorium endivia.* It is not known where endive originated, although it possibly evolved in the eastern Mediterranean region. Endive was known to the ancient Egyptians, was cultivated by the Greeks and Romans, and had appeared in central Europe by the sixteenth century. The plant is now grown throughout the world, including the tropics.

Endive leaves are normally used fresh in salads although they may be blanched to reduce bitterness. Its nutritional value is very much like that of lettuce. Two groups of cultivars can be recognized, (1) curled endive with curled pinnatifid leaves (4)—the usual form utilized, and (2) the group 'Escarole' with broad, almost entire, rather flat leaves. Red anthocyanin pigment may be present.

The species is an annual, sometimes biennial, herb which, if left to flower, produces a branching, leafy, flowering stem (50–150 cm in height) with pale-blue flower-heads about 4 cm across. The 'seed' is 2–3.5 mm in length with a minute pappus.

CHICORY (5–6) *Cichorium intybus.* The wild species is found throughout Europe (except the north), western Asia, and central Russia. Because of confusion between endive and chicory, there is dispute concerning the time or origin of the cultivated forms of chicory but it could have been as late as the sixteenth century.

Chicory is utilized in a number of ways. The root is a substitute for coffee (see p. 118). Broad-leaved forms (5) are used in salads or cooked as a vegetable. Some cultivars are blanched to give *chicons* (6B) rather like 'cos' lettuces. To obtain *chicons*, large first-year roots of selected cultivars are dug up, the root tips and the foliage (except for the basal 3 cm) are cut off. The remaining structures are replanted indoors in the dark at an optimum temperature of about 18 °C. The resulting blanched leafy growths are the *chicons* which are eaten raw or cooked. 'Witloof' is a well-known cultivar (developed in Belgium) used for *chicons.* Nutritionally speaking, chicory is similar to lettuce and endive.

Chicory is a perennial herb with leaves (6) covered in short hairs (endive is hairless). It produces a branching flowering stem, up to 1.5 m in height, bearing bright-blue flower-heads (6A). The 'seed' is crowned by a pappus of short scales.

In France there is some confusion regarding the vernacular names. 'Witloof' chicory is called *endive*, endives are known as *chicorée.*

PLANTS × ¼ FLOWERS LIFE SIZE

1 **CABBAGE LETTUCE 'WEBB'S WONDERFUL'** 1A Flowers and fruits
2 **LETTUCE 'BUTTERCRUNCH'** 3 **COS LETTUCE 'ST. ALBANS'** 4 **ENDIVE**
5 **CHICORY 'ROSSA DI VERONA'** 6 **CHICORY 'SUGAR LOAF'** 6A Flowers 6B Blanched head

Crucifer (Brassicaceae) salad plants

WATERCRESS (1–2) *Nasturtium officinale* and *N.microphyllum* × *officinale*. This is an aquatic hairless perennial with fine white roots at the base and dark-green, pinnate leaves. The small, white flowers produce fruits (13–18 mm in length) with seeds in differing arrangements according to the species.

Nasturtium officinale is green watercress, the leaves of which remain green in the autumn but it is susceptible to frost damage in winter and spring. Its fruit (a siliqua) bears two distinct rows of seeds in each cell. This is one of the two commercial species. The other is brown or winter watercress (*N. microphyllum* × *officinale*), which is a hybrid between one-rowed watercress (2) (*N. microphyllum*, with seeds in one row in each cell of the siliqua) and *N. officinale*. The leaves of brown or winter watercress turn purplish-brown in the autumn but it is relatively frost-hardy, thus producing a crop in the winter. It does not really produce viable seed but spreads by vegetative propagation.

Watercress is found in the wild from Britain through central and southern Europe to western Asia. Although it has probably been gathered for food from ancient times, commercial cultivation only started in the nineteenth century. Now it is grown in many parts of the world if clear, clean, running water is available, giving up to 10 crops a year. Watercress is included in salads, garnishes, and may be cooked as a vegetable or included in soup. As other leafy salad plants, it contains small amounts of sugars and fat, the usual range of B vitamins and minerals, also vitamin E, but the amounts of protein (3 per cent), iron (2 mg/100 g), carotenes (a high content of β-carotene), and vitamin C (62 mg/100 g) are relatively high.

WHITE MUSTARD (3) *Sinapis alba* syn. *Brassica hirta*. The seed of this species is included in table mustard (see p. 142). The seedlings (young plants showing the first pair of seed leaves or cotyledons and a few centimetres in height) may be included in 'mustard and cress', used in salads and garnishes. The crop is produced commercially all around the year, it is also easily grown at home on wet flannel or suitable substitute. Very often nowadays marketed 'mustard and cress' is actually rape (*Brassica napus*, see p. 28), the seedlings of which are said to be hardier. It is sometimes stated that rape cotyledons are a more intense shade of green than those of white mustard, but this type of distinction is often difficult to apply in practice.

If allowed to go to seed, the difference between the two species is apparent in several characteristics, especially in the fruits, which are bristly, hairy, with a flattened sabre-like beak in *Sinapis alba* (**3A**), and non-hairy with a slender, tapering beak in *Brassica napus*.

CRESS (4) *Lepidium sativum*. Cress seedlings may be included in 'mustard and cress'. As its seed takes several days longer than mustard to germinate, allowance must be made in sowing the seed if both types of seedlings are to be harvested at the same time. On the continent of Europe the plant is grown to near maturity and used in salads and soups. Cress is probably a native of the Near East.

Cress seedlings have deeply three-lobed cotyledons and are therefore easily distinguishable from white mustard and rape. The plant is an annual, growing to a height of 15–30 cm, its single or branched stem bearing, at the base, leaves with long stalks and, further up the stem, pinnate or bipinnate leaves (sometimes crisped). Its white or reddish, minute flowers develop into flattened, elliptical pouch-like fruits (**4A**), notched at the apex. Its fruit, about 5 mm in length is, botanically speaking, a 'silicula' and usually contains one seed per cell.

'Mustard and cress' contains a high content of carotenes and vitamin C (33 mg/100 g).

WINTER CRESS OR LAND CRESS *Barbarea verna*. This is a native of central Europe but is now found in many lands. It is a useful salad plant because, if protected, its leaves can be picked in the winter, also in the early spring, but it is rarely cultivated now. The plant is a biennial, usually grown as an annual, which produces a flowering stem, up to 1 m in height, with pinnately lobed leaves, bright-yellow flowers, and a narrow, cylindrical, short-pointed siliqua, 3–6 cm in length.

ROCKET (5) *Eruca sativa*. This is a native of the Mediterranean region and has been used as a salad plant in a number of European countries since classical times. It now appears in supermarkets in the United Kingdom. The plant is grown for oilseeds (*jamba* or *taramira*) in India, Pakistan, and Iran. Rocket is an annual with an erect stem up to 60 cm in height and bearing pinnately lobed leaves with a large terminal lobe (like radish or turnip). The flowers are white or yellowish, with deep-violet or reddish veins. The fruit is rather like that of white mustard, but lacking bristly hairs—a cylindrical siliqua, with a broad, flattened, sabre-like beak.

LIFE SIZE

1 **CULTIVATED WATERCRESS** 2 **WILD WATERCRESS**
3 **MUSTARD** seedlings 3A Flowers 4 **CRESS** seedlings 4A Flowers and fruits
5 **ROCKET**

Oriental leaf vegetables

The regions of eastern Asia have their own distinctive vegetables. Because of the great interest in Chinese and other types of Asian cuisine, some of these vegetables have become familiar in western countries and are cultivated locally. The genus Brassica *(Cruciferae/Brassicaceae) provides a wide range of oriental vegetables.*

PAK-CHOI (1) *Brassica chinensis.* This is one of the Chinese cabbages, although it is more nearly related to turnip or turnip rape than to the European cabbage. It evolved in China and its cultivation was recorded in that country in the fifth century AD. Pak-choi is widely cultivated in the Far East and South-East Asia, also it is gaining popularity in western countries. It is used in soups and stir-fried dishes but is seldom eaten raw. The leaves are arranged in a spiral and do not form a heart; also, the leaf-stalks or petioles are not winged. These characters distinguish pak-choi from pe-tsai. The flowers and fruits are typical of the Cruciferae.

PE-TSAI (2,3,4,6) *Brassica pekinensis.* This is another of the Chinese cabbages, a native of China, and is grown all over the world. It is well known in western countries, sometimes described as Chinese leaf. Some forms (**2**, **3**, **4**) form loose rosettes but the form (**6**) with a compact head is best known in western countries. The petioles are winged. Pe-tsai is used as a salad plant, a vegetable, in soups, and in pickles.

A number of other forms of *Brassica chinensis*, also *B. juncea* (see p. 142) are utilized as vegetables. Chinese kale (*Brassica alboglabra*) is related to European cabbage.

GARLAND CHRYSANTHEMUM, TANGHO, OR SHUNGIKU (5) *Chrysanthemum coronarium.* This is a member of the Compositae (Asteraceae) family and said to be a native of the Mediterranean region, although it is best known as a food plant in the Far East. The leaves, which may be lobed or dissected, are utilized as a vegetable or salad plant. They contain almost 3 per cent protein, carotenes, and possibly up to 45 mg/100 g vitamin C. Sometimes the leaves have a very strong taste.

6 PE-TSAI × ½

QUARTER LIFE SIZE

1 **PAK-CHOI**
2 **PE-TSAI** 3 **WONG BOK** 4 **CHIHLI**
5 **SHUNGIKU**

The European brassicas (1)

(varieties of Brassica oleracea*)*

WILD CABBAGE (1) *Brassica oleracea.* This is a native of the Mediterranean region and south-western Europe, extending northwards to southern England. It grows on seaside cliffs. It is a glabrous biennial or perennial plant, with a rather woody, more or less decumbent stem. Its glaucous, blue-green leaves are few, compared with most of its cultivated relatives. The lower leaves are stalked and fairly large, with irregularly wavy margins and often with a few small lobes near the base. The flower and fruit features are characteristic of the Cruciferae: four pale-yellow petals, six stamens (two outer short and four inner long), and the fruit (5–10 cm in length) is a siliqua with a short, usually seedless, beak.

The wild species has evolved into a number of varieties where different parts of the plant (stem, leaves, buds, and flowers) have become the edible constituents. Opinions regarding the date of first cultivation of cabbage forms vary from a few hundred to thousands of years BC. Although essentially temperate, *Brassica oleracea* forms are now grown in other regions all over the world.

KALES, BORECOLES (WINTER GREENS), OR COLLARDS (2) These are non-heading Brassicas and are probably the closest to wild cabbage in that they have an erect stem bearing large leaves. They are very hardy and can overwinter to produce young leaves and shoots in the spring. The curly kales (2A), with curled and crimped leaves, are more popular for human consumption. Marrow-stem kale (a thickened stem up to 10 cm in diameter) and 1000-headed kale (growing to over 1 m in height) are cultivated for livestock feed and forage. Ornamental kales with white or pink markings on the leaves are often cultivated as curiosities.

HEAD CABBAGES (3) These very familiar plants are characterized by having a short stem and a greatly enlarged terminal bud. They are thought to have originated in Germany at least by the twelfth century. 'Summer cabbages' (pointed or round heads) are planted out in April or May and ready for harvest from June to August; 'autumn cabbages' are planted out in June and are mature by September and October; 'winter whites' are planted out in June, mature in November and December, and possess white heads.

Sauerkraut, very popular in Germany and some other European countries, is fermented shredded cabbage leaves.

RED CABBAGE (4) This has anthocyanin pigment in addition to chlorophyll, hence the red colour. It is used for pickling.

SPRING CABBAGE (5) The term is applied to two types: (1) spring greens grown for their fresh, loose, leafy heads, and (2) spring hearting cabbage, which is hardy and grows slowly through the winter to mature in spring.

SAVOYS (6) These are cabbages with wrinkled leaves, generally hardy and frost resistant, planted in early July and harvested from October to March. They tend to be milder in flavour than the smooth-leaved winter cabbages and make excellent shredded salad—'coleslaw'.

BRUSSELS SPROUTS (7) This may have been the last major form of *Brassica oleracea* to have developed—the earliest records are from Belgium in the eighteenth century. Dense, compact axillary buds (like miniature cabbages) are borne close together all along a tall, single stem. They are very popular for deep-freezing. The terminal bud is sometimes available as 'sprout tops'. F_1 hybrids are often available for cultivation.

'FLOWER CABBAGES' (8) These are ornamental cabbages, introduced from Japan, attractive in garden borders and popular for floral arrangements. The head is loose, with the inner leaves bright pink or pale yellow and the outer leaves beautifully variegated.

PLANTS × ⅛ DETAILS × ½

1 **WILD CABBAGE** leaf and flower details 2 **KALE** 2A **CURLY KALE** leaf detail
3 **ROUND CABBAGE** 4 **RED CABBAGE** 5 **SPRING CABBAGE** 5A Leaf detail
6 **SAVOY** 6A Leaf detail 7 **BRUSSELS SPROUTS** 7A Detail of sprout
8 **FLOWER CABBAGES**

The European brassicas (2)

(varieties of Brassica oleracea*)*

CAULIFLOWER (1) This is thought to have originated in the eastern Mediterranean region and appeared in Italy at the end of the fifteenth century. Forms suitable for hot and humid tropical climates developed in India over the past 200 years. Cauliflowers produce a single stem bearing a large, swollen, roundish flower-head (the edible part) consisting of a tightly packed mass of under-developed white or creamy-white flower buds (sometimes purple, yellowish-green, or orange)—the 'curd'. In the United Kingdom, summer and autumn cauliflowers are grown all over the country, but winter cauliflowers mature during the winter in mild areas such as Cornwall, also they are imported from countries such as France, Israel, Italy, and Spain. Mini-cauliflowers are now available with heads about 8 cm across.

Cauliflower is sometimes served raw, with or without a French dressing, but usually boiled with a white (including *béchamel*) sauce or *au gratin*; it is also found in various pickles.

SPROUTING BROCCOLI (2) This became popular in northern Europe in the eighteenth century. It is similar to cauliflower in the structure of its flower-head, but instead of producing a single head it produces a rather loose terminal cluster of flower-heads (2A) on one or several branches, and a large number of smaller heads in the axils of the leaves lower down the branches. Purple and white cultivars are available. Sprouting broccoli, sown in April or May, is an overwintering annual ready in early spring, although some cultivars may be available in December.

GREEN SPROUTING BROCCOLI OR CALABRESE (3) This is presumably of Italian origin and was introduced into the United States of America by Italian immigrants in the early twentieth century. It differs from the previous vegetable (2) in that the curd is green and the terminal head is fairly dominant, although side branches may develop, particularly if the terminal head is removed. Also, it is not overwintering but matures in the summer. Flowering stems are often fused together, a situation described as 'fasciation'. It is boiled for consumption and, in addition, deep-frozen and canned.

KOHLRABI (4) Kohlrabi first appeared in Europe in the fifteenth century. It is now grown in many parts of the world and is important in China and Vietnam. The turnip-like globe (5–12 cm in diameter) is not a root but the swollen base of the stem (leaves are attached) and may be white, green, or purple. It is a fast grower, maturing in 40–70 days. Kohlrabi is boiled for human consumption and is fed to livestock.

Brassica vegetables are consumed in enormous quantities throughout the world and are important in human nutrition. Their leaves have little starch, sugar, or fat, but their fibre is useful. There is some vitamin E and a range of B vitamins. Their content of carotenes is significant, although it is related to the concentration of chlorophyll, for example the outer leaves of a cabbage may contain 50 times as much carotene as the inner leaves, and white cabbage contains only a trace of carotene. Vitamin C is present at a range of 35–115 mg/100 g and the concentration of protein is almost 4 per cent.

The flavour, often mustard-like, of these vegetables, is related to substances known as glucosinolates which, in certain situations, can be toxic (see p. 207), but there have been claims that they can offer some sort of protection against cancer.

PLANTS × 1/8   DETAILS × 1/2

1 **CAULIFLOWER** 1A Details of inflorescence
2 **PURPLE SPROUTING BROCCOLI** 2A Detail of flowering stem
3 **CALABRESE** detail 4 **KOHLRABI** 4A Detail of stems

Other leaf vegetables

SPINACH (1) *Spinacea oleracea.* Spinach belongs to the goosefoot family (Chenopodiaceae) and originated in Iran and neighbouring areas. It spread to China about AD 600, from there to Korea and Japan in the fourteenth to seventeenth centuries; the Arabs introduced it to Spain in the eleventh century, but widespread cultivation in Europe only began in the eighteenth century. Spinach is now cultivated in temperate regions and cooler parts of the tropics all over the world.

The plant is an annual herb, with its crinkled leaves forming a rosette (1) in the young plant. Two main types of spinach are cultivated: (1) round or summer spinach with rounded 'seeds' (actually fruits); (2) prickly or winter spinach with fruits bearing two or three prickles. This type, compared to the former, has a more spreading habit and the leaves are more triangular than rounded. The theory that prickly spinach is hardier than rounded is now discounted; indeed, most modern cultivars are of the latter type—plant breeding has increased their hardiness. Seed can be sown in spring, summer, and late summer to give pickings at frequent periods throughout the year. Spinach may 'bolt' to give a leafy inflorescence (**1A**) up to 60 cm in height, bearing unisexual flowers (**1B**) on separate plants. Modern cultivars have a decreased tendency to bolt.

Spinach can be eaten raw, boiled (in little or no water), in soups, quiches, soufflés, green pastas, *oeufs Florentine*, also frozen and canned. Its pleasant acidic taste is due to oxalic acid which is not considered toxic in the usual amounts consumed. Nutritionally it is most useful, with about 3 per cent protein, fibre, a range of minerals, 3.5 mg β-carotene/100 g, vitamin E and the range of B vitamins, and 26 mg/100 g vitamin C.

SPINACH-BEET (2) *Beta vulgaris.* This is a form of the garden beetroot and sugar-beet (see p. 16) but it is grown solely for its succulent leaves which are used as a green vegetable, like spinach, which it resembles in flavour. The whole leaf can be eaten, including the long, green stalk. It may be sown in spring, for picking in summer and autumn; or in late summer, for picking in late winter and early spring. There is good evidence that by the first century AD leafy beet forms were being cultivated.

SEAKALE-BEET OR CHARD (3) *Beta vulgaris.* This is very closely allied to spinach-beet and is used in the same way. It differs mainly in having a broad, white leaf-stalk, up to several centimetres across, which is often eaten as a separate vegetable, cooked and served in the same way as seakale, while the green blade is used like spinach. Some cultivars have reddish-purple leaf-stalks and blades. Both spinach-beet and seakale-beet belong to the family Chenopodiaceae.

ORACHE (4) *Atriplex hortensis* Orache is also a member of the family Chenopodiaceae and is a native of western Asia and south-eastern Europe. It is of ancient cultivation and was widely grown until the eighteenth century but is not now of importance. Orache can be utilized like spinach. It is an annual herb growing to 2 m in height, with triangular leaves 8–12 cm in length. Male and female flowers are one the same plant. Most female flowers have two large, sepal-like bracts instead of a perianth, others have a perianth but not bracts. Red and yellow forms are often grown as ornamentals.

NEW ZEALAND SPINACH (5) *Tetragonia expansa.* This is a member of the ice-plant family, Aizoaceae, and is a native of Australia, Tasmania, New Zealand, the Pacific Islands, Japan, China, and Taiwan. Captain Cook brought the plant to Europe from New Zealand in the eighteenth century and used it as a source of vitamin C on his voyages. It can be employed as a spinach substitute with the added advantage that it can tolerate dry soils and hot weather. The plant is usually consumed boiled to reduce the saponin content (see p. 209) and is rich in minerals and vitamins, for example β-carotene (4 mg/100 g) and vitamin C (25–50 mg/100 g).

It is a fast-growing annual with sprawling stems (35–50 cm in length) bearing fleshy leaves (5–12 cm in length). The small, yellowish-green flowers produce green top-shaped fruits (about 8 mm in length) crowned with four small thorns.

AMARANTHUS SPINACH (6) *Amaranthus* spp. Several *Amaranthus* species are important tropical leafy vegetables in South-East Asia, Africa, and the Caribbean area. In addition to cultivated species, wild plants are also utilized. *Amaranthus* vegetables are good sources of protein, minerals, β-carotene (4–8 mg/100 g) and vitamin C (60–120 mg/100 g). Certain *Amaranthus* species (e.g. *A. caudatus*) are traditionally grown as pseudo-cereal crops for their grain in Central and South America.

PLANTS × ¼ FLOWERS × 1 DETAILS × 3

1 **SPINACH** 1A Male flowers 1B Male and female flower details
2 **SPINACH-BEET** 2A Flowers 2B Detail 3 **SEAKALE-BEET (CHARD)**
4 **RED ORACHE** 4A Flowers 5 **NEW ZEALAND SPINACH**
6 **AMARANTHUS SPINACH** 6A Flowers

Some plants grown for their young stems and leaf stalks

RHUBARB (1) *Rheum raphonticum* syn. *R. rhabarbarum.* Rhubarb is sometimes described as a hybrid (*Rheum* × *cultorum*) although of uncertain origin. It probably evolved in northern China and eastern Siberia. This and other *Rheum* species have been used in medicine in China for at least 2000 years. Rhubarb was introduced into Europe in the sixteenth century but only used for medical reasons. Its use as a fruit substitute began in the nineteenth century. Rhubarb is now cultivated in many countries with a temperate climate. The plant part utilized is the leaf-stalk or petiole, which is not, of course, a fruit botanically speaking but it is in culinary terms a fruit, being included, with added sugar, in pies, preserves, tarts, crumbles, and wine. Rhubarb contains very little protein, fat, sugar, vitamin C (6 mg/100 g), or β-carotene (0.05 mg/100 g). The leaf-stalk contains citric, malic, and oxalic acids, which contribute to the sharpness of rhubarb which is often tempered in food products with added sugar. Cases of poisoning have been associated with consumption of the leaf-blade because of high concentrations of oxalic acid (see p. 207) and an anthraquinone.

Rhubarb is a perennial with a rootstock consisting of short rhizomes bearing large leaves (maybe more than 1 m in length) and very large inflorescences (up to 2 m in height) bearing small, white flowers. For the best leaf-stalks for eating, the rootstocks are unearthed and exposed to the winter frosts, then planted in warm, dark conditions to force growth. Rhubarb belongs to the buckwheat family—Polygonaceae.

SEAKALE (2) *Crambe maritima.* This occurs on shingle and sandy coasts of the Baltic, Atlantic, and parts of the Mediterranean. Today it is not cultivated extensively. The blanched leaf-stalks (2) (ideally about 20 cm in length) have a pleasant but rather bitter flavour and are boiled like asparagus, then served in a white sauce or butter. The blanched leaf-stalk is terminated by a very reduced leaf-blade. It contains very little protein, fat, or sugar, and the vitamin C in boiled material is 18 mg/100 g. Blanching is carried out by covering young plants with a pot or box; wild plants have been blanched with a covering of shingle.

Seakale belongs to the Cruciferae (Brassicaceae) family. It has large (up to 30 cm in length), basal, bluish-grey leaves (2A) and flowering stems (up to 60 cm in height) with white four-petalled flowers. The roundish fruit, which does not split at maturity, contains a single seed—an unusual condition for the family.

ASPARAGUS (3) *Asparagus officinalis.* This is a member of the family Liliaceae and is a native of central and southern Europe, North Africa, and western and central Asia. It was cultivated by the ancient Egyptians, Greeks, and Romans but lost its popularity in the Middle Ages, then regaining it in the seventeenth century. Today it is cultivated all over the world, with the United States as the greatest producer (a distinct subspecies, ssp. *prostatus* is a rare plant found in parts of the United Kingdom and Europe). The part eaten is the young shoot or 'spear', which grows from the rootstock between late April and early July and is cut when 20–30 cm in height. Young, unemerged shoots are sometimes harvested as white asparagus. Green asparagus is one of the most delicious vegetables, being lightly cooked and served with butter, white sauce, or vinaigrette; much of the crop is deep-frozen or canned. Fresh green asparagus contains about 3 per cent protein, little fat, 2 per cent sugar, 0.3 mg/100 g carotene, and 12 mg vitamin C/100 g. White asparagus contains smaller amounts of vitamins and minerals.

The shoots that are left on the rootstock elongate to form much-branched stems about 1.5 m in height, bearing clusters of needle-like 'cladodes' (modified branches that function as leaves) in the axils of scale (reduced) leaves. The small yellowish or pale-green male and female flowers are normally borne on separate plants, singly or in groups of two or three, at the junctions of the branchlets. Occasionally hermaphrodite flowers are produced. The fruit is a small, round berry, red when ripe.

BAMBOO SHOOTS (4) Bamboo shoots, as an article of food, are the thick, pointed shoots that emerge from the ground under a bamboo plant and which would, if left, develop into a new stem or culm (4A). The usual practice is to cover the bases of the plants in winter with mud and manure and to cut the new emerging shoots, about 15 cm in length, in the spring. After removing the leaf sheaths, the stems are boiled for about half an hour to remove any bitterness (cyanogenic glycosides) but to retain their crisp texture. Many species of bamboo, including those of *Phyllostachys*, *Bambusa*, and *Dendrocalamus*, are utilized. Bamboo shoots are a popular food item in eastern Asia and in many other countries because of the popularity of Chinese cuisine. They are often canned. Bamboo shoots contain about 3 per cent protein, little fat, about 5 per cent carbohydrate, but only 4 mg/100 g vitamin C. Their crispy texture adds to the acceptability of a meal. Bamboo species belong to the grass family (Gramineae/Poaceae).

SHOOTS AND LEAF-STALKS × ½ PLANTS × 1/12

1 **RHUBARB** leaf-stalks 1A Plant 2 **SEAKALE** blanched leaf-stalks 2A Plant
3 **ASPARAGUS** shoots 3A Plants 4 **BAMBOO** shoot 4A Culms

Globe artichoke, cardoon, okra

GLOBE ARTICHOKE (1) *Cynara scolymus.* The young flower-heads or 'chokes' have numerous large scales or bracts with fleshy bases (1B). This fleshy base is the part usually eaten. The flower-head may be baked, fried, boiled, stuffed, and served hot with various sauces (e.g. *sauce hollandaise*) or melted butter, or served cold with vinaigrette. Also, the receptacle (1C) or 'heart' is eaten. Particularly in Italy, small and immature flower-heads are cooked and preserved in olive oil. Globe artichokes can be canned. They contain about 3 per cent protein and 3 per cent carbohydrate, little fat, and a small amount of vitamin C.

Globe artichoke probably evolved in the Mediterranean region, maybe from a wild form of *Cynara cardunculus* (2). It was known as a food plant to the Greeks and Romans. The plants can be grown in many parts of Europe but they require protection against frost. California is a major growing region. There are many cultivars available and these are best raised from suckers which should be planted in the spring. Plants raised from seed are variable. Renewal of globe artichokes takes place every 3–4 years.

The globe artichoke is a thistle-like plant belonging to the Compositae (Asteraceae) family. It is an herbaceous perennial, 80–180 cm in height, usually grown as a triennial and with greyish-green, deeply-lobed leaves, up to 75 cm in length. The globose flower-heads (1A) (5–10 cm in diameter) consist of broad, fleshy, green scales or bracts surrounding the central violet-blue florets, both bracts and florets being attached to a receptacle or 'heart'. According to cultivar, the bract has a blunt or notched apex, although occasionally it bears a sharp spine.

CARDOON (2) *Cynara cardunculus.* Cardoon seems to be well known in certain European countries (e.g. France, Italy) but it is not of importance as a food plant in the United Kingdom, although it is attractive as an ornamental. It is very closely related to globe artichoke, and the wild cardoon may have been the ancestor of both species. The blanched leaf-stalks (superficially resembling celery) are the food items produced from cardoon. They may be boiled as a vegetable and served with a white or cheese sauce, or breaded and fried, or served raw with *bagna cauda* (a hot anchovy and garlic dip). Cardoon contains very little protein, fat, carbohydrate, or vitamin C (2 mg/100 g).

It is larger (up to 2.5 m in height) than globe artichoke and is cultivated as an annual.

OKRA, OKRO, LADY'S FINGERS, GUMBO, OR BINDI (3) *Hibiscus esculentus.* It is a member of the cotton family, Malvaceae, and is probably a native of tropical Africa, although it is now cultivated in many tropical and subtropical countries (e.g. India, Brazil, Thailand, Turkey, and Spain). The plant is an annual herb (1–2 m in height) with palmate leaves, and flowers with yellow petals, each having a red spot at the base. Its fruit or 'pod' is the part eaten but for this purpose it is harvested at a tender, immature stage (5–12 cm in length). Okra fruits may be boiled, fried, dried, deep-frozen, canned, or pickled. They are highly mucilaginous and are used to thicken soups and stews. Raw okra contains about 3 per cent protein, very little fat, 3 per cent carbohydrate, and reasonable amounts of carotene and vitamin C (21 mg/100 g).

ROSELLE OR RED SORREL *Hibiscus sabdariffa.* This is a tropical plant. Its succulent calyx is used to make a drink, jellies, sauces, chutneys, and preserves. The young shoots and leaves are eaten raw or cooked as vegetables.

PLANTS × 1/12 DETAILS × ½

1 **GLOBE ARTICHOKE** 1A Immature flower-head 1B Cooked flower-head 1C Receptacle
2 **CARDOON**
3 **OKRA** 3A Flower 3B Fruits

Onions and related crops (1)

The genus Allium *contains about 500 species, a number of which have been most important as food plants for a very long time. Based on the evidence of archaeology, literature, and illustrations, it is very clear that onion, garlic, and leek were cultivated in ancient Egypt around 3000* BC. Allium *species are included in the family Alliaceae. They have underground bulbs which may produce aerial stems bearing flowers of varying colours, according to species.*

Onions, garlic, and their allies contain substances known as 'alliins' *(see p. 206)* *which break down to give a number of volatile sulphur-containing chemical compounds which are responsible for the characteristic odour of the plants and are the lachrymatory principles of onion. Onions and garlic have been used in many traditional medical practices (e.g. to treat intestinal worms, gastroenteritis, high blood pressure, and as antibacterial, antifungal, and antitumour agents). Some epidemiological and clinical investigations have claimed that the characteristic chemical substances of these plants can reduce heart disease. 'Odour-free' garlic products are now available.*

ONION (1–4) *Allium cepa.* As described above, the common onion is of ancient cultivation, although its origin is not clear. It is possible that it evolved in south-western Asia. The Romans took the plant throughout Europe and it became popular in the Middle Ages. Columbus introduced the onion to America. Nowadays it is an important crop in many parts of the world. Some large producers are China, Russia, India, and the United States of America.

The culinary uses of onions are extraordinarily numerous. They are eaten raw, fried, boiled, and roasted; in soups, sauces, stews, curries, and a great variety of other savoury dishes; and they are a main ingredient of many pickles and chutneys. Dried-onion products (rings, flakes, and powder) are produced for the food processing industry, also onion oil by distillation.

The onion, although a biennial, is grown as an annual. Its bulb (1), composed of fleshy enlarged leaf-bases, is the edible part. Bulbs (2, 3, 4) vary in skin colour (brown, yellow, white, red, or purple), shape (globe or spindle), size (spring onions (4) are harvested when immature), and pungency (Spanish onions (2) are mild in flavour).

Raw onions have restricted amounts of protein, fat, carbohydrate, minerals, and vitamins. Probably their main value in a diet is the flavour.

CHIVES (5) *Allium schoenoprasum.* As a native plant, chives are widespread in the northern hemisphere, from arctic Russia to Japan, from northern Europe to parts of the Mediterranean region, and in North America. They occur also in Asia Minor and the Himalayas. In the United Kingdom they are found wild in some rocky pastures and occasionally become established in other places where they have escaped from cultivation. The plant is 15–40 cm in height with small, white bulbs (1–3 cm in length) and narrow tubular leaves. Its flowers are pale purple or pink. The plant part used is the leaves, raw and chopped as a garnish in soups, omelettes, salads, and sandwiches (frozen leaves are sometimes available). As a garnish, the leaves do not constitute a significant source of nutrition, although they do contain reasonable quantities of carotene and vitamin C (45 mg/100 g). They seem to have been cultivated in Europe since the sixteenth century, in gardens for their leaves and as an ornamental. With the present interest in herbs, there is production on the market-garden scale.

WELSH ONION, JAPANESE BUNCHING ONION, OR CIBOULE (6) *Allium fistulosum.* Despite its name, the Welsh onion is not a native of Wales, nor has it ever been cultivated extensively in that country. 'Welsh' is a corruption of the German *welsche* meaning foreign, applied when this onion was introduced into Europe from central Asia towards the end of the Middle Ages. It is not known in the wild state but possibly evolved in eastern Asia from *Allium altaicum.* The crop has been known in China and Japan since ancient times and was the principal onion grown but, in Japan, it has been overtaken by the common onion (*A. cepa*). It is sometimes grown in the United Kingdom.

The plant has an indistinct bulb with laterals; hollow, cylindrical leaves (hence the species name *fistulosum*); and a flowering stem, up to 80 cm in height, with yellowish-white flowers. In the East, the green leaves may be included in salads or used as a flavouring in soups and other dishes. Also, sometimes the whole plant is cooked and the lower part of the plant may be blanched. Its pungency is not strong, with the green leaves containing about 2 per cent protein, very little fat, about 5 per cent digestible carbohydrate, carotene, the vitamin B complex, vitamin C (33 mg/100 g), and a range of minerals.

TWO-THIRDS LIFE SIZE ONION PLANT × 1/8

1 **ONION** diagram 1A Plant
2 '**SPANISH**' **ONION** 3 **ONION** '**RED BLOOD**' 4 **SPRING ONION** '**WHITE LISBON**'
5 **CHIVES** 6 **WELSH ONION**

Onions and related crops (2)

SHALLOTS (1) These are considered to be a variety of onion, belonging to the Aggregatum group. Shallots differ from the common onion in that the bulbs multiply freely, producing several lateral bulbs (1). Propagation is usually vegetative, by dividing the cluster of bulbs and replanting them singly, but seed may be produced by a few cultivars. It is said by some that the first reliable record of shallots dates back to twelfth-century France, although it has been stated by others that they were used by the ancient Greeks and Romans. They are now found in many countries. Shallots can be eaten raw or cooked (their smaller size sometimes makes them more convenient than the common onion) and are very useful for pickling. The nutrient composition of the bulb is very similar to that of common onion. The potato onion or multiplier onion also has lateral bulbs but they are enclosed within a common skin.

TREE ONION, EGYPTIAN ONION, AND CATAWISSA ONION These forms all belong to the Proliferum group of onions, in which the inflorescence produces a cluster of 'bulbils' or small bulbs instead of flowers. Sometimes both bulbils and flowers are borne in the same inflorescence. They are of little commercial value.

LEEK (2) *Allium ampeloprasum* var. *porrum* syn. *Allium porrum.* This is not known in the wild state and probably evolved from the wild *A. ampeloprasum* found around the Mediterranean, in the islands of the Azores, Canaries, Cape Verdes, and Madeira, also much of central and northern Europe, including the United Kingdom. It was certainly cultivated in ancient Egypt from 2000 BC onwards and was known in Europe in the Middle Ages. Leek is essentially a European crop, although it is sometimes grown in the tropics at high elevations. The plants are grown from seed and their lower parts are blanched (with soil or other coverings) to give the best-quality leeks. These blanched regions are the parts consumed in soups, stews, and as a separate boiled vegetable. Nutritionally they resemble other *Allium* species.

The leek has flat leaves folded sharply lengthwise, their long bases encircling each other to form an elongated, cylindrical bulb or 'pseudostem' (**2C**). The flowering stem (40–150 cm in height) is stout, cylindrical, terminating in a more-or-less globose inflorescence (**2A**) which is at first enclosed within a papery spathe with a long beak. The flowers are pale purple, with exserted stamens (**2B**).

The leek is one of the traditional emblems worn by the Welsh on St. David's Day. It is supposed to relate to a Welsh victory over the Saxons in AD 640 when the Welsh soldiers wore leeks to distinguish themselves from the enemy.

KURRAT This plant is closely related to the leek, and similar to it but smaller, with narrower leaves. It is cultivated in the countries of the Near East for its leaves which are utilized like chives.

GARLIC (3) *Allium sativum.* This does not grow wild but probably evolved from *Allium longicuspis* (wild garlic) of central Asia. It was cultivated from at least 2000 BC in Egypt and Mesopotamia. Garlic is now known all over the world. The bulb is used as a flavouring (fresh or dried and powdered), or as a vegetable in its own right (Provençal '40-clove chicken'). Nutritionally speaking, raw garlic contains about 8 per cent protein, 15 per cent starch, small amounts of fat and sugars, a large amount of potassium (620 mg/100 g), and 17 mg/100 g of vitamin C. In some countries the chopped leaves are also consumed. Garlic is considered of great importance in many countries, although there are some people who find the resulting breath odour offensive. Its possible medical advantages are discussed on page 206.

The garlic bulb, unlike other *Allium* species, develops entirely underground (3). It consists of a number of segments or 'cloves' within a white, pinkish, or purple skin (**3B**). The leaves are flat and rather slender, the flowering stem (30–60 cm in height) smooth and solid. The whitish flowers are usually mixed with bulbils. The young flower-head is enclosed in a papery, long-beaked spathe, which is soon shed (**3A**). Single cloves are used to propagate the plant.

ROCAMBOLE This name is often applied to forms of *Allium sativum* with coiled stems but also to *Allium scorodoprasum* (sand leek), which has been used in place of garlic.

CHINESE CHIVES *Allium tuberosum.* A well-known vegetable in the East, these are available in Chinese outlets in western countries. They have conspicuous rhizomes but little-developed bulbs. The young leaves and flower-stalks have a garlic-like flavour and are used for seasoning soups, noodle dishes, and omelettes.

RAKKYO *Allium chinense.* This is well known in China and Japan. Its bulbs are mainly used for pickles.

PLANTS × ⅛ INFLORESCENCES AND BULBS × ⅔ DETAILS × 3

1 **SHALLOT** plant 1A Inflorescence 1B Flower detail 1C Bulbs
2 **LEEK** plant 2A Inflorescence 2B Flower detail 2C Elongated bulb
3 **GARLIC** plants 3A Inflorescence 3B Composite bulb

Salad roots

BEETROOT (1) *Beta vulgaris.* As explained earlier in the book (p. 16), the various forms (leaf beets, see p. 170, and root beets) of *Beta vulgaris* evolved from the wild sea-beet, a common seashore plant with a wide distribution in Europe and western Asia. The root forms include the red beetroot, sugar-beet, mangels or mangolds, and fodder beets. Beetroots are boiled and then eaten hot or cold and may be pickled. They are the basis of the famous Russian and Polish soup, 'borsch', and of home-made beetroot wine. The beetroot contains 6–10 per cent sucrose. The red pigment is a nitrogen-containing anthocyanin known as betanin which is used as a food colourant. Beetroot is grown widely in Europe and America.

Beetroot belongs to the goosefoot family (Chenopodiaceae). It is grown as an annual or biennial. The root (the upper part of which is the 'hypocotyl'—an intermediate region towards the stem) is usually spherical but other shapes can be found. The leaves are variable in shape and colour, but are often rhomboid–ovoid near the base of the plant and dark green or reddish. The small, green, hermaphrodite flowers are borne in cymes, arranged in a tall, branching, spike-like inflorescence. The fruits are one-seeded, but two or more are usually joined together by the swollen perianth bases to form a 'seedball' which, when sown, can be expected to produce more than one seedling.

Probably the leaf beets were developed before the root beets but the red beetroot was cultivated by the Romans.

RADISH (2) *Raphanus sativus.* This is unknown in the wild state but probably evolved in the eastern Mediterranean region. It was cultivated in Egypt before 2000 BC and spread to China by about 500 BC and to Japan around AD 700. Radishes are now grown all over the world and their fleshy roots are usually consumed. They may be divided into four groups: (1) western or small radishes (2, 3, 4); (2) oriental radishes (5); (3) leaf radishes; and (4) rat-tailed radish. Western radishes are utilized raw as relishes or appetizers because of their pungency (glucosinolates). They contain a reasonable amount of vitamin C (around 25 mg/100 g). Winter radishes (3, 4) are so-named because their solid firm-fleshed roots can be stored for winter use without becoming hollow. Oriental radishes (5) have mild-flavoured roots and, in the East, are used in soups, sauces, or cooked with meat. These roots, which can be very large (up to 20 kg in weight), are sold in supermarkets in the United Kingdom as 'mouli' or 'rettich'. Leaf radishes are cultivated for fodder and the rat-tailed radish (sometimes known as *Raphanus caudatus*) is cultivated in Asia for its young pods (up to 30 cm or more in length), consumed raw, cooked, or pickled.

Radish is a member of the Cruciferae (Brassicaceae). The four petals (3A) are white, lilac, or pinkish and the fruit is a fleshy siliqua with a pointed beak. Its leaves are lobed and irregularly toothed, with a large terminal segment and smaller, paired, lower segments.

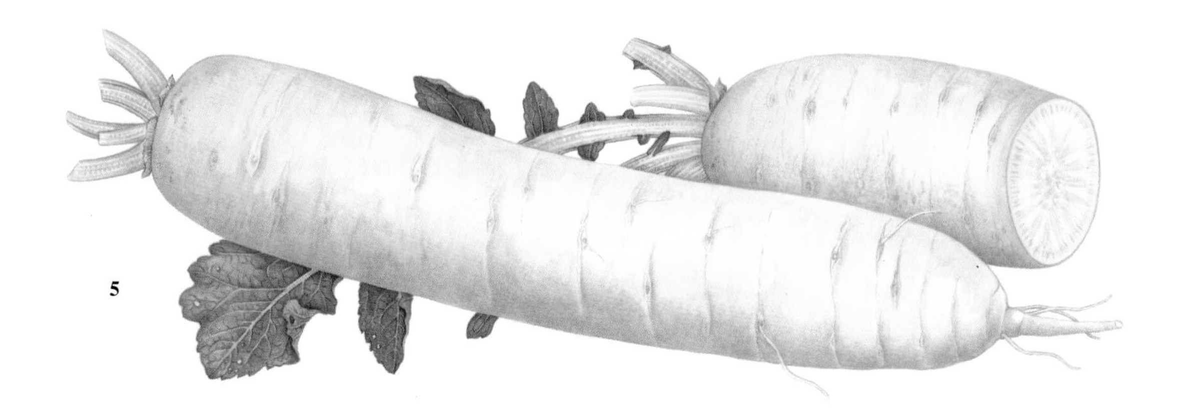

5 **JAPANESE MOOLI RADISH** × I

HALF LIFE SIZE

1 **BEETROOT**
2 **RADISHES 'SCARLET GLOBE', 'FRENCH BREAKFAST', SPARKLER', ICICLE'** 2A Flowers and fruits
3 **WINTER RADISH 'ROUND BLACK SPANISH'** 3A Flower 4 **WINTER RADISH 'CHINA ROSE'**

Crucifer and composite root crops

TURNIP (1) *Brassica campestris* syn. *Brassica rapa*. This is the 'root' form of the turnip rape oilseed crop (see p. 28). It is grown primarily for its swollen 'root', used in soups and stews, but also for its leafy tops—'spring greens'. Turnip was described in the period of Alexander the Great and it must have spread from his empire, which included the Middle East and Persia, to eastern Asia. It is therefore an ancient crop but its exact origin is not known. The 'root' contains about 5 per cent sugar and 17 mg/100 g vitamin C; the leaf has a reasonable amount of carotene (4.6 mg/100 g) and a large amount of vitamin C (139 mg/100 g).

Turnip belongs to the Cruciferae (Brassicaceae) family. The swollen 'root' consists mainly of the hypocotyl, that is the intermediate region between the true root and the stem. 'Roots' are usually round in European cultivars but some Japanese forms have long, carrot-like 'roots'. The flesh of the turnip 'root' may be white or yellow, the skin may be yellow, white, green, or purple-topped. The plant is biennial. Turnip can be distinguished from swede and cabbage by the way in which its open yellow flowers are raised above the unopened buds (**1A**). The fruit is a siliqua.

SWEDE OR RUTABAGA (2) *Brassica napus*. This is the 'root' form of oilseed rape (colza) (see p. 28). The plant evolved as a hybrid between turnip (*B. campestris*) and cabbage or kale (*B. oleracea*), such hybridization has been carried out artificially in modern times and it is not known when the original hybridization took place. It has been stated that swede was known to the ancient Greeks and Romans, but it has also been claimed that the first descriptions appeared in Europe in the seventeenth century. Nevertheless it seems reasonable to assume that the plant is of European origin. Swedes are grown both for human food and for feeding to livestock. They are used mainly in stews or served 'mashed' as a separate vegetable. Swedes may contain up to 5 per cent sugar but little protein and fat and 25–30 mg/100 g vitamin C. As the swollen 'root' is composed of both the hypocotyl and the base of the leafy stem, a swede can be distinguished from a turnip by the presence of a swollen 'neck' bearing a number of ridges, the leaf-base scars. Swedes may be purple, white, or yellow, with yellow or, less commonly, whitish flesh. The plant is usually a biennial. When in flower it may be distinguished from a turnip by its open flowers (**2A**) not being raised above the unopened buds. It belongs to the family Cruciferae (Brassicaceae).

SCORZONERA OR BLACK SALSIFY (3) *Scorzonera hispanica*. The characteristic root with a black skin and white flesh is eaten as a boiled vegetable. It is sweet and contains the carbohydrate inulin, which is made up of fructose units, and is suitable for diabetics. Another use is as a coffee substitute, in the same way as chicory (see p. 160). The young leaves are sometimes eaten in salads.

It belongs to the family Compositae (Asteraceae) and is native to central and southern Europe where it grows wild. Scorzonera was probably first cultivated in Spain and Italy in the sixteenth century, appearing somewhat later in England, although it has always been more popular on the continent of Europe. It is actually a perennial but is usually grown as an annual or biennial; the root may be lifted in the first autumn although it will continue to increase in size if left in the ground for another year. The dandelion-like yellow flower-heads (**3A**) are borne on long stalks, on flowering stems about 50 cm in height. The 'seed' bears a pappus of feathery hairs.

SALSIFY OR OYSTER PLANT (4) *Tragopogon porrifolius*. This also belongs to the family Compositae (Asteraceae) and is better known on the continent of Europe than in the United Kingdom. Its white roots are boiled and eaten with melted butter, also with cream and cheese. The cooked roots have a nutty flavour hardly reminiscent of oysters although their glossy appearance may explain the name 'oyster plant'. The leaves may be used in salads.

It is a native of the Mediterranean region and was probably brought into cultivation about the same time as scorzonera. The plant is usually a biennial, 50–100 cm in height, with terminal, long-stalked, purplish flower-heads (**4A**). The 'seed' bears a pappus of both feathery and simple hairs.

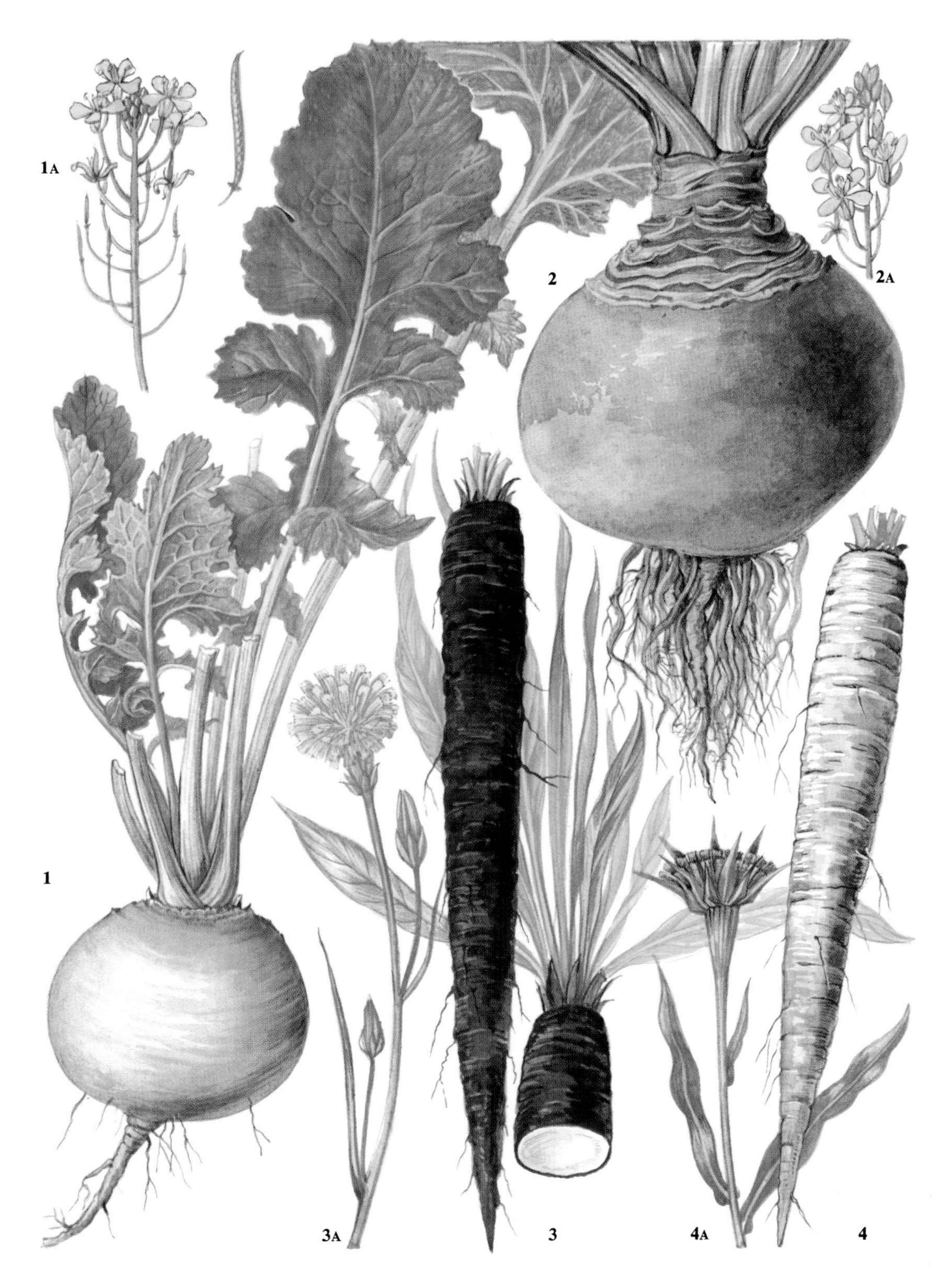

HALF LIFE SIZE

1 **TURNIP** 1A Flowers and fruit 2 **GARDEN SWEDE** 2A Flowers
3 **SCORZONERA** 3A Flower-head 4 **SALSIFY** 4A Flower-head

The umbellifer (Apiaceae) root crops

CARROT (1) *Daucus carota.* The species includes a number of cultivated and wild forms or subspecies which are found in Europe, south-western Asia, Africa, and America. The wild carrot, which often grows near the sea, has a comparatively small, tough, pale-fleshed tap root bearing little resemblance to the thick, fleshy, orange or red root of the cultivated carrot. It has been suggested that the well-known cultivated carrot originated in Afghanistan from forms with roots coloured purple with anthocyanin pigment. Some of these forms were yellow mutants, devoid of anthocyanin. The Afghanistan forms spread westwards and eastwards (Asia Minor, tenth/eleventh centuries; Arab-occupied Spain, twelfth century; continental north-western Europe, fourteenth century; England, fifteenth century; China, fourteenth century; Japan, seventeenth century). Carrots cultivated in north-western Europe before and during the sixteenth century were all purple or yellow, with long roots. The yellow roots were preferred because cooked purple roots released the anthocyanin pigment into the surrounding sauce or soup. In the seventeenth century selection was carried out, in The Netherlands, for carrots with denser orange (carotene) pigment and from these forms the modern cultivars have developed.

Carrots are of enormous importance as a vegetable, being eaten raw, cooked (e.g. in soups stews, infant and invalid foods), or processed into juice. They are the best plant source of provitamin A, containing about 7 mg/100 g of β-carotene. Carrots possess about 7 per cent of sugar (glucose, fructose, sucrose). They are low in other nutrients, for example the vitamin C content is 6 mg/100 g.

The plant is a biennial. The roots of the cultivated carrot are of varying shapes (**1B–1D**). If allowed to grow to a second year, it bears a terminal compound umbel of white flowers (**1A**) subtended by ternate or pinnatifid bracts, the central flower of each umbel often red or purple.

PARSNIP (2) *Pastinaca sativa.* Wild parsnip is found throughout southern and central Europe and was introduced into the United Kingdom and northern Europe. The cultivated parsnip, which has a thicker and more succulent root (**2B**) than the wild form, is grown in temperate regions all over the world. Its root is used as animal fodder or as a cooked vegetable. Parsnip was cultivated in Roman times but good, fleshy forms were not developed until the Middle Ages. The root contains about 6 per cent starch and the same amount of sugar; exposure to frost is supposed to increase the conversion of starch to sugar. The vitamin C content is 17 mg/100 g. Parsnip wine is sometimes made. The plant has a characteristic smell, hollow, furrowed stems, and large, simple, pinnate leaves with ovate and toothed leaflets. The small, yellow flowers (**2A**) are borne in an umbel up to 10 cm across.

CELERIAC (3) *Apium graveolens* var. *rapaceum.* This is closely related to celery (see p. 158) but whereas the leaf stalk is the edible part of celery, the part of the celeriac that is eaten is the swollen base of the stem, and possibly the upper part of the root. Unlike celery, celeriac leaf stalks are little swollen and of moderate length; also they are bitter and therefore unfit for use in salads. The swollen edible part is globose in shape, 8–12 cm in diameter, and with a transversely wrinkled brown surface layer containing a whitish flesh. It may be eaten boiled as a separate vegetable or included in soups and stews, also raw in salads. The flavour is similar to that of celery. It contains about 14 mg/100 g of vitamin C.

Celeriac was probably brought into cultivation after celery.

PARSLEY, TURNIP-ROOTED, OR HAMBURG PARSLEY (see p. 156)

CHERVIL, TURNIP-ROOTED *Chaerophyllum bulbosum.* This plant should not be confused with chervil (*Anthriscus cerefolium*) (see p. 156). Turnip-rooted chervil has large (12–15 cm in length) grey to almost black thickened roots with yellowish flesh. The root is cooked as a vegetable but it is not of any commercial importance. It is a native of southern Europe but has become naturalized in much of central and northern Europe.

PLANTS × ⅛ FLOWER-HEADS AND ROOTS × ½

1 **CARROT** plant 1A Flower-head
1B, 1C, 1D Roots of intermediate main-crops, forcing and stump-rooted varieties
2 **PARSNIP** plant 2A Flower-head 2B Mature plant 3 **CELERIAC** plant 3A Root

Potatoes

POTATO (1) *Solanum tuberosum.* This is one of the most important crops of the world, coming fourth in food production following wheat, maize, and rice; but less important as a source of calories because of its water content. It is cultivated in about 150 countries all over the world, except in the lowland tropics. Major producers include Russia, China, Poland, Germany, and India. There are seven cultivated species—the one of universal importance is the 'Irish' or 'European' potato (*Solanum tuberosum*). Some 230 species are wild and are distributed from the south-western United States, Mexico, through Central America, and into South America, with a strong concentration in the Andes.

Remains of wild potatoes (dated 11 000 BC) have been found in southern Chile; remains of cultivated material from at least 5000 BC. A number of sixteenth-century accounts, written by the Spanish conquerors, describe the use of potato tubers by the Indians in the Andes of South America. Potatoes were brought to Europe late in the sixteenth century but the details are obscure. The stories that attribute their introduction to Sir Francis Drake and Sir Walter Raleigh are considered inaccurate and mere legends. It is probable that the first potatoes were brought from the port of Cartagena in Colombia to Spain. These were of the *andigena* type and 'short-day adapted', able to tuberize only under 12-hour day length or less. Such types produced small tubers in November and December and therefore were only able to grow in the milder areas of such countries as Spain, Italy, southern France, and Ireland. As the short-day adaptation was bred out of them, potatoes were adopted as a crop throughout Europe, giving earlier maturing and heavier cropping cultivars. In England the crop was not universally adopted until the mid-eighteenth to early nineteenth centuries. Potatoes were first received into the North American colonies in about 1620 from Bermuda (these had come from England not South America). India, China, and Japan acquired the crop in the late seventeenth century. In 1845 and 1846 in Ireland the potato crop was largely destroyed by the fungus (*Phytophthora infestans*) blight. This led to famine and a mass emigration of the population to England and America.

As a freshly cooked vegetable, potato tubers may be served in a great variety of ways—boiled, steamed, fried, baked, roasted, or as an ingredient of soups, stews, pies, and other dishes. They can be processed into crisps (known as chips in North America), chips (known as French fries in North America), potato flour, dehydrated or dried potato. Potatoes are canned or frozen (the frozen potato, or *chuño,* is an ancient form of preservation in the high, cold mountains of South America). Individual products of potatoes are starch, alcohol (the basis of vodka and schnapps), glucose, and dextrin. The tubers are also used as an animal feed. Potatoes contain about 80 per cent water, 2 per cent protein, 18 per cent carbohydrate (most of which is starch), a range of minerals, and they are a good source of vitamin C (21 mg/100 g in a freshly dug potato), although this decreases during storage. As a crop cultivated per unit area and compared to cereals, potatoes produce more carbohydrate (hence more energy) and more vitamin C, although not such much protein, but this is richer in lysine than cereal protein. These features explain, at least in part, the world importance of the potato.

There are hundreds of potato cultivars (**2–9**). They are classified according to the date when they are normally marketed into early (new) and main. The cultivars show variation in tuber shape and skin colour—some have a deep-purple skin. They can vary according to their value in the different types of cooking and processing.

The potato is a perennial herb (**1**) of the Solanaceae family, with rather weak, straggling, or more or less erect, branching stems, 0.3–1 m in height. It has odd-pinnate leaves with three or four pairs of ovate leaflets, with smaller ones in-between. The flowers (**1A**) are white to purplish, about 2.5 cm across, with yellow anthers joined laterally to form a cone-shaped structure which conceals the ovary. The fruit (infrequently produced) is a tomato-like green or yellowish berry, 1.5–2 cm across. The plant has fibrous roots and many rhizomes (underground stems) which become swollen at the tip to form the edible tubers. These tubers are normally lifted and may be stored in clamps. The fresh tubers, which are planted in spring, are usually obtained from certified disease-free stocks grown in favourable areas (particularly Scotland and Ireland). Potatoes used for planting are called 'seed potatoes' although botanically they are tubers, not seeds. The young shoots (**1B**) develop from the buds or 'eyes' of the seed potato.

All green parts of the plant, including potato tubers which have been exposed to light, contain poisonous glycoalkaloids (solanines) (see p. 207) so it is advisable to avoid eating tubers with green patches.

TUBERS × ⅔ PLANT × ⅛ FLOWER DETAIL × 1

1 **POTATO** plant 1A Flower detail 1B Seed potato
2 **'ARRAN PILOT'** 3 **'KING EDWARD'** 4 **'MAJESTIC'** 5 **'CRAIG'S ROYAL'**
6 **'RED CRAIG'S ROYAL** 7 **'HOME GUARD'** 8 **'RECORD'** 9 **'DR MCINTOSH**

Other tubers

JERUSALEM ARTICHOKE (1) *Helianthus tuberosus.* This has crisp-fleshed, underground stem tubers (1A), white to yellow or red to blue in colour, 3–6 cm in thickness and 7–10 cm in length, usually irregular and knobbly in shape. They can be eaten boiled or baked. The tubers contain the carbohydrate 'inulin' which is made up of fructose units, which are released on hydrolysis. Fructose, as a replacement for glucose, is tolerated by diabetics.

The species belongs to the family Compositae (Asteraceae) and is closely related to the sunflower (it is in the same genus). Jerusalem artichoke is a native of North America where it was probably eaten by the Indians and taken to Europe in the early 1600s. The plant grows to a height of 3 m but only flowers in the United Kingdom and northern Europe after a long, warm summer.

OCA (2) *Oxalis tuberosa.* Its stem tubers (2A) are white, yellow, or red and cylindrical with a series of grooves and bulges. In the Andean highlands (from Venezuela to Argentina) they constitute a staple food for the local people, second only in importance to the potato. Their nutrient composition is at least as good as that of the potato and they may be boiled, baked, fried, or eaten fresh. Some cultivars are acid (oxalic acid) but this can be removed by sun-drying or traditional freeze-drying. Oca is also cultivated in Mexico and New Zealand.

The plant belongs to the wood sorrel family (Oxalidaceae) and has been grown as an ornamental in Europe.

ULLUCO (3) *Ullucus tuberosus.* This is another staple crop of many highland areas of the Andes. Its edible stem tubers are yellow, pink, red, or purple and shaped like small potatoes or long and curved (2–15 cm in length). The tubers (*papa lisa*) are sometimes sold in modern packaging in supermarkets of some of the South American cities. Ulluco tubers are boiled or pickled. Canned ulluco is imported from Peru to the United States. Fresh tubers contain about 14 per cent starch and sugars, 1–2 per cent protein, and 23 mg/100 g vitamin C. The leaves are also consumed.

The plant is a member of the family Basellaceae.

YSAÑO OR MASHUA (4) *Tropaeolum tuberosum.* This is another stem tuber crop of the Andes. It is a climber (up to 2 m in height) and the tubers are white to yellow. The plant is hardy. Because of their sharp flavour (glucosinolates—see p. 207) the tubers are not eaten raw but boiled, baked, or fried. Dry tubers may contain 14–16 per cent protein, almost 80 per cent carbohydrate, about 9 μg/100 g β-carotene, and almost 480 mg vitamin C/100 g.

Ysaño is closely related to the garden nasturtium and belongs to the family Tropaeolaceae.

TIGER NUT OR CHUFA (5) *Cyperus esculentus.* This is not a nut in the accepted sense but is a small underground stem tuber belonging to a plant of the sedge family (Cyperaceae). The 'nuts', which are cultivated in West Africa, can be eaten raw or roasted. They have been used to make non-alcoholic beverages in Africa and Spain. Tubers contain about 4 per cent protein, 24 per cent fat, 30 per cent starch, and 16 per cent sucrose. Material has been found in Egyptian tombs (2400–2200 BC).

5 TIGER NUTS × 1

TWO-THIRDS LIFE SIZE Fig. 1 × 1/12

1 **JERUSALEM ARTICHOKE** 1A Tuber 2 **OCA** 2A Tubers
3 **ULLUCO** 4 **YSAÑO (TROPAEOLUM)** 4A Tubers

Starchy rooted plants used in making tapioca and arrowroot

CASSAVA, MANIOC, OR TAPIOCA (1) *Manihot esculenta* syn. *Manihot utilissima.* This is the fourth most important source of calories in the human diet in tropical regions. It originated in the Americas; although the exact place is in dispute, Brazil, Colombia, Venezuela, Paraguay, and Mexico have all been suggested. Cassava was being cultivated by tribal groups in the Amazon basin by 2000 BC and, in the sixteenth and seventeenth centuries, it was taken by the Spanish and Portuguese to Africa, India, and South-East Asia. Today it is cultivated in almost all subtropical and tropical countries. The plant is a shrub growing to a height of 3 m. It has woody stems, sections of which are used as cuttings to propagate the crop, and swollen tuberous roots (**1A**)— the part eaten, although the leaves are also utilized as vegetables. The roots are usually harvested 9–12 months after planting. Cassava contains the glycoside linamarin which breaks down under enzyme action (see p. 208) to give poisonous prussic acid (HCN). Cultivars have been divided into 'bitter' (potentially 1350 mg HCN/kg) and 'sweet' (potentially 30–100 mg/kg), although there are intermediates and the glycoside content is affected by the environment. The various forms of food processing (peeling the root, washing, boiling, toasting, and fermentation), if carried out correctly, remove toxicity.

Cassava is a good famine reserve plant because it can tolerate adverse conditions and its mature tubers may be left in the ground for 2 years, although mature tubers, if lifted, can deteriorate within 24–72 hours unless properly stored. The plant is almost immune to locust attack.

Raw cassava contains about 35 per cent starch but little protein, about 1–3 per cent. The vitamin C content is 30 mg/100 g. Cassava food products are of importance in various countries, for example *gari* in West Africa, *farinha* in Brazil, and *gaplek* in Indonesia. Tapioca, made in Malaysia and India by washing out the starch and then drying it, is exported to temperate countries for use in puddings, biscuits, and confectionery. In recent years there has been a considerable export trade from Thailand to Europe of dried cassava chips for use as an animal feed.

The species belongs to the family Euphorbiaceae.

ARROWROOT (2) *Maranta arundinacea.* This is a tropical herbaceous perennial belonging to the family Marantaceae and possessing swollen starchy (over 20 per cent starch) rhizomes (**2A**). The starch is in very fine grains which are easily digestible and is thus suitable for invalid and infant diets. St. Vincent (West Indies) is well known for the cultivation of this plant, although there is some production in China and Brazil. The plants are propagated by rhizome tips ('bits') and dug up when mature at 10–12 months, and the starch is extracted by processes of crushing, washing with water, and drying.

Other plants used for arrowroot include 'Queensland arrowroot' (*Canna edulis*), 'Indian arrowroot' (*Curcuma angustifolia*), and 'East Indian arrowroot' (*Tacca leontopetaloides*).

TARO (3) *Colocasia esculenta* syn. *Colocasia antiquorum.* This is known in the Pacific islands as 'taro'; in the West Indies as 'eddoe' or 'dasheen'; and in West Africa 'old cocoyam'. It originated in South-East Asia and was probably cultivated before rice. Taro is widely cultivated in southern and central China and is a staple food in many Pacific islands. The part eaten is the corm (**3A**) which is roasted, baked, or boiled. If it is not sufficiently cooked, irritation is caused to the mouth by calcium oxalate crystals. Also consumed are the subsidiary corms (cormels), leaf stalks, and blades (**3**). Raw taro corm contains about 25 per cent starch, little protein, and up to 13 mg/100 g vitamin C. The raw leaves contain about 7 mg/100 g carotene and 52 mg/100 g vitamin C. The plant belongs to the arum-lily family (Araceae) and grows to a height of 2 m. It rarely flowers and the leaf stalk is attached near the centre of the blade.

TANNIA, YAUTIA, OR NEW COCOYAM (4) *Xanthosoma sagittifolium.* This originated in the New World and was cultivated in tropical America and the West Indies in pre-Columbian times. The plant is taller than taro and the leaf stalk is attached to the edge of the blade at the notch, giving an arrow-shaped leaf. The corm is processed for consumption in very much the same way as taro and its nutritional composition is also similar, but its starch is more difficult to digest. In West Africa the corms are often preferred to taro for the preparation of *fufu.* The species belongs to the family Araceae.

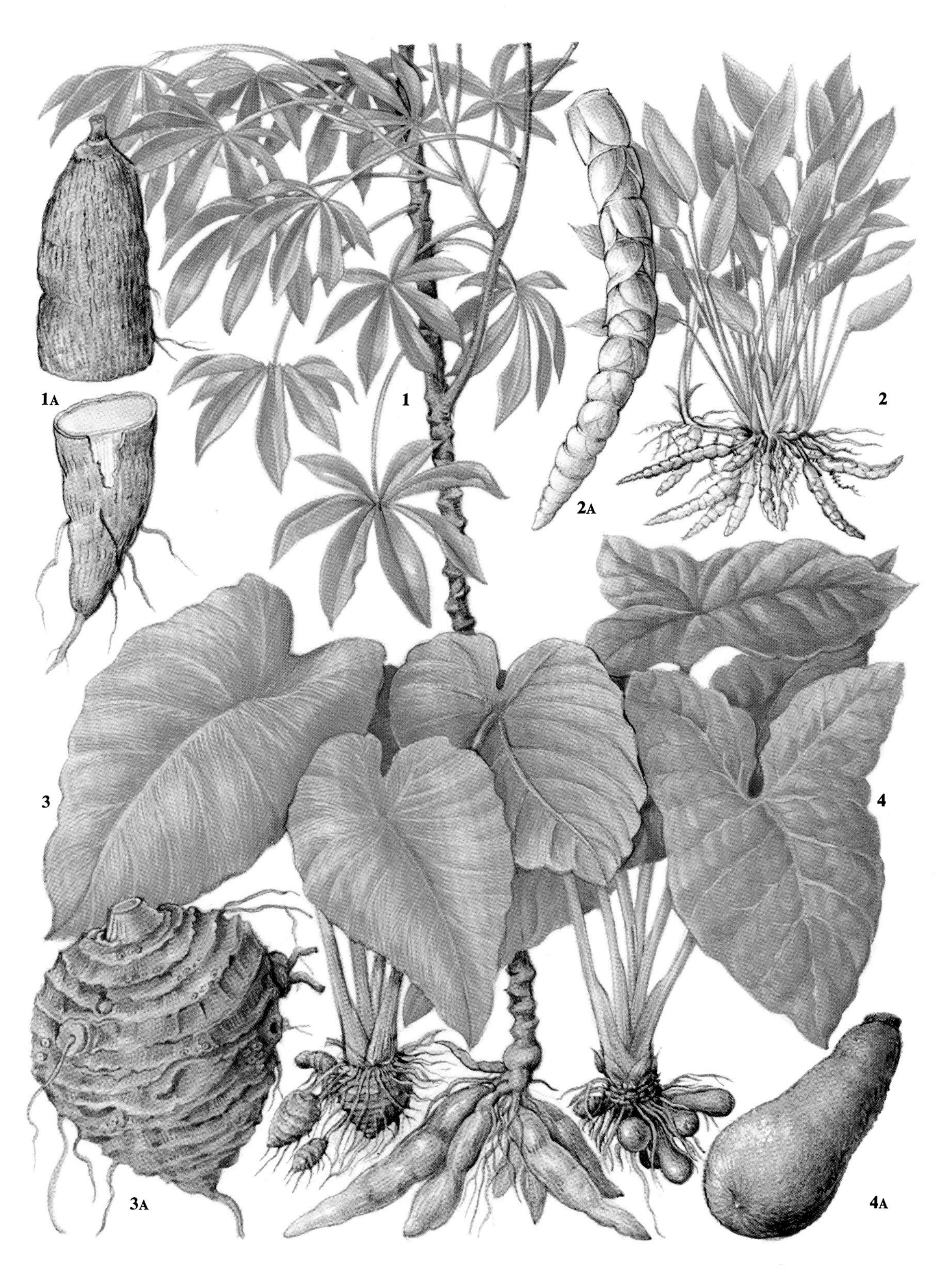

PLANTS × 1/12 TUBERS, etc. × ¼

1 **CASSAVA** 1A Root 2 **ARROWROOT** 2A Rhizome
3 **TARO** 3A Corm 4 **TANNIA** 4A Tuber

Sweet potato and yams

SWEET POTATO (1) *Ipomoea batatas.* This should not be confused with the ordinary potato (*Solanum tuberosum*). Sweet potato belongs to the morning glory family—Convolvulaceae. It is not known in the wild state and it is normally accepted that it originated in tropical America, possibly from the wild Mexican *I. trifida.* At a very early date it was cultivated in Mexico, Central and South America, and the West Indies, and in some way was transported to Polynesia and New Zealand. On the return from his first voyage to the New World, Columbus took the plant to Spain. Somewhat later the Spaniards and the Portuguese transported the sweet potato to Asia and Africa. It is now an important crop in Asia and Africa with some cultivation in the Americas and Europe.

The tuberous roots (**1**, **1A**, **1B**) are the main edible parts of the plant. They vary in shape (elongated to nearly globular); skin colour (white, tan, yellow, or red); and flesh colour (white, yellow, or orange). The tubers can be eaten boiled, baked, or candied. They are also canned and are a source of starch and alcohol. Sweet potatoes contain about 16 per cent starch and quite a high percentage of sugars (6 per cent). The cultivars with yellow or orange flesh are good sources of carotene (4 mg/100 g). Sweet potatoes contain an appreciable amount of vitamin C (23 mg/100 g). Aboveground, trailing or twining stems (1–5 m in length) are produced with, in tropical regions, purple flowers (**1C**). The tender parts of these aerial stems have been used as human food in Africa and Asia, also given to livestock. In the tropics sweet potatoes are propagated by stem cuttings; in temperate regions by sprouts or slips obtained from small tubers. Harvested tubers do not store well.

YAMS (2) *Dioscorea* spp. The term 'yam' is normally applied to cultivated *Dioscorea* species but in the United States of America it has been used for sweet potatoes with orange flesh. Most world production takes place in West Africa, particularly Nigeria, but there is also cultivation in South America, the Caribbean region, and South-East Asia. Five *Dioscorea* species are commonly cultivated for food:

1. *Dioscorea alata* (greater yam), originated in South-East Asia but is now found throughout the tropics.
2. *Dioscorea cayensis* (yellow Guinea yam) is a native of West Africa where it is widely cultivated. It was taken to the West Indies in the sixteenth century. The species is not cultivated in Asia.
3. *Dioscorea esculenta* (lesser yam) is of ancient cultivation in the East but is not widely grown outside Asia and the Pacific region.
4. *Dioscorea rotundata* (white Guinea yam) originated in, and is the most important cultivated yam in, West Africa.
5. *Dioscorea trifida* (cush-cush yam) is the only food yam originating in the New World and is cultivated in the Caribbean region.

Wild yam tubers have been utilized in times of famine but it is usually necessary to remove their toxic alkaloids by boiling or soaking in water.

The edible part of the yam is the underground stem tuber (**2**, **2A**, **2B**—*D. rotundata*). According to species, a plant may possess one or several tubers. Yams are prepared for consumption by boiling, frying, or roasting, and in West Africa, where in some parts they constitute a staple, they are processed into a food known as *fufu* (this is also prepared from plantains, cassava, or cocoyams). The tuber contains about 28 per cent starch and no carotene unless it possesses yellow flesh. Vitamin C is present at about 5 mg/100 g. Yams store well and are easy to handle. They were carried as food supplies on ships engaged in long voyages in the Indian and Pacific Oceans in the pre-European era, later by the Portuguese, and also on slave ships between West Africa and the New World. Their vitamin C content, although restricted, would have been useful in preventing scurvy.

They are climbing plants (**2B**) which twine clockwise or anticlockwise according to species. Yams belong to the family Dioscoreaceae.

YAM-BEAN (3) *Pachyrrhizus erosus.* This belongs to the family Leguminosae (Fabaceae). It grows wild in Mexico and Central America and was cultivated there in pre-Columbian times. It is now also grown elsewhere in the tropics (e.g. India and South-East Asia). The species is a climbing plant (**3A**) with simple or lobed tubers (**3**) containing about 10 per cent starch and 20 mg/100 g vitamin C. The tuber may be thinly sliced and eaten raw with garnish or in salads, also cooked or pickled. In the United States it has been used as a substitute for water-chestnut in oriental cuisine. The young pods may be eaten but the mature seeds are poisonous.

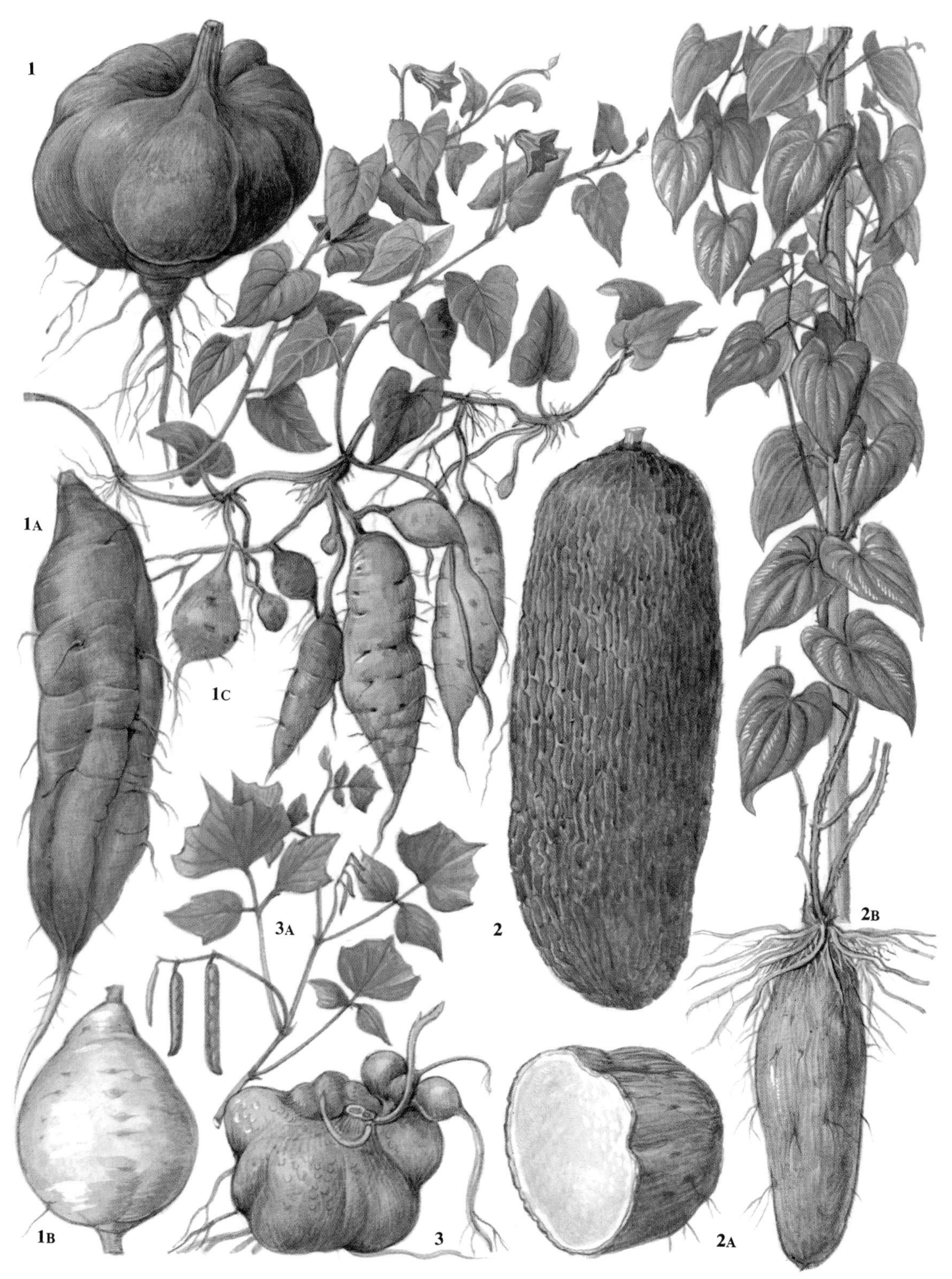

TUBERS × ¼ FOLIAGE × ⅛

1, 1A, 1B **SWEET POTATO** tubers 1C Flowering plant with tubers
2, 2A **YAM** tubers 2B Plant with tuber
3 **YAM-BEAN** tuber 3A Bean-bearing shoot

Seaweeds

(For general information on non-flowering plants, see p. xix)

*Certain seaweeds, often dried, may be used directly as food (boiled as vegetables or used as garnishes and seasonings, or in salads and soups) in many parts of the world, but the major usage seems to be in the Orient, particularly Japan, where some 50 species from 29 genera are employed. Usually the red and brown algae are involved but the green sea-lettuce (*Ulva lactuca*) has been used locally in Scotland in soups and salads. In Europe, France supplies seaweeds as food to supermarkets. Carbohydrates known as phycocolloids (agars, alginates, and carrageenans) are extracted from seaweeds and are used as thickeners and stabilizers (see p. 201) in a vast array of foods, including canned commodities, confectionery, ice-cream, jellies, soups, and sauces. As regards their nutritional properties, there is variation between species but, generally speaking, there is little protein and fat, no starch, and traces of sugar as known in flowering plants. The protein has an amino acid profile similar to that of legumes. Of the minerals, the quantity of iodine can be very high but this relates to the species. Seaweeds contain carotene, vitamins E and C. Compared to other plants, seaweeds are unusual in that some reports state that they contain a little vitamin B_{12} in addition to the other B vitamins, although the presence of B_{12} is open to debate. Seaweeds in a diet provide fibre.*

LAVER (1) *Porphyra umbilicalis.* This is the red seaweed involved in the production of laver bread, particularly popular in South Wales although it is also collected and eaten elsewhere on the western coast of the United Kingdom and Ireland. To produce laver bread, the seaweed is washed to remove sand, boiled for 8–12 hours in salty water, and minced to give a dark brown or black product which, as a traditional dish, is coated in oatmeal before being fried and served with bacon and eggs. It is also eaten with potatoes and butter.

Laver is common on rocks and stones all round the coasts of the British Isles and other temperate North Atlantic countries. It is rosy purple turning to olive-green or brown. The thin, flexible, wavy-edged frond is attached to the rock on which it grows by a small, disc-shaped holdfast.

Porphyra is popular in China, Korea, and Japan (where it is known as *nori*). The Japanese cultivate *Porphyra* by sinking bundles of bamboo canes or brushwood offshore. When a good crop of the young seaweed has become established, the bundles are transferred to less salty water where they will grow better than in undiluted sea water. In the Orient *Porphyra* is used as a food in the ways described at the beginning of this page.

DULSE (2) *Palmaria palmata* syn. *Rhodymenia palmata.* This is a red seaweed. As human food it is consumed in the ways already described. Also, as other seaweeds, it is used as a masticatory, an animal fodder, and fertilizer. The dark-red, rather leathery, wedge-shaped frond is usually divided dichotomously or palmately, and old fronds often have rows of smaller, thinner 'leaflets' along their margins.

CARRAGEEN OR IRISH MOSS (3) *Chondrus crispus.* This is collected commercially in Canada for carrageenan extraction. Small quantities are harvested in Ireland and France and then dried. They may be sold whole or ground for use as a health food, a thickener in cooking, or in cough mixtures. It is a red seaweed, being attached to rocks by a disc-like holdfast with a stalk of varying length, branching repeatedly into a series of flattened segments, more-or-less fan-like in appearance.

KNOTTED WRACK (4) *Ascophyllum nodosum.* This is a brown seaweed and is common in temperate Atlantic countries. It is harvested in Ireland, Scotland, and Norway for alginate extraction, also meal for human and animal consumption. The plant has conspicuous bladders.

OTHER SEAWEEDS A number of species of *Laminaria* (e.g. *L. hyperborea*, *L. digitata*, and *L. saccharina*), a brown seaweed, are utilized for animal and human food, also as a fertilizer and source of alginates. In Japan, *Laminaria* species are known as ***kombu.*** Other seaweeds utilized include *Gelidium sesquipedale* (source of agar), *Furcellaria lumbricalis* (source of carrageenan), *Phyllophora truncata* (source of carrageenan), and *Undaria* spp. (***wakame***).

'Seaweed' utilized in Chinese cuisine in western countries is often cabbage (*Brassica oleracea*).

LIFE SIZE

1 **LAVER** 2 **DULSE** 3 **CARRAGEEN**
4 **KNOTTED WRACK**

Mushrooms, truffles, and other edible fungi

Of the commercially cultivated mushrooms Agaricus bisporus *is the most important, with about 60 per cent of world mushroom production. It is produced widely in Europe, the United States of America, and some Far Eastern countries. Others of commercial importance are Japanese black forest mushroom or shiitake (*Lentinus edodus*), Chinese or straw mushroom (*Volvariella volvacea*), oyster mushroom (*Pleurotus ostreatus*), and winter mushroom (*Flammulina velutipes*). Mushrooms add flavour and texture to a meal. The nutrient composition will vary somewhat according to species but taking the common mushroom (*Agaricus bisporus*) as a representative example, there is about 2 per cent protein (which contains all the essential amino acids), 0.4 per cent unsaturated fat, about 5 per cent carbohydrate, a range of minerals with a large amount of potassium (320 mg/100 g), no carotene, some vitamin E, a range of B vitamins but no B_{12}, and about 4 mg vitamin C/100 g. A microfungus now being produced commercially for food is* Fusarium graminearum *('Quorn'). It has a similar texture and eating quality to meat; it contains a large amount of protein compared to other fungi—about 12 per cent.*

TRUFFLE (1) *Tuber aestivum.* This fungus, which grows underground in woods, especially beech woods, is irregularly globose, 2.5–10 cm across, dark brown and warty, its flesh permanently solid, white, soon turning buff, with a network of white veins. This, the best flavoured of British truffles, is regarded as inferior to the French Périgord truffle (*T. melanosporum*) which is used in *paté de foie gras.*

CHANTERELLE (2) *Cantharellus cibarius.* Common in woods, in summer and autumn, Chanterelles are rather firm-fleshed and need cooking longer than mushrooms. The funnel-shaped cap is egg-yellow, with paler flesh, having a faint odour reminiscent of apricots. The pale pinkish-buff spores are produced in narrow folds.

MOREL (3) *Morchella esculenta.* This grows in spring, often in woodland clearings. Morels have a distinctive appearance, their caps criss-crossed with irregular, pale-brown ridges between which are darker brown hollows in which the spores are produced. The stalk is whitish, becoming yellowish or reddish when old. There are other European species.

FIELD MUSHROOM (4) *Agaricus campestris.* This wild *Agaricus* species is found in Europe and is the one primarily collected in the United Kingdom. Found in meadows and pastures in summer and autumn, it has a white cap, when young connected to the stem by a membrane (partial veil) which tears as the cap expands, its remains persisting for a time as a narrow ring around the stem. The gills are white at first, soon turning pink, and finally dark purplish-brown. The spores are dark brown.

Agaricus bisporus is distinguished microscopically from *A. campestris* by its spores, being produced in pairs instead of fours. This is the common cultivated mushroom, which is white and is sold in different developmental stages, for example 'button'—young; 'flat'—older and expanded. 'Chestnut' mushrooms are brown forms of the species. Large quantities of mushrooms are canned.

BLEWITS (5) *Lepista* spp. These have a pleasant smell and are good to eat, cooked like mushrooms. *Lepista saeva* is found in open grassland, in autumn. It is mushroom shaped, with a greyish or brownish cap, tinged with lilac or purple, white-fleshed when young. The stem is whitish with bluish streaks. The wood blewit, *Lepista nudum,* grows mainly in woods in late autumn and is wholly lilac or purple.

OYSTER MUSHROOM (6) *Pleurotus ostreatus.* This is rather flavourless, but as it often grows in colonies it is easy to collect and it dries and keeps well. The colour of the cap varies with age, from dark bluish-grey to pale brown. the widely spaced, yellowish-white gills merge into a very short stem which attaches the side of the cap to the host tree—often a beech. It is now one of the commercially produced mushrooms.

CEP (7) *Boletus edulis.* A member of a large genus, most (but not all) of which are edible, being fried when fresh or dried and used in casseroles and soups. It is common in woods (especially beech woods) in summer and autumn. It has a brown, smooth, moist, shining cap, its flesh white, often tinged with pink. Beneath is a spongy mass of vertical tubes, white at first, becoming yellowish green, in which the brown spores are produced. The stalk is stout, pale brown, with a fine network of raised white veins towards the top.

SHAGGY PARASOL (8) *Macrolepiota rhacodes.* This is a member of a large genus. It has a cap, up to 18 cm across, covered with large, yellowish or brownish scales, except for a smooth brown disc in the centre. The long, stout, whitish stem is smooth. The flesh is white, turning red when cut. It grows in open spaces in woods and gardens, usually in rich soil. **Parasol mushroom** (*M. procera*) is another well-known species.

FAIRY-RING CHAMPIGNON (9) *Marasmius oreades.* This forms the well-known 'fairy-rings' on lawns and short-turfed pastures. Its cap is brownish, often tinged with pink, paler when dried, slightly domed in the centre. The gills and the slender tough stem are pale buff. The flesh is white and has a pleasant, mushroom-like flavour but the tough stem should be discarded. This fungus is easy to dry and keeps its flavour well when reconstituted by soaking. It can also be pickled and used for making a ketchup.

GIANT PUFF-BALL (10) *Calvatia gigantea.* This grows in woodland and pastures and is large. It can be eaten only when young, white, and firm-fleshed. With age it turns yellowish and then brown. A mature puff-ball contains an astronomical number of powdery, olive-brown spores. There are several other edible puff-balls. The illustration here is about one-tenth actual size.

TWO-THIRDS LIFE SIZE

1 **TRUFFLE** 2 **CHANTERELLE** 3 **MOREL**
4 **FIELD MUSHROOM** 5 **BLEWIT** 6 **OYSTER MUSHROOM**
7 **CEP** 8 **SHAGGY PARASOL** 9 **FAIRY-RING CHAMPIGNON** 10 **GIANT PUFF-BALL**

Some wild plants

In many parts of the world, wild plants are still utilized as food and in other parts there has been a resurgence of interest (Mabey 1972). It is not always possible or easy to find nutrient analyses of these plants but when the cultivated types are very close to the wild forms some information can be obtained—see in previous sections of this book: hazel-nut (p. 30), wild crab (p. 52), rose (p. 68), sloe (p. 72), blackberry (p. 84), cloudberry (p. 84), dewberry (p. 84), bilberry (p. 88), cranberry (p. 88), strawberry tree (p. 88), chicory (p. 118), dandelion (p. 118), juniper (p. 146), hop (p. 146), wormwood (p. 146), caraway (p. 148), fennel (p. 148), peppermint (p. 150), spearmint (p. 150), marjoram (p. 150), tansy (p. 154), chamomile (p. 154), samphire (p. 156), sweet cicely (p. 156), lovage (p. 156), mustard (p. 142), watercress (p. 162), rocket (p. 162), asparagus (p. 172), sea kale (p. 172), chives (p. 176). A very good modern example of a scientific study of the nutritional properties of wild plants in Canada is that by Kuhnlein and Turner (1991).

ELDER (1) *Sambucus nigra.* Elder flowers (actually the corollas detached from the stalks) are used to make wine (still or sparkling) and cordial, also a preserve with gooseberries and, with flour and egg, fritters. Elder berries are used to make a well-known wine but can also be added to apple pies or apple and blackberry jellies. An interesting product once made from elder berries was Pontack Sauce.

Elder belongs to the honeysuckle family—Caprifoliaceae. It is a deciduous shrub or small tree, common in woods, hedgerows, and waste places. Its branches have brownish-grey, corky bark and there is a large proportion of soft, light, whitish pith. The creamy-white flowers are bisexual and the fruit is purplish black.

Studies on North American species have indicated that the leaves, bark, roots, and seeds are poisonous because of glycosides but the berries processed in the ways already described are obviously acceptable.

BARBERRY (2) *Berberis vulgaris.* Barberry is not now as common as in the past in hedgerows and bushy places because, being an intermediate host of the black rust fungus (*Puccinia*) of cereals, it has been systematically eradicated. The fairly acid red berries make jelly without additional pectin and have been candied, pickled, and used for garnishing.

Barberry belongs to the family Berberidaceae. It is a spiny shrub, 1–2 m in height, with bisexual flowers, borne in a pendulous raceme and succeeded by bright red fruits.

GOOD KING HENRY (3) *Chenopodium bonus–henricus.* This was formerly cultivated in medieval and Elizabethan times as a green vegetable (like spinach) and remains have been found in neolithic sites. It is now found growing wild in Europe, western Asia, and North America. The plant belongs to the goosefoot family (Chenopodiaceae), which includes two other species formerly of importance as vegetables—orache (*Atriplex hortensis*) and fat hen (*Chenopodium album*).

Good King Henry is a perennial herb, 30–50 cm in height. It bears rather fleshy, triangular, arrowhead-shaped leaves and the long stigma sticks out of the small, green flower. The plant contains almost 6 per cent protein, a range of B vitamins, and good quantities of carotene and vitamin C.

STINGING NETTLE (4) *Urtica dioica.* Stinging nettle has had a number of food uses. The young tops, gathered when about 15 cm high, can be used as a green vegetable, usually in the form of a purée like spinach—the older leaves are bitter in taste. Stinging nettle is said to be rich in vitamin C (75 mg/100 g). In Scotland there is a recipe for nettle pudding or haggis, nettles with leeks or onions, broccoli or cabbage, and rice or oatmeal, served with butter or gravy. Nettles have been used to make soup, beer, tea, and herbal preparations.

The plant belongs to the family Urticaceae, and has stinging hairs. *Urtica dioica* is a perennial with the blades of the lower leaves longer than their stalks and the small, green male and female flowers borne on different plants. The less common small nettle (*Urtica urens*) is an annual. The blades of its lower leaves are shorter than their stalks and the unisexual male and female flowers are borne on the same plant.

SORREL (5) *Rumex acetosa.* This is a common wild plant which was used as food in ancient Egypt and by the Romans. It is sometimes cultivated, particularly in France where improved forms have been developed. Sorrel can be utilized as a vegetable, in soups, salads, or sauces. The sharp flavour is due to oxalic acid.

It belongs to the family Polygonaceae and is a perennial herb, 30–100 cm in height. The arrowhead-shaped leaves are acid-flavoured and have downward-pointing basal lobes. The male and female flowers are borne on different plants. Other *Rumex* species sometimes used a food are *R. scutatus* (round-leaved sorrel) and *R. patientia* (herb patience).

TWO-THIRDS LIFE SIZE

1 **ELDER** 2 **BARBERRY**
3 **GOOD KING HENRY** 4 **STINGING NETTLE** 5 **SORREL**

Nutrition and health

Role of plants in food supply

Food provides all the energy to maintain life in humans and other living organisms, the materials for growth, repair, secretions, and reproduction, and essential vitamins and minerals required for all these processes. For the world population as a whole, food plants supply the great majority of these nutrients and only in the richer countries do animal foods provide a significant proportion. For example, the proportion of total energy in the diet from plant foods ranges from 63 per cent in the United Kingdom to 96 per cent in Bangladesh, the remainder coming from animal foods (see Table 1, p. 214).

The main characteristics of plant versus animal foods, remembering that all animal nutrients are ultimately derived from the plants the animals consumed, are as follows: plant foods contain a higher proportion of carbohydrates, and uniquely provide complex carbohydrate including dietary fibre; generally lower levels of protein; and lower levels of fat, which is usually mainly unsaturated. Vitamins or their precursors specifically related to plant foods are carotenes (related to vitamin A), and vitamin C, while retinol (preformed vitamin A), and vitamin B_{12}, are specific to animal foods. These characteristics can be altered by traditional and commercial food-processing methods. Minerals are widespread in both plants and animals but some, such as iron, are less easily absorbed from plant foods unless foods are properly combined. However, specific foods in both categories can flout these general rules. For example, legumes are relatively high in protein, coconut fat is mainly saturated, and fermented soya is rich in vitamin B_{12}.

These general characteristics mean that the composition of the food supply in different countries is variable, although general patterns can be distinguished, largely in relation to the level of affluence. These are discussed later (p. 211).

Food components

See Tables 2–10, pp. 215–223. Plant foods contain nutrients, other biologically active substances, inert substances, and some natural toxicants and pollutants. The main nutritional components are the so-called proximal nutrients: carbohydrate, fat, and protein (and alcohol processed from plants). These form the bulk of the foods, apart from water, and provide all the energy, which is their common denominator. Vitamins and minerals are present in very small quantities but are equally essential for specific physiological functions. Essential nutrients are those which must be supplied from the diet as they cannot be formed from other substances in the body. Non-essential nutrients are those which exist in foods but can also be made in the body. Vitamin C is an example of a nutrient that is essential for humans but non-essential for most other animals.

Water

Water is the largest component of most foodstuffs, both plant and animal. The human body is 70 per cent water, as is meat. The water content of plant foods ranges from about 5 per cent in nuts and oilseeds to 10 per cent in grains and pulses, 80 per cent in potatoes, and 95 per cent in water-melon.

The low water and fat content of cereal grains accounts for their importance in the world food supply as they can be most easily stored and transported without spoilage and with calorific efficiency, as well as supplying carbohydrate, protein, and fat in the proportions most closely compatible with physiological needs.

Energy

Energy is needed for activity, but most of the energy consumed from food is used for basal metabolism, the chemical processes that keep the heart beating, the blood circulating, and the brain functioning. Energy is measured in kilocalories or kilojoules (1 kcal = 4.18 kJ) and adults require approximately 1500–3000 kcal/day, depending on sex, body size, and activity level. This is obtained from the major components of food: carbohydrate, fat, and protein, as well as from alcohol. Vitamins and minerals provide no energy. Pure carbohydrate provides approximately 4 kcal/g; fat, 9 kcal/g; protein, 4 kcal/g (and alcohol 7 kcal/g). Plant foods contain a mixture of the first three in varying proportions. However, the energy value of most fruits and vegetables is generally less than 1 kcal/g, as plants contain a considerable amount of water, ranging from 10 per cent for grains to 95 per cent for water-melon. They also contain dietary fibre which yields little energy. Exceptions are the grains and pulses, which have an energy value of about 3.5 kcal/g because of their low water content before cooking, and oilseeds and nuts, which contain little water and much fat, resulting in an energy value of approximately 6 kcal/g.

Carbohydrates

Carbohydrates supply most of the energy intake of the majority of the world's population. These are derived almost exclusively from plant foods, as the carbohydrate content of most foods of animal origin is negligible, except for milk sugar. Green plants synthesize carbohydrate from carbon dioxide and water in sunlight, thereby producing oxygen. Sugars, which are soluble in water, are formed first and transported throughout the plants, some are then linked together into two main types of polysaccharides: starch, which is stored in the plant cells: and non-starch polysaccharide (NSP), including polysaccharides that constitute the cell wall (cellulose) as well as other storage forms. NSP is the principal component of the more commonly called dietary fibre. Dietary carbohydrates are classified according to the complexity of single sugar units (monosaccharides) that are linked together.

MONOSACCHARIDES Monosaccharides normally contain three to seven carbon atoms with hydrogen and oxygen attached, and are named, respectively, trioses, tetroses, pentoses, hexoses, and heptoses. The hexoses and pentoses are the most important in the diet, and can exist either as straight chains or in a ring structure.

Hexoses include glucose, which is not abundant in free form in natural foods, but is found in small amounts in fruits and vegetables, particularly grapes and onions. It is one of the main constituents of honey and can be manufactured from starch. Fructose is found in fruits, vegetables, and honey. Galactose is found in milk as a component of the disaccharide lactose, and in the storage polysaccharides of some plants. Two important pentose sugars are ribose and deoxyribose which form the genetic materials RNA (ribonucleic acid) and DNA (deoxyribonucleic acid).

DISACCHARIDES Combinations of two simple sugars are the disaccharides. The main ones is sucrose, which is the most commonly used sugar, extracted from sugar-beet and sugar-cane. It is formed from glucose and fructose. Table sugar is 99.9 per cent pure sucrose and is the main source of sucrose in the diet. Sucrose is also present in fruit and vegetables. It is broken down into glucose and fructose as it is digested in the intestine. Lactose is formed from glucose and galactose, but is only found in milk, not in plant foods. Maltose, consisting of two molecules of glucose, is present in sprouted (malted) wheat and barley, from which malt extract is produced for brewing and for malted food products.

Trehalose, or mushroom sugar, is another form composed of two molecules of glucose, constituting up to 15 per cent of the dry matter of mushrooms. A specific enzyme exists in the body to break down this disaccharide for digestion, indicating that mushrooms were in the past a more important part of the human diet than now. The mono- and disaccharides are sweet, but glucose is only half as sweet as sucrose.

OLIGOSACCHARIDES Short chains of sugars are called oligosaccharides. These include raffinose, stachyose, and verbascose, which are found mainly in legumes such as peas and beans, and fructans, found in cereals, onions, garlic, asparagus, and Jerusalem artichokes which contain the longer-chain fructan, inulin. They are not broken down by digestive enzymes and so pass into the large intestine, where they are fermented by bacteria, producing flatulence.

SUGAR ALCOHOLS Sugar alcohol is the chemical name of specific forms of sugars found in nature and prepared commercially, not to be confused with the drinking alcohol prepared by the fermentation of various carbohydrates. Sugar alcohols include sorbitol, found in fruits such as cherries; mannitol, extracted from a seaweed; and inositol, present in many foods, in particular cereal bran. Sorbitol is used commercially as a sweetener in foods and drinks for diabetics as it is absorbed from the intestine more slowly and so has less effect on blood glucose levels than sucrose. It is only 60 per cent as sweet as sucrose. It is also used in sweets and chewing gums to reduce the effect of sugar on tooth decay. The bacteria in the mouth also use this form less readily and so produce less of the acid which causes decay. Inositol combines with phosphate to form phytic acid, present in many plant foods, which is important nutritionally as it impairs the absorption of calcium and iron in the intestine.

POLYSACCHARIDES The polysaccharides are not sweet. Starch is the major polysaccharide and carbohydrate in the human diet and the main storage form of energy in cereals, root crops, and plantains. It consists of two types of chains of glucose molecules: amylose is an unbranched chain; amylopectin is highly branched. Starch granules from different plants contain varying proportions of amylose and amylopectin. Starch is insoluble in water and therefore not easily digested until cooked, which causes it to swell. Recooling reduces digestibility, forming 'resistant starch'. Other polysaccharides include cellulose and gums.

DIETARY FIBRE Dietary fibre is a term that has been used loosely to describe the plant material that is resistant to digestion in the human intestine. This could include lignin, a woody substance that is not a carbohydrate, some forms of starch, and other components. However, to characterize it chemically and allow a better understanding of the effects of various components on digestion and metabolism, it is now generally accepted to define it as non-starch polysaccharide (NSP). NSP includes cellulose (insoluble) which is the main component of plant cell walls, and non-cellulose polysaccharides (mainly water soluble), including: pectin, which is extracted commercially and used as a gelling agent in jams; β-glucans, of which cereals such as oats and barley are a good source and which may explain the cholesterol-lowering effect of oat bran; gums such as guar gum and gum arabic, which are extracted commercially and used in the food industry as emulsifiers, stabilisers, and thickeners; mucilages such as alginates, carrageenans, and agar, which are found in seaweeds and other algae, and are used as thickeners in dairy products and confectionery.

The published values for dietary fibre content of foods are very variable because of different analytical techniques using a variety of digestive enzymes. 'Crude fibre' values are found in older publications. This is the material left after extraction with petroleum, dilute sodium hydroxide, and dilute hydrochloric acid, under strictly specified conditions. This technique is no longer used in human studies, but it is still used for the analysis of animal feedstuffs. It measures mainly cellulose and lignin and therefore greatly underestimates the dietary fibre in food. 'Total fibre' gives much higher values as it includes a broad range of non-digestible substances. These values are found in most food tables and on most food labels. 'Non-starch polysaccharide' (NSP) values are considerably lower and are included along with total fibre in recent editions of British food tables, as the currently accepted technique (see pp. 215–223, Tables 2–10).

Wholegrain cereals are especially rich sources of NSP, mainly insoluble in wheat, maize, and rice, whereas a significant proportion is soluble in oats, barley, and rye. In vegetables the soluble and insoluble fractions are approximately equal, but in fruits the proportions are widely variable. Intakes of NSP in developed countries range between 10 and 18 g/day, with approximately 12 g/day in the United Kingdom, whereas in developing countries intakes are considerably higher. The equivalent intake of total fibre in the United Kingdom is about 21 g/day, compared to China, for example, where the country average intake is 33 g/day with over 70 g/day average in some regions.

Dietary fibre, especially insoluble fibre such as wheat bran, decreases intestinal transit time and increases faecal bulk. For this reason lack of dietary fibre is a major factor in several gastrointestinal diseases common in industrialized societies, such as constipation and colorectal cancer. Populations with higher intakes of dietary fibre have generally a lower incidence of diseases such as coronary heart disease and gallstones, related to the blood cholesterol level. Soluble NSP (pectins, gums, etc.) are effective in reducing cholesterol in the blood and other tissues, but insoluble NSP is not. Soluble NSP prevents the reabsorption by the intestine of bile acids synthesized in the liver and secreted into the intestine via the gall-bladder to facilitate the absorption of fat. The bile acids, which are formed from cholesterol, therefore have to be replaced from the cholesterol pool, so reducing blood cholesterol levels.

Cholesterol synthesis in the liver may also be altered by the short-chain fatty acids (see below) derived from the colonic fermentation of the soluble fibre.

DIGESTION AND ABSORPTION Digestion of carbohydrate occurs mainly in the small intestine where it is broken down by various enzymes to monosaccharides that are absorbed, glucose being absorbed faster than fructose. After a meal the blood glucose rises to a maximum in about half an hour and returns to fasting level in about 2 hours. In diabetes blood glucose levels tend to be raised and so it is important to control the peak and rate of return to fasting values in order to stabilize blood glucose levels. The rate of glucose absorption from the breakdown of carbohydrate in various foods in comparison with pure glucose is called the glycaemic index which is a useful indicator for the choice of foods in dietary control for diabetics. For example carbohydrate from legumes raises blood glucose less than half as much as pure glucose which has a glycaemic index of 100. The glycaemic index is 30–40 for beans, 40 for apples, 65 for wholemeal bread, and 70 for white bread.

Fat

TYPES In the diet of the United Kingdom and other developed countries, approximately 40 per cent of the energy comes from fats and oils, but this can be as low as 10 per cent in some developing countries. The term 'fat' is commonly used for solid greasy substances, and oils for those that are liquid at room temperature, but chemically they are very similar or identical. Lipids is the chemical term for substances that are insoluble in water but soluble in solvents such as alcohol and chloroform; this includes the fats and oils, and cell membrane lipids. The latter include cholesterol, which is found only in animal not plant membranes and other tissues, dietary cholesterol therefore coming almost entirely from animal foods. They also include the polar lipids, phospholipids in animals, and glycolipids in plants. Polar lipids are miscible in both lipids and water and are therefore important to stabilize emulsions, both in the body, for example bile salts in the digestion of fat, and in foods, for example lecithins in foods such as chocolate, processed cheese, and mayonnaise. Lecithins (phosphatidylcholine) are present in egg yolk and are prepared commercially from soya bean, peanuts, and maize. Lipoproteins are combinations of lipid and protein, found in animal systems, that can transport lipids in a water-based liquid such as blood.

TRIGLYCERIDES Dietary fats and oils are chemically mainly triglycerides, molecules composed of three fatty acids attached to a glycerol core. Triglycerides are an important form of energy storage in both animals and plants. The fatty acids are composed of carbon chains of varying length with hydrogen atoms attached to the bonding sites. Fatty acids are classified as saturated, monounsaturated, and polyunsaturated, referring to the extent to which hydrogen fills the available carbon bonds. Saturated fatty acids have all bonds filled, monounsaturates have one bond unfilled, so that there is one double bond between the carbon atoms in the chain, and polyunsaturated fatty acids have several unfilled. These groups have different properties and functions. The longer the chain length and the more saturated, the harder the fat. For example, lard contains a higher proportion of long-chain saturated fatty acids and cooking oils a higher proportion of unsaturated. The extent of saturation is also important for susceptibility to rancidity. The unsaturated fats react more readily with oxygen and become rancid. Saturated fats are more stable for storage.

In general animal fats contain mainly saturated (40–60 per cent) and monounsaturated fatty acids (30–50 per cent), with little polyunsaturated (<10 per cent), while plant foods contain fats that are mainly polyunsaturated (40–60 per cent) and monounsaturated (30–40 per cent) with smaller proportions of saturated fats (approximately 20 per cent). However, there are a few exceptions such as coconut oil and palm oil which are highly saturated (90 per cent, 50 per cent), while poultry and game have more polyunsaturated fats, especially turkey (35 per cent). Milk fat contains a high proportion of short-chain saturated fatty acids (see Table 11, p. 224).

Unsaturated fatty acids exist in two different forms, *cis* and *trans*. Most occur only in the *cis* form but *trans*-fatty acids are produced in the commercial processing of oils, such as in the production of margarine, and are also found in the milk of ruminant animals. There is some controversial evidence that at high intakes *trans*-fatty acids may adversely affect blood cholesterol levels.

HEALTH EFFECTS Dietary fats have been implicated in several aspects of health. If the amount of fat in the diet is very low, for example less than 10 per cent of the energy, malnutrition may occur in young children. Fat is the food component that provides the most energy for the same weight or volume (9 kcal/g) and so reduces the bulk of the food required. Young children require a relatively large amount of energy for their body size and they cannot eat enough food to satisfy their needs if the fat content is too low. Another problem with low intake is the poor absorption of the fat-soluble vitamins such as A, E, and D, and deficiencies of these can occur.

High fat intakes, especially saturated fat, are epidemiologically associated with a higher incidence of atherosclerosis and coronary heart disease and certain cancers such as bowel, prostate, and breast cancers. Blood cholesterol levels are a risk factor for heart disease, in particular the LDL fraction (low-density lipoprotein) which carries most of the blood cholesterol and is responsive to diet. The smaller-fraction HDL (high-density lipoprotein) is less responsive to diet but increases with exercise and is protective against heart disease. It is thought that if the LDL becomes oxidized by free radicals it is scavenged by specialized cells (macrophages) in the cell wall, converting them to 'foam cells' and leading to the development of atherosclerotic lesions. Antioxidants (see p. 207) can inhibit this effect and protect against heart disease.

Blood cholesterol is affected to only a small extent by cholesterol in the diet as intakes are only about 10–20 per cent of the amount that is produced by the body itself. The main dietary factor is saturated fat. As most saturated fats in the diet come from animal products, diets of predominantly plant-based foods result in reduced blood cholesterol levels.

If diets are virtually fat free, as has occurred with infant formulas, clinical deficiency of the 'essential fatty acids' occurs, resulting in flaky skin and other impairments. This dietary factor was originally called vitamin F until identified as the essential fatty acids linoleic and linolenic acids. These essential fatty acids are long-chain polyunsaturated acids that are required for the rapid brain development occurring in infancy and for the production of prostaglandins that regulate a variety of functions, including inflammation, blood clotting, and blood vessel dilation and contraction. Very small amounts are required to prevent deficiency (about 1 per cent of dietary energy). Care is now taken to ensure that infant feeds contain all essential nutrients including 'essential fatty acids'.

Protein

STRUCTURE Protein constitutes 10–15 per cent of the energy in almost all human diets. It is also important in the structure of all cells in the body, as well as forming enzymes, molecules that transport substances in the blood, and some hormones. There is a large variety of proteins in both food and the body, with different compositions and functions. They are composed of about 20 different amino acids, which all contain nitrogen (about 16 per cent), carbon, oxygen, and hydrogen. Some also contain small amounts of sulphur and phosphorus. The body is able to synthesize some amino acids in the liver (non-essential amino acids) but others have to be obtained from food (essential amino acids). There are eight essential amino acids: valine, leucine, isoleucine, phenylalanine, tryptophan, threonine, methionine, and lysine, plus arginine and histidine for infants as their capacity for synthesis is low and insufficient to meet the requirements for growth. Others can be essential under certain conditions, such as cysteine and tyrosine when their amino acid precursors methionine and phenylalanine are not in abundance in the diet.

Amino acids can be linked by 'peptide' bonds (between the nitrogen-containing amino group of one and the acid group of another). Peptides range from dipeptides, with two amino acids, to polypeptides with up to thousands. Proteins are made up of polypeptide chains folded into structures with different shapes and functions. Examples of some plant proteins are gluten, gliadin, and zein, and of body or animal proteins: collagen, haemoglobin, and insulin. Ingested proteins are degraded in the intestine by specific enzymes and absorbed as free amino acids and small peptides into the blood to be carried to the tissues for processing into specific proteins and other substances required for body structure and function, and any surplus is oxidized to supply energy.

PROTEIN QUALITY The 'quality' of a protein is influenced by its amino acid composition and its digestibility. In each food the content of substances such as dietary fibre, digestive enzyme inhibitors, protein structure, and processing affect protein digestibility.

The body requires different amounts of each essential amino acid, and the closer the pattern of amino acids in a particular type of protein is to the ideal pattern, the better it will be retained in the body. If it is not close, the amino acids present in relative excess will be used only to provide energy and excretion products. Animal proteins, including milk and egg, have a composition similar to that of body requirements and have traditionally been used as a reference or 'ideal' protein. Plant proteins are relatively low in certain amino acids. The lowest essential amino acid in relation to the quantity in the reference protein is termed the 'limiting amino acid'. This limiting amino acid varies between plant food groups. For example in grains the limiting amino acid is lysine; in legumes, methionine and cysteine.

Many plant-breeding programmes have been undertaken to improve the amino acid pattern of food crops. However, the theoretical 'quality' of individual proteins is not so important in practice as all diets, except the very poorest, contain a mixture of plant proteins, each with different amino acid patterns that compensate for each other even without animal protein in the diet. In addition, if the total amount of food consumed provides the energy required, there is usually more protein than is required and so adequate quantities of even the limiting amino acid. Protein deficiency is likely to occur only when total food intake is too low, or where the diet, especially a weaning diet, consists almost entirely of plant food with a particularly low protein content in relation to total calories, such as cassava and plantain. This can be expressed as the calories from protein as a percentage of total calories: cassava contains 2.4 per cent and plantain 3.6 per cent, whereas other staple foods contain higher amounts, such as: wheat, 13–15 per cent; maize, 10.0 per cent; rice, 7.1 per cent; and potato, 6.7 per cent. Protein-rich foods such as animal products and legumes contain 20–30 per cent: beans and peas, 22 per cent; milk, 22 per cent; eggs, 33 per cent; chicken and beef, 26 per cent. For comparison, recommended dietary intakes for protein energy in relation to total calories is in the order of 7 per cent for adults and 4 per cent for toddlers.

Soya beans are higher in protein than most other legumes (37 per cent of the calories) and soya foods have been used for centuries as an important part of the diet in Asia, and in recent decades by the food industry in the West. Although the limiting amino acids are the same as in other legumes, methionine and cysteine, the level is higher, so the quality of soya protein is greater. One advantage over animal proteins is that it does not increase the excretion of calcium and therefore may partly explain the lower rates of osteoporosis (thinning of the bones after middle age) in Asian countries. Soya beans also contain isoflavonoids (see pp. 206–7) which may inhibit bone resorption. Soya protein also lowers blood cholesterol and soya isoflavonoids may suppress the oxidation of LDL-cholesterol, both risk factors in heart disease (see p. 202).

DIGESTION AND ABSORPTION Before absorption from the intestine, dietary proteins have to be broken down to their component amino acids. The first stage of this process occurs in the stomach, where the strong acid causes the folded structures to uncoil. This exposes the bonds between the amino acids to specific enzymes, collectively called proteases, that break the chains into smaller segments, so that by the time they pass out of the stomach into the small intestine they are already mostly broken into strands of two, three, or more amino acids (dipeptides, tripeptides, polypeptides), or single amino acids. In the small intestine the acid is neutralized by alkaline juice from the pancreas which allows other enzymes such as trypsin and chymotrypsin to act to complete digestion to amino acids which are absorbed into the intestinal cells, and released into the bloodstream. Occasionally minute quantities of whole proteins or large polypeptides manage to cross the intestinal wall and can cause an immune response and play a part in food allergies. Some substances in food can inhibit the action of proteases and so affect digestion (see p. 207–8).

Alcohol

Alcohol is not a component of any natural plant food; however, in almost all cultures it is produced from whichever plants are grown in abundance and provides a considerable source of energy in the diet. In the United Kingdom the national average is 6 per cent of the total dietary energy. Alcohol is produced by yeast fermentation of carbohydrates from a wide variety of sources: vodka from potato; rum from sugar; palm wine from the sweet sap of palm trees; whisky from barley; and wine from grapes.

Vitamins

Vitamins are substances required in very small amounts, µg or mg per day, to maintain normal metabolism. The body is unable to synthesize them and so they must be obtained from food.

NAMES OF VITAMINS The names of the vitamins result from the history of their discovery. At the beginning of the twentieth century it was known that fats, carbohydrate, protein, and minerals were required in the diet but animals fed on these purified constituents did not grow unless milk was added to the diet. Studies showed that two factors were essential: factor A, fat soluble, in the cream; and factor B, water soluble, in the fat-free milk. Chemical tests identified factor B as an amine. From these 'vital amines' the term 'vitamine' was coined by Dr Casimir Funk in 1913, but the final 'e' was later dropped since not all were amines.

Vitamin B was later shown to be a mixture of substances with different functions and these were given numbers in the sequence with which they were discovered. B_1 was later identified as thiamin, B_2 as riboflavin. What would have been B_3 was used for pantothenic acid and sometimes quite wrongly for niacin. The latter was already known as a chemical 'nicotinic acid' before its physiological effect was investigated, and so the number was not used. There are other gaps in the sequence as compounds thought to be vitamins were given numbers but subsequently shown not to be vitamins or already described by other researchers under different names. B_6 describes not one but six vitamins which have similar biological action. Similarly, niacin is a generic term used for two compounds, nicotinic acid and nicotinamide, both of which have the same biological activity. The only other B vitamin for which the number is still used is vitamin B_{12}, identified as cobalamin.

Vitamins C, D, and E were named in the order of their discovery. The terms vitamin F, G, and H are not used as F turned out to be the essential fatty acids, G to be what was already known as vitamin B_2. H is called biotin. Vitamin K was named not by sequence but from the Danish 'koagulation' because of its function in blood, discovered by Henrik Dam in Denmark. Dietary deficiency of each vitamin causes specific signs and symptoms.

ESSENTIALITY By definition, vitamins are essential nutrients. However, two are essential only under specific conditions as they can be synthesized to a certain extent in the body. These are vitamin D and niacin. Vitamin D is formed in skin exposed to sunlight and is essential only if sunlight is inadequate. It is therefore considered by some to be a hormone rather than a vitamin. Niacin can be formed from the essential amino acid tryptophan and it is now thought that the amount formed is as important as the preformed niacin obtained from the diet.

Vitamins can be toxic in high doses, especially the fat-soluble vitamins that can accumulate in the liver and fatty tissues. However, when consumed in plant foods there is no risk of toxicity and the problem can only arise from the excessive consumption of vitamin supplements or from the consumption of liver from animals in which high levels of vitamin A have accumulated from their food.

FUNCTIONS The functions of all vitamins cannot be covered in this summary. Vitamins are therefore listed in Table 12 (p. 225) with the main effects of deficiency and an indication of the prevalence of clinical deficiencies. Some vitamins are widely spread in many foods, such as vitamin E and pantothenic acid, and so deficiencies are rare; others are concentrated in certain foods, such as vitamins A and C, and so deficiencies are more frequent. The vitamins of particular relevance to plant foods are discussed briefly in this section.

FAT-SOLUBLE VITAMINS Vitamins are classified into fat soluble (A, D, E, and K) and water soluble (B vitamins and C). Deficiencies of the fat-soluble vitamins can occur if the level of fat in the diet is so low that absorption is impaired. Conversely, toxic effects of high doses are likely to occur mainly with this group as they can be stored and accumulate in the body. Vitamin A covers two groups of compounds: retinol, or preformed vitamin A, which is found only in animal foods; and the carotenes, from which retinol can be formed in the body, which are found in green-, yellow-, and orange-coloured fruits and vegetables. Carotenes are converted to retinol with varying efficiency. β-carotene has the highest vitamin A activity, 6 mg is equivalent to 1 mg retinol. For others such as α-carotene and cryptoxanthine, 12 mg is equivalent to 1 mg retinol. Retinol is stored in the liver of animals and excess liver consumption can lead to toxicity. However, this risk does not exist with the precursor carotenes as their conversion to retinol is regulated.

Vitamin A deficiency is one of the most widespread micronutrient deficiencies worldwide, but only in less-developed countries. The well-known function of vitamin A is in the metabolism of the visual pigments of the eye. Deficiency of vitamin A leads first to night blindness and subsequently to complete blindness. Another extremely important function is in the control of cell differentiation so that deficiency leads to abnormalities of the skin, mucosa, and of growth, and resistance to respiratory and gastrointestinal disease.

Carotenes also have a separate function as antioxidants (see p. 207). They are therefore important in resistance to cell damage by the many substances that produce 'free radicals'. These are highly reactive molecules that result in the oxidative damage associated with ageing, heart disease, and cancer.

Vitamin D is obtained only from animal foods but its absorption is affected by the phytate content of plant foods and so deficiencies occur in communities where people, especially women and children, do not go out in the sun and the diet consists largely of unrefined cereals and contains few animal products. In the United Kingdom rickets was common at the beginning of the twentieth century in industrial cities such as Glasgow, but has been virtually eliminated except for a few cases in Asian communities where the requisite conditions exist.

Vitamin E (the name given to eight tocopherols) is a highly effective antioxidant (see p. 207) and has an important role in protecting polyunsaturated fatty acids and other components of cell membranes from oxidation by free radicals. It may have other functions such as an anti-inflammatory action and stimulation of the immune response. Vegetable oils are major sources of vitamin E in the diet. The quantities required to prevent overt deficiency with signs of damage to cell membranes, including muscle and nerve cells, are less than the intake needed to reduce the risk of cancers, coronary heart disease, and other degenerative diseases.

WATER-SOLUBLE VITAMINS On a worldwide basis, diseases due to deficiencies of B vitamins that were once common, such as pellagra and beriberi, have been almost eliminated through improved diets, better food processing, and fortification, although during the 1980s they have reappeared in the expanding population of refugees in the world due to the use of emergency food supplies unsuitable for long-term feeding.

Beriberi, a disease caused by thiamin deficiency, was epidemic in Asia during the nineteenth century, in populations consuming a monotonous diet consisting mainly of polished rice. Dr Casimir Funk, the originator of the term 'vitamine', isolated the antiberiberi factor from rice polishings in 1911 and the chemical structure was later established. The disease is characterized by extreme muscle weakness, multiple nerve damage, and in some cases oedema and shortness of breath. Epidemics of this scale no longer exist but clinical deficiency is still seen in Asia in poor individuals consuming mainly rice. It is also seen in chronic alcoholics and in patients with chronic disease, intestinal malabsorption, and anorexia. It can also occur when there is increased demand for thiamin, such as during refeeding of malnourished patients with excessive carbohydrate.

Pellagra, caused by niacin deficiency, is another disease related to particular types of food plants and was first recognized to occur in poor people on a corn-based diet. It also occurs with poor, sorghum-based diets. Corn (maize) contains little niacin which is in a form that is not readily available. The disease was prevalent in Europe and the southern United States after Spanish explorers introduced corn from Central America. It is classically characterized by the three Ds: dermatitis, diarrhoea, and dementia. The tongue is inflamed and a red rash occurs in areas of the skin exposed to the sun. In 1914 Goldberger showed in mental hospital patients in the United States that pellagra could be treated by substituting mixed grains for corn in the diet. He also showed that corn-based diets devoid of animal protein resulted in pellagra, leading to the discovery that the amino acid tryptophan can substitute for the vitamin niacin in the diet. Pellagra does not exist where maize is prepared in the traditional way by soaking in lime, which increases the bioavailability of niacin. Other cereals such as oatmeal, rice, and wheat provide moderate amounts of niacin. Pellagra still occurs in parts of India and Africa associated with the consumption of a poor, sorghum-based diet, and is occasionally found in malnourished alcoholics.

The B vitamins that remain problematic are vitamin B_2 in communities that do not consume rich sources such as milk, and folic acid and vitamin B_{12} which both result in megaloblastic anaemia. The anaemia is characterized by large, immature blood cells, in contrast to the more common iron-deficiency anaemia where the blood cells are smaller than normal. As vitamin B_{12} is present in animal but not plant foods, obtaining adequate amounts of this vitamin can be problematic amongst strict vegetarians, unless supplements are taken or fermented foods such as yeast products are incorporated into the diet, as microorganisms also form vitamin B_{12}. However, most vitamin B_{12} deficiency is not due to its lack in the diet but to a pathological condition which prevents its absorption in the intestine. The metabolism of folic acid requires vitamin B_{12} and so the two vitamins are closely linked. Vitamin B_{12} deficiency results in secondary folic acid deficiency. Supplementation with folic acid would treat the resultant megaloblastic anaemia but would mask the other effect of vitamin B_{12} deficiency, degeneration of the spinal cord. It is therefore important to eliminate vitamin B_{12} deficiency as a cause of megaloblastic anaemia before treating with folic acid.

Dietary deficiency of folic acid is not uncommon, especially in developing countries, despite its presence in most fresh foods, both animal and plant, as it is easily oxidized. Pregnancy increases the requirement and, if the mother's stores are low, neural tube defects, such as spina bifida, can occur in the infant. Such cases also occur in the United Kingdom and other developed countries.

Vitamin C deficiency was a common problem in the past at the end of winter after months without fresh fruit and vegetables. It also afflicted sailors, often fatally, on long sea voyages in previous centuries until the naval surgeon James Lind advocated the use of citrus fruit in his *Treatise of the Scurvy,* published in 1753. It took the admiralty 42 years to introduce a daily ration of lime juice, abolishing scurvy from the navy and giving British sailors the nickname 'Limey'. Modern transport, storage, and commerce have made fresh produce available the year round and so signs of deficiency are uncommon but do occur in groups such as the elderly or alcoholics, where intake is low.

Vitamin C is an antioxidant, as is β-carotene and some other food components, and it plays a role in the prevention of diseases such as coronary heart disease and certain cancers. Being a water-soluble vitamin it is excreted once the body tissues are saturated, at an intake of 70–100 mg/day, and therefore there is no good evidence for the very high doses claimed to be beneficial in the treatment or prevention of cancer, nor of the common cold. It is also important in aiding the absorption of iron from the diet.

Minerals

All nutritionally important minerals have to be provided in the diet and several are known to be essential for specific functions. The main characteristics of groups of minerals and specific features of minerals especially relevant to plant-based diets are discussed in this section.

Those required in larger amounts (g/day) are calcium, magnesium, and phosphorus, which form the structure of bones and teeth as well as having other metabolic functions, and sodium, chloride, and potassium, which are important in the maintenance of the normal composition of fluids outside and inside the cells of the body.

Others required in smaller amounts (μg or mg/day) are associated with specific enzymes: copper with energy metabolism enzymes; iron in haemoglobin, which carries oxygen in the blood cells, and in myoglobin in muscle; molybdenum with a small number of enzymes; selenium in antioxidant enzymes; zinc in over 100 varied enzymes and in cell receptors for some hormones and vitamins; chromium in glucose tolerance factor, possibly related to some forms of diabetes; iodine in thyroid hormones; magnesium with enzymes related to genetic functions; and manganese with several enzymes, including antioxidants. Others are known to be essential but their function is not yet known: silicon, vanadium, nickel, and tin. Cobalt forms part of the structure of vitamin B_{12}.

Some have beneficial effects but are not established as essential: fluoride, which alters the structure of bone and dental enamel, reducing susceptibility to dental decay; and lithium, which is not essential nor has beneficial effects in healthy people but is effective as a pharmacological agent in manic-depressive disease.

In general, deficient mineral intake occurs where there is a particular deficiency in the soil and the population consumes only locally grown products. In practice this is now rare, except for iodine deficiency, with extensive commerce and mixed diets. Many minerals are toxic in excess but this is unlikely from food, except where crops are grown on soil with very high selenium levels. Problems more commonly arise with inappropriate supplements or contamination of food or water. Fluorosis, causing bone and teeth defects, is common in areas where the fluoride content of the water supply is considerably greater than 1 p.p.m., the level considered optimal for health.

The main mineral deficiencies are caused by excessive losses and poor absorption. Iron, iodine, calcium, and zinc are the most problematic worldwide. Iron deficiency is mainly due to loss of blood at a rate greater than it can be replaced by absorption from the diet as a result of intestinal parasites, especially hookworm, and heavy menstrual losses. It is therefore particularly common in women. In foods, iron is in the form of haem in meat and inorganic iron salts in plants. Haem iron is better absorbed. Absorption of inorganic iron is facilitated by vitamin C and other antioxidants if consumed at the same time, as well as some proteins, but it is hindered by other food components including phytate, dietary fibre, calcium, and tea.

Iodine is required by the thyroid gland to form the hormone thyroxine which controls metabolism. Deficiency leads to goitre and, if severe, cretinism. It is particularly prevalent in isolated mountainous areas where the soil is leached of iodine and which are far from the sea so that fish and other iodine-rich marine products are not consumed. In many areas of Europe and other developed countries, such as Derbyshire in the United Kingdom, and Switzerland, iodine deficiency has been virtually eliminated by food trade and fortification of salt, but it remains an endemic problem in many parts of the world, including Europe. Some plants, such as brassicas (see pp. 162–9), contain goitrogens, substances that prevent the normal metabolism of thyroid hormones and are another cause of goitre.

Calcium requirements are especially high during rapid growth in infancy, adolescence, pregnancy, and lactation. In many countries the main source of calcium is milk products, but in others milk is not consumed after childhood and so calcium intakes are lower, the main sources being cereals and sometimes soft bones as in fish. The absorption of calcium from the intestine is closely regulated and depends on vitamin D but also on the availability of dietary calcium. Some food substances form complexes or insoluble calcium salts that cannot be absorbed. The most important is phytic acid found in cereal bran and some nuts and pulses. When bread is leavened the enzyme phytase in the yeast partially breaks down the phytate, improving the availability of calcium and other minerals. However, populations that consume unleavened wholegrain bread, such as chapatis and tortillas, absorb a lower amount of calcium and other minerals.

Rickets and osteoporosis are two health problems related to calcium metabolism. Calcium deficiency reduces the growth rate of children but calcium deficiency alone does not seem to result in rickets, a condition in which bones fail to calcify properly and become distorted. However, it may contribute to rickets when vitamin D status from food and sunlight is low. Nor does calcium deficiency seem to be the cause of osteoporosis, the progressive loss of bone density with age leading to brittle bones. The main cause is inactivity and reduced levels of sex hormones after middle age. Bone reaches its maximum density (peak bone density) at about the age of 30 and a high density can protect from the effects of osteoporosis in later life. Therefore adequate calcium and vitamin D up to this age is important but high intakes of calcium have no beneficial effect once the peak bone density has been achieved.

Zinc deficiency, resulting in impaired growth and sexual development, has been identified in population groups in the Middle East who consume diets low in rich sources of zinc, such as animal products, and high in phytates, such as unleavened bread. It may exist in other communities where the staple food is chapatis and tortillas and may partially account for the reduced stature of such populations. Zinc is also implicated in immune function and therefore resistance to disease.

Sodium, chloride, and potassium are important for the balance of extracellular and intracellular fluids, and the ratio of sodium to potassium intakes is related to blood pressure. Excessive intakes of sodium are associated with hypertension. Sodium and chloride are not naturally found in high concentrations in foods but are added to many during processing and as table salt. Potassium is particularly abundant in vegetables, potatoes, fruit (especially bananas), and juices.

Other biologically active substances

Apart from nutrients and dietary fibre, food contains substances that are biologically active and can be beneficial in terms of health or subjective effects. Some of these substances are difficult to classify as they may be beneficial in the quantities found in some plants but could also be considered to be toxicants in other circumstances (see p. 207). These include: large families of polyphenols, which share a similar chemical structure, are broadly referred to as tannins, include the flavonoids, and are generally highly coloured and astringent; alkaloids, of which the stimulant caffeine is one example; plant steroids which are hormone-like substances including the phytoestrogens (see below); and many other types of compounds. Little scientific attention has been paid to their nutritional functions in the past but recently there has been extensive research into their protective effects against some common degenerative diseases. Some of this research explains the use of particular foods and herbs to promote health in parts of the world such as China, where, traditionally, less distinction has been made between food and drugs. Plant substances that protect against cancer, cardiovascular, or other diseases are loosely called 'phytochemicals'.

Some of the evidence for the beneficial effects of different plant foods comes from epidemiological studies in which the prevalence of certain diseases in various population groups is correlated with the level of consumption of specific foods. These can indicate a relationship but cannot prove that the food causes the effect on health. Further studies in animals and people are required to show the physiological mechanisms that could explain the relationship with the food or with its specific phytochemical components. The number of these chemicals is vast and their classification by chemical structure complicated. Only some examples of substances with beneficial effects are given in this section.

ALLIINS (see p. 176) Garlic and allied species contain a family of thioallyl compounds, alliins, that may protect against cancer and cardiovascular disease. Raw garlic contains 1–2 per cent of these sulphur compounds which can lower blood levels of cholesterol and triglycerides, and reduce blood clotting in animals and possibly in humans. They have also been shown to suppress cancer initiation and development in animal studies, consistent with epidemiological evidence that high garlic consumption is associated with low cancer incidence.

PHYTOESTROGENS Phytoestrogens are steroid substances derived from plants that have been reported to reduce the levels of the biologically active free oestrogens (female hormones) in the body and so lower the risk of oestrogen-responsive cancers, such as certain types of breast cancer. Most of the research has involved breast cancer but these phytochemicals also affect other related cancers, such as ovarian cancer, as well as the metabolism of male hormones and prostate cancer. One type, the lignans, are diphenolic compounds (mainly enterolactone and enterodiol) derived from the bacterial digestion of polyphenols. Many oilseeds and grains (see p. 28), such as flaxseed, soya bean, rapeseed, and wheat are rich sources of lignans or their precursors. The incidence of breast cancer is low in countries that have high intakes of such foods, although lignans are not the only factor as these foods also contain many other phytochemicals, such as flavonoids and other phenolic compounds, which also have anticancer properties. Another class of phytoestrogens are the coumestans, found in clover and therefore relevant to the nutrition of animals but not of humans, for whom the isoflavonoids are the other main class. Daidzein and genistein are the major isoflavonoids found in many plant foods but soya bean is a particularly rich source.

The phytoestrogens stimulate the production, in the liver, of sex hormone binding globulin (SHBG), on which a large proportion of the hormones are carried in the blood. The higher the level of SHBG the more the hormones are bound, resulting in less of the biologically active free hormone. They also have other effects that are cancer protective. Some reduce the proliferation of cells that respond to oestrogens, such as breast and uterus cells, as they compete with oestrogens for the binding sites but have only a weak action; others inhibit enzymes associated with cell proliferation; and others are antioxidants which have anticancer effects.

ANTIOXIDANTS Antioxidants are substances that reduce the rate of oxidation of substances that are susceptible, such as polyunsaturated fatty acids (see p. 202), by neutralizing free radicals. These are highly reactive molecules produced in the body in various chemical reactions in normal metabolic processes but also from external sources including infections, smoking, exposure to sunlight, and other onslaughts that cause tissue damage. They increase the rate of ageing, and of degenerative diseases, including heart disease (see p. 202), cataract, and cancer. Antioxidants include: some vitamins (vitamin E, vitamin C, and β-carotene); some trace elements that are components of antioxidant enzymes (including selenium, copper, zinc, and manganese); some non-nutrients, such as ubiquinone (coenzyme Q) and phenolic compounds (for example phytoestrogens, flavonoids, phenolic acids, and butylated hydroxytoluene (BHT), used as a food preservative); and some antinutrients (such as phytic acid) (see below). Phenolic acids such as chlorogenic, caffeic, gentisic, ferulic, and vanillic acids are abundant in whole grain, especially in the bran layer.

FLAVONOIDS Flavonoids are naturally occurring, phenolic, water-soluble, antioxidant substances widely distributed in vegetables, fruits, and plant-based beverages such as tea and wine. They have an important effect on colour and flavour. They include hundreds of different substances but all have a common chemical structure, based on the phenolic compounds, flavans (with two six-carbon rings joined by a three-carbon chain) combined with a sugar (glycoside). Flavonoids have antioxidant properties and therefore potentially important effects on health by preventing the oxidation of several cellular components (including DNA and low-density lipoprotein (LDL)) and reducing blood clotting, both implicated in coronary heart disease. The most widely distributed is quercitin. Others include kaempferol, myricetin, catechin, apigenin, and luteolin. In a Dutch study it was found that 95 per cent of total dietary flavonoids were from two compounds: 63 per cent from quercitin and 32 per cent from kaempferol. The major dietary contributor of flavonoids intake was black tea (61 per cent), onions (13 per cent), and apples (10 per cent). The risk of mortality from heart disease was substantially reduced in people with the highest flavonoid intakes.

TANNINS 'Tannins' is the term used to describe collectively the polyphenols responsible in part for the astringent flavour of wine and tea. Thirty per cent of the dry weight of the leaves of the tea plant consists of polyphenolic substances, which are oxidized during fermentation. Almost all are flavonoids, mainly catechin or its derivatives, which interact with the bitterness of caffeine to provide the astringency. Red wines have up to 800 mg/litre of catechins, compared to 50 mg/litre in white wine. These antioxidants are postulated to explain the 'French paradox' that death rates from coronary heart disease are less than a quarter of the rates in England and Wales, despite a diet high in fat, and prevalent smoking.

ALKALOIDS Alkaloids are chemically compounds with nitrogen in a complex molecular ring structure. They have an extremely bitter taste and many have highly undesirable effects on animals and man, and would therefore be classified under toxins (see below). Others have pharmacological effects that are useful or desirable to man. Nicotine, atropine, emetine, and quinine are all alkaloids, as is the methylxanthine caffeine, and related compounds such as theophylline and theobromine. These stimulants are found in coffee, tea, and cocoa, and increase alertness, heart rate, and urine production.

GLUCOSINOLATES Glucosinolates are a group of pungent compounds contained in cruciferous vegetables including horseradish, radish, and mustards, and to a lesser degree in the brassicas, such as cabbage, Brussels sprouts, and kale. When cut, grated, or chewed they are broken down by enzymes in the food and hydrolysed to give the pungent and volatile isothiocyanates and other products, unless cooked before these reactions can occur. Glucosinolates are thought to be protective against cancer and therefore beneficial. However, glucosinolates can cause goitre and from this aspect should be classed as a toxin (see p. 208).

Natural toxicants and antinutrients

The public is often concerned about the hazards of chemical additives and contaminants and assumes that natural products are safe. However, plants contain natural toxins. We need to distinguish between 'toxin' and 'hazard'. Almost any substance, including water and essential nutrients, are toxic in high doses but few represent a hazard in the quantities normally consumed. Some examples of natural toxicants that can be a hazard under certain conditions are described in this section. These toxins exist in plants partly as defence mechanisms against insects.

METAL BINDERS Phytate is a common constituent of plants, especially wholegrain cereals, that is of great importance in nutrition. Chemically it is a cyclic compound (inositol, a dietary essential, a vitamin for microorganisms and some animals but not human beings) containing six phosphate molecules. It has the ability to bind minerals such as calcium, magnesium, zinc, copper, and iron to form an insoluble complex that is not readily absorbed from the intestine. It is therefore implicated in mineral deficiencies in areas that rely on wholegrain cereals as the main component of the diet, especially if unleavened products are used. The fermentation process in leavening bread helps to decompose the phytate and render the minerals more absorbable.

Oxalic acid is another calcium-binding substance that is present in certain plants, such as rhubarb and spinach, although only very high levels of consumption (as in rhubarb leaves) are likely to cause any problems of calcium absorption.

ENZYME INHIBITORS Solanine is an alkaloid (see above), present mainly in the sprouts and skin of green potatoes, which inhibits the enzyme cholinesterase, important in the metabolism of the neurotransmitter acetylcholine. The clinical symptoms of poisoning include gastrointestinal disturbances and neurological disorders. The content in potatoes depends on the cultivar, and so all existing and new cultivars are monitored for alkaloid content. The upper acceptable limit for safety is generally 20 mg/100 g fresh weight, and a dose of 3 mg/kg body weight is considered the toxic level for man. The average person in the United States consumes about 10,000 mg/year, which would be lethal if taken in one dose. Solanine is not destroyed by cooking.

Protease inhibitors exist in many plants, including cereals and potatoes, but are particularly common in legumes. They inhibit the action of enzymes important in the digestion of protein, such as trypsin and chymotrypsin (see p. 203). However, most are destroyed by heat during cooking.

NEUROTOXIC AMINO ACIDS Lathyrism is a disease caused by the consumption of large quantities of the seeds of certain *Lathyrus* species (leguminous drought-resistant plants, including *Lathyrus sativus* (grass-pea or chickling vetch) see p. 50) which are widely grown in Asia and North Africa). *Lathyrus* species have many unusual amino acids. Lathyrism is a public health problem in India, especially in poor areas of the country when food is scarce and *L. sativus* forms a large part of the diet for a period of several months. It also affects horses and cattle. In man it is manifested by the sudden onset of spastic paralysis of the legs as a result of lesions of the spinal cord caused by neurotoxic amino acids in the seeds. In extreme cases it can lead to death. The sale of *L. sativus* has been banned in many states in India but it is still grown where suitable alternative arid crops are not available. There is no cure for the disease. The toxin can be largely removed by soaking or cooking in an excess of water, the main disadvantage being the loss of B vitamins. However, when used in the normal method of preparing chapatis there is no destruction of the toxin.

FAVISM Favism is a haemolytic disease caused by an inborn error of metabolism which renders the subjects susceptible to substances in the faba (fava) or common broad bean *Vicia faba* (see p. 44). Consumption of beans or inhalation of pollen of the plants results in symptoms associated with haemolysis (breakdown of the red blood cells), including pallor, fatigue, breathlessness, nausea, pain in the abdomen or back, fever, and chills. In more severe cases jaundice and dark urine from the destroyed blood cells can occur, with subsequent anaemia. Symptoms can start within minutes of exposure, especially to pollen, but in most cases they occur several hours or even days after consumption of the beans. The acute phase lasts for 1–2 days, with spontaneous recovery. However, it can be fatal in young children. The chemical identity of the toxic principle in fava beans has not been definitively established, but possible candidates are (the non-sugar) part of the glycosides vicine and convicine. These are hardly true toxins as they affect only a minority of people.

Favism occurs in populations that are genetically susceptible because they are deficient in a particular enzyme which catalyses a special biochemical pathway essential in blood cells for the regeneration of a blood antioxidant needed to deal with the challenge of fava and also certain drugs. Favism is very prevalent around the Mediterranean, including North Africa, and in the Middle East. It is also prevalent in China, and affects some 100 million people worldwide.

CYANOGENS Trace amounts of cyanide are widespread in plants, mainly as cyanogenetic glycosides. Relatively high concentrations occur in certain grasses, pulses, root crops, and fruit kernels. Four of these are of practical importance in human consumption: amygdalin, dhurrin, linamarin, and lotaustralin. Amygdalin is present in bitter almonds and other seed kernels from the fruits of *Citrus*, *Malus*, and *Prunus* species. Dhurrin, which is closely related, occurs in sorghum and other grasses. Linamarin (phaseolunatin) and lotaustralin (methyllinamarin) are the glycosides of pulses, especially lima or butter-bean, and of the root crops cassava, sweet potato, and yam. Cassava (see p. 190) is an important staple food in many parts of Africa and Asia. The glycoside releases cyanide when acted on by enzymes released from cells in the cassava. Fatal cyanide formation can occur if the cassava is not adequately processed, especially the bitter cassava which has a higher cyanide content. The toxicity of cassava has been recognized for hundreds of years and traditional processing methods have been developed by peeling, grating, and washing in running water or soaking for several days, by which time most of the cyanide is lost in fermentation. Any remaining free hydrogen cyanide is readily volatilized on boiling.

Human, and more frequently animal, poisoning has been recorded in relation to lima beans, cassava, millet, and fruit-stone kernels. Chronic poisoning can also occur and is linked to chronic neurological conditions, including loss of vision and blindness (due to damage to the optic nerve) and ataxia (loss of locomotor control), where nutritional status is also poor especially in relation to cystine and the B vitamins cobalamin, niacin, and riboflavin, which are involved in cyanide detoxification.

HAEMAGGLUTININS (LECTINS) Lectins are proteins found in many plant groups, including fungi, potatoes, and legumes. They are found mainly in seeds but also in tubers and plant saps. The first discovered, at the end of the nineteenth century, was the protein, ricin, in castor beans, found capable of causing the red blood cells to stick together (agglutinating). Many edible pulses contain lectins. The only common characteristic is that they are all proteins, and have a special affinity for certain sugar molecules to which they attach in cell membranes. Some lectins, such as those in the garden pea, act on all types of blood groups, others, such as the lima bean lectins, act on specific blood groups. Most research has been conducted on red kidney beans (see p. 40), which have a particularly high content. If eaten raw or undercooked, nausea, vomiting, and diarrhoea can occur. The toxin is destroyed by vigorous boiling for 10 minutes but may persist if cooked at low temperatures in a slow cooker.

GLUCOSINOLATES (THIOGLYCOSIDES, GOITROGENS) Glucosinolates (also called thioglycosides) are responsible for the pungent flavours of horse-radish and mustard and contribute to the flavours of turnip, cabbage, and related vegetables. The glucosinolates in these cruciferous vegetables have protective qualities (see p. 207) but some can also produce goitre when consumed in large quantities, and from this aspect they should be classed as toxins. The glucosinolates are broken down by enzymes in the food when it is cut or grated, and hydrolysed to give volatile isothiocyanates. These can spontaneously form a ring-shaped molecule (goitrin) that interferes with the uptake of iodine in the thyroid gland and is therefore goitrogenic. Goitre is the enlargement of the thyroid gland in an attempt to capture iodine which is required in the production of the hormone thyroxine (see p. 206). Most goitres are caused by a lack of iodine in the diet but a small percentage may result from the inhibition by goitrin of iodine uptake into the thyroid gland.

SAPONINS Saponins are a group of glycosides (substances that have a chemical linking sugar and other components, often pigments) that occur in a wide variety of plants, especially legumes. Foods particularly rich in saponins are soya beans, chickpeas, haricot beans, and kidney beans. They have a bitter taste and detergent properties so that they foam in water, causing bloat in animals. They can haemolize red blood cells but taken orally have low toxicity. They also bind cholesterol, reducing plasma cholesterol levels in experimental primates, indicating the potential to reduce the risk of coronary heart disease in humans.

GOSSYPOL Gossypol pigments are polyphenolic substances in cottonseed, the oil of which is used in edible products such as margarine and salad oil. The toxic pigment is therefore detrimental in this use and also in the use of the meal as a protein supplement. Symptoms of gossypol toxicity in humans and animals are cardiac irregularities, depressed appetite, and weight loss. It reduces the oxygen-carrying capacity of the blood and causes haemolysis of red blood cells. The mechanisms of tissue damage are not well known but may involve changes in cell permeability as fluid accumulates in the body cavities. However, there are few reports of gossypol toxicity in human beings. In China gossypol has been used as an antifertility pill for men as it blocks sperm formation. It was evaluated to be almost 100 per cent efficient and side-effects were reported to be mild and infrequent.

ALLERGENS Allergens are not strictly toxins. True toxins are constituents that have an undesirable effect on anyone who consumes them, the effects being proportional to the quantity eaten. Allergens are usually normal food constituents that are generally innocuous except to individuals who have an abnormal reaction, an allergy. The intensity of the reaction depends not on the quantity consumed but the degree of sensitivity of the person. Symptoms are variable and can affect any tissue of the body, but the most commonly affected are the skin and the respiratory tract, with eczema, itchy skin, runny nose, and asthma. Almost any food can produce an allergic reaction but some are common, others rare. In plant foods common allergens are found in cereals, legumes, and nuts. Apart from milk products and egg, wheat is the most common allergen, the causative fraction probably being the protein albumin. Maize and barley, including malt in flavourings and beverages, are also common allergens. These are frequently replaced in the diet of allergic subjects by polished rice or oats as they are less frequently allergenic, in the case of rice probably because the causative fraction is removed in the polishing. Amongst the legumes the peanut is the most allergenic. Strawberries commonly produce a skin rash, but washing the fruit with scalding water and chilling can remove the factor, either by heat destruction or by washing off a surface contaminant. Nuts and seeds are the most highly allergenic food group, occasionally causing life-threatening reactions, including coconut, cashew nuts, and the seeds of cotton, mustard, sesame, poppy, etc.

CARCINOGENS Many substances in food are potentially carcinogenic. These include contaminant mycotoxins such as aflatoxin (see p. 210), cycasin, *Senecio* alkaloids, bracken fern, and safrole. Cycasin is a glycoside that occurs in cycad plants in the tropics and subtropics. These are resistant to drought and hurricanes, and the high levels of starch in the roots, seeds, and stems provide a source of food in an emergency. Populations using cycads for food have long been aware of its toxicity and have developed traditional methods of preparation, including fermentation, heating, water extraction, and sun drying that detoxify the final product.

Senecio alkaloids are found in the genus *Senecio*. Livestock grazed on pastures in which the plants were growing have been noted to die from liver and lung lesions. The alkaloids may also be involved in human liver diseases, occlusion of the veins, and kwashiorkor. Humans can also be affected indirectly if residues remain in the milk or meat of animals consumed. *Senecio* alkaloids are also used in folk medicine and their use has been associated in the past with veno-occlusive disease in children in Jamaica.

Bracken fern (*Pteridium aquilinum*) is widely distributed and has long been recognized to cause the poisoning of livestock. There are two types of symptoms: one of thiamin deficiency caused by the presence of the enzyme thiaminase that destroys the vitamin, the other includes the production of benign and malignant tumours in the intestine and bladder. The fern is used for human consumption in several parts of the world, including Japan, the north-eastern United States, and Canada. Indirect exposure to the toxin could also come from the milk of cows fed the fern. However, epidemiological evidence relating consumption to human cancer incidence is not available.

Safrole (with a benzene ring structure) is a component of many essential oils, such as the oils of star anise, camphor, sassafras, mace, ginger, bay leaf, and cinnamon leaf. It was used as a flavour in soft drinks until it was discovered that it caused liver cancer in animals.

OESTROGENS Many plants, including cereals, legumes (in particular soya beans), and oilseeds, contain isoflavones that have weak oestrogenic activity, which has certain beneficial effects (see p. 206). However, toxicity can occur in unusual circumstances, such as in The Netherlands during the food shortages of the Second World War when tulip bulbs were eaten as a famine food. Their high concentration of oestrogens resulted in abnormalities of the menstrual cycle in women.

STIMULANTS, DEPRESSANTS, AND HALLUCINOGENS The alkaloids in certain plants can be useful in medicine and for other social purposes (see p. 207) but can also be the cause of poisoning in humans and livestock. For example seeds from the genus *Datura* have been used as a medicinal and hallucinogenic brew in several developing countries and have been the cause of many cases of poisoning. These have occurred particularly in children and livestock, sometimes through the contamination of bread and forage with the seed, which contains the alkaloids atropine, hyoscyamine, and scopolamine.

Nutmeg also contains a toxic substance, myristicin, which constitutes 4 per cent of the volatile oils, and has been used as a remedy for a variety of ailments, including toothache, dysentery, rheumatism, and to induce abortions. It is also hallucinogenic and can have other toxic effects such as nausea, constipation, tachycardia, and stupor.

The pressor amines (serotonin, noradrenaline, tyramine, tryptamine, and dopamine), which constrict the blood vessels and cause increased blood pressure, occur in fairly high levels in some plant foods, such as pineapple, banana, plantain, and avocado. They are normally detoxified in the liver by specific enzymes. However, some antidepressant drugs inhibit the enzyme (monoamine oxidase) and lead to heightened sensitivity to such foods, resulting in increased blood pressure and severe headache. Some fungi contain mycotoxins, including those that are used to produce hallucinations, such as *Amanita muscaria*. The toxicity and inedibility of many fungi is well known; however, some species considered edible have recently been associated with testicular damage in man, and sensitivity to alcohol (*Coprinus*) as well as cancer of the lung, liver, and intestine (*Gyromitra*).

FLATUS-PRODUCING SUBSTANCES One of the main factors limiting the human consumption of legumes is their ability to produce gas in the intestinal tract, causing physical and social discomfort associated with the flatulence, nausea, cramps, and diarrhoea. The oligosaccharides raffinose and stachyose are the main causative factors. The human intestine does not have the enzymes that can break the chemical linkages between the component sugars and therefore cannot digest them. The intact oligosaccharides therefore pass into the lower intestine, where they are fermented by bacteria with the production of the gases carbon dioxide, hydrogen, and methane. Traditional bean foods such as soya-bean curd (tofu) and fermented soya bean (tempeh) have little flatus activity as the oligosaccharides are probably removed during preparation. Cooking in water causes some reduction in the flatus activity of other beans, and cultivars are developed by selective breeding that have low contents of raffinose and stachyose. Legumes are not the only foods that contain flatus-producing factors. Wheat and certain fruits and juices, especially raisins, bananas, apple juice, and grape juice, also tend to increase human intestinal gas production.

Pollutants/contaminants

The moulds that grow on dry foods such as nuts and bread produce several toxins that are potent carcinogens. One example is aflatoxin produced by *Aspergillus flavus*. The main source in the United Kingdom, the United States of America, and West Africa is mouldy nuts (see p. 26). Aflatoxin may cause liver damage and may lead to liver cancer. Ergot is a toxin produced on mouldy grain and causes neurological effects described as St. Vitus's dance. Serious outbreaks occurred in the past in the United Kingdom and still occur in other countries (see p. 4). Patulin is a mycotoxin associated with mouldy apples.

The residues of agricultural chemicals in foods are considered to represent no significant hazard to consumers in the United Kingdom when current regulations are adhered to. However, the existence and enforcement of such regulations are not universal in all countries. Many consumers prefer organic produce on which no modern agricultural chemicals have been used.

Effects of processing and storage

Changes occur in food nutrient content or the availability of nutrients in foods during storage, in cooking, and in other kinds of processing (see Table 13, p. 225).

Cooking

Some plant foods can be eaten raw but many need processing, such as soaking to remove toxic substances, and cooking to break down cell walls. This improves the digestibility and the availability of nutrients and may inactivate toxic substance and enzymes that can destroy some vitamins. Disintegration of the plant cell wall allows the enzyme phytase to hydrolyse phytates, thereby making minerals more available. For example phytase in yeast reduces the phytic acid in bread dough. However, excessive heat will destroy the phytase. Moist heat also causes starch to swell and form a gel which is more digestible. For example, potato starch is not digestible unless the cell walls are broken down by cooking.

There are also losses during cooking depending on the cooking method. Cooking in water causes soluble vitamins to be leached into the water and therefore lost if the water is discarded. Some can also be partially destroyed by heat and oxidation. The loss of vitamins in cooking and heat processing depends on the acidity or alkalinity of the medium, exposure to air and to light. The most sensitive vitamins are vitamin C, folates, and thiamin, but under specific conditions losses of other vitamins also occur.

Vitamin C losses in processing and cooking can be very large as the vitamin is easily oxidized in alkaline conditions and if exposed to the air, especially in the presence of copper. Adding bicarbonate to keep the colour of vegetables during cooking is therefore detrimental to vitamin C content. When plant tissues are cut or mashed an enzyme is released which oxidizes vitamin C but it can be inactivated at temperatures of 70 °C therefore 'blanching' fruits and vegetables in boiling water for a short time minimizes the enzyme activity, although there is some extraction of the vitamin into the blanching water. Vitamin C is also slowly destroyed during frying. Thiamin is also water soluble and sensitive to alkali, and significant losses occur in cooking water, especially if the water is slightly alkaline. Folates are very susceptible to oxidation in neutral or alkaline conditions.

Roasting, baking

The dry heat involved in roasting and baking damages vitamins unstable to heat, especially thiamin, on the surface of the product, and produces a brown crust caused by the so-called Maillard reaction between protein and carbohydrate, making the lysine unavailable.

Blanching

Blanching is the process of heating foods at a high temperature for a short time to inactivate enzymes that cause deterioration. It is an essential step before freezing, drying, or canning vegetables. It also reduces bulk, expels gases, and helps to maintain the colour. Blanching usually involves immersion in boiling water for one to several minutes, or in hot air, steam, or irradiation. There is some destruction of nutrients due to water extraction and oxidation, especially of vitamin C, which is reduced by one-third to a half, but the extent is very variable depending on the surface area and other factory conditions, including temperature, time, equipment, and the degree of maturity of the food. There may be considerable loss of folic acid and other water-soluble vitamins such as niacin, riboflavin, and thiamin. Carotenes are more stable. Losses are greater in leaf vegetables, such as spinach, than those with less surface area, such as asparagus. In general, losses are less with steam blanching. There is also loss of other water-soluble nutrients such as mineral salts, and, to a small extent, proteins and carbohydrates.

Drying and canning

The losses in both methods are mainly in blanching and so there is little difference between them. In drying, vitamin C is the most unstable. In canning, apart from the effects of leaching if the liquor is discarded, there is also some chemical destruction depending on the temperature, residual oxygen in the can, and a metallic surface. There is greater loss of vitamin C in lacquered cans than plain as in the latter corrosion uses up residual oxygen. Vitamin E and most of the B vitamins are stable in the canning process, except thiamin which is reduced by about half, others to a lesser extent.

Freezing

The principal losses in freezing are in the preliminary blanching. At the freezing stage there are no losses. Some losses can occur on thawing. The vitamin content of frozen foods on the plate can be higher than that of fresh foods as fruits and vegetables are frozen immediately after harvesting, whereas 'market fresh' produce may be stored at ambient temperatures for some days with some loss of nutrients, mainly vitamin C.

Other processes

These include radiation, which is used to destroy microorganisms, to inhibit the sprouting of potatoes, and to disinfest wheat from insect pests. Losses are similar to other methods of processing. Fermentation is used in several cultures, such as in the Far East where fermentation of cooked soya beans with moulds and bacterial cultures produces such products as miso and tempeh. The fermentation process increases the content of B vitamins.

Harvesting and storage

Mechanical harvesting of fruits and vegetables such as potatoes, carrots, peas, and tomatoes can cause bruising which results in some destruction of vitamin C. If harvesting of fruits such as tomatoes is done while they are unripe, to be matured off the vine, there is a reduction in the vitamin C content of approximately one-third compared with the fruit harvested ripe.

Losses in storage depend on storage conditions. When leafy vegetables wilt, the enzyme ascorbic acid oxidase comes in contact with vitamin C and there is a rapid destruction of the vitamin. Under tropical conditions 90 per cent of vitamin C can be destroyed in 24 hours. Storage of dried foods results in losses of vitamin C dependent on the moisture content and the temperature: in dried tomatoes with 1 per cent moisture there is no loss at 4 °C but losses increase considerably at higher moisture contents and temperatures. After canning, further vitamin losses of thiamin, pantothenate and niacin occur in storage. Other B vitamins are stable in storage. Vitamin E is stable in processing but is reduced by long storage. Freezing is the best method of preservation, particularly at temperatures of about -20 °C and below, but cannot be applied to fruits and vegetables. Uncooked produce can be stored under gas: carbon monoxide to destroy enzymes, then ethylene dioxide to destroy microorganisms. This leads to a loss of vitamin C. Storage of cereals for long periods results in only very small losses of vitamins when the moisture content is less than 10 per cent.

Milling

As the highest concentration of vitamins and minerals in grain are in the germ, the aleurone layer, and the endosperm, a proportion is removed during milling of wheat and husking of rice, depending on the extraction rates. Cereals are important sources of thiamin and niacin in most diets and so milling losses are serious; in many countries legislation requires nutrients to be restored to white flour (see below). Similarly with rice, machine milling leads to refined rice with high losses of B vitamins which can lead to the thiamin deficiency disease beriberi if the diet is heavily dependent on rice. After milling, the lipids in rice are more susceptible to oxidation and rancidity as the milling process destroys the natural antioxidants and exposes the lipids to air.

Parboiling

Parboiling is a process used to improve the keeping qualities of rice. It also limits the losses of vitamins in the milling of rice. The unhusked rice is soaked in water and steamed or boiled and then dried before husking. Part of the B vitamins migrate to the inner part of the grain, the husk is removed more easily with less loss of germ, and the grain is partially gelatinized and so less easily damaged in milling. The main disadvantage of this process is that it changes the colour and flavour of the rice compared to the polished white product. The commercial equivalent of this domestic process is 'converted rice'.

Enrichment

The terms enrichment, fortification, supplementation, nutrification, and restoration refer to the addition of vitamins and other nutrients to foods. They have specific meanings but are often used interchangeably. Enrichment refers to nutrients that are added to legally defined standards. In the United Kingdom enrichment of cereal flours was introduced after the Second World War and the regulations remain in force to restore the levels of iron, thiamin, and niacin in white flour to the level found in 85 per cent extraction flour, and to add calcium. In the United States riboflavin is also added to white flour. Fortification refers to the addition of nutrients either not present before processing or present in only very small amounts. Supplementation or nutrification refers to the addition to snack foods to provide a special source of nutrients in the diet. Restoration restores nutrients lost in processing to the previous level.

International dietary patterns

In poor countries carbohydrate, mainly starch, provides up to 80 per cent of the energy intake, whereas in the richer countries of North America, Europe, and Australia carbohydrates account for less than 50 per cent of the calories, with about half coming from simple sugars.

Food consumption and preferences are moulded by many geographic, economic, and social factors, but certain trends can be discerned internationally. As people become more affluent they change their consumption away from coarse grain cereals such as millet and sorghum to wheat and rice, they then increasingly replace cereal and root crop staples with animal products, especially meat. Other additions to the diet are refined sugar and more expensive fruits and processed foods (see Figure, p. 226).

These changes in types of food consumed result in progressive changes in the proportion of dietary calories coming from each major constituent. The percentage of calories from protein in all national diets is fairly constant, between 10 and 15 per cent, although with increasing affluence animal sources increasingly replace vegetable sources. Differences in the percentage of calories from carbohydrate are therefore reflected by reciprocal differences in fat calories. Populations in poor countries may receive on average only 10 per cent of their calories from fat, while those in the rich receive approximately 40 per cent from this source.

This level of fat is considered too high for health, being implicated in heart disease, obesity, and some cancers. Dietary guidelines in most of these affluent countries therefore recommend reducing fat intake to approximately 30 per cent of the calories and increasing carbohydrates to approximately 60 per cent, while limiting extracted sugar to no more than 10 per cent. In contrast, the very low levels of fat in the diets of some poor countries can be problematic for the consumption of adequate energy, particularly in children.

These general patterns describe food consumption under normal circumstances. In times of food shortage plants that are not normally desired, or even toxic plants, may be consumed. These famine foods can include coarse grains, root crops resistant to drought and other environmental disasters, various wild plants, tulip bulbs, and even ground date stones. In some cases such foods are nutritionally sound but carry a social stigma as they are not preferred by taste or custom and are associated with poverty and distress.

Sources of information about national diets

Information about the food and nutrients available and consumed in a country come from a variety of sources.

Food Balance Sheets

Each country has data from Food Balance Sheets which are prepared from estimates of the population and of all the food available for human consumption from data on crop and animal production, imports, exports, and losses in harvesting, storage, to animal feed, and to seed. This is calculated as the food available/person/day in terms of quantity and calorie content. The data presented in Table 1 p. 214 are derived from Food Balance Sheets prepared by FAO, the United Nations Food and Agricultural Organization. Such data are useful for international comparisons but indicate the food available for consumption, not necessarily that consumed. They tend to overestimate national consumption and cannot provide information about variations within each country.

Household surveys

These surveys of nationally representative samples are conducted at regular intervals by many governments and collect data for food purchased for home use over a specified period of time. Some also include an inventory of larder stores at the beginning and end of the survey period. These surveys can be used to compare consumption patterns of households in various groups such as by region, economic status, or ethnic origin, but not of individuals. They sometimes also include foods purchased outside the home.

For example, in the United Kingdom household surveys have been conducted annually since the 1950s. The Ministry of Agriculture, Fisheries and Food conducts the National Food Survey, of domestic food consumption and expenditure over a reporting period of 1 week in which the quantities of, and expenditure on, individual types of foods are recorded as purchases. Reporting weeks are spread over the year to account for seasonal variation and the survey results are published annually, with food consumption expressed as expenditure, quantities consumed, and nutritive value. Since 1992 foods consumed outside the home have been included to enable evaluation of total consumption and expenditure on food.

In addition, the Department of Employment conducts annual Family Expenditure Surveys in which household recording lasts 14 consecutive days. These surveys serve several purposes but the main use is to monitor trends in household food consumption.

Other countries have similar data collection, such as the United States' Nationwide Food Consumption Survey (NFCS), although usually at less frequent intervals.

Nutrition surveys

These obtain information about the dietary intake, and often measure the nutritional status, of individuals, and so can be used to monitor potential problems by age and sex group as well as by region, socio-economic, and ethnic group. Intakes may be estimated prospectively by weighing all food consumed or by measuring with household measures, or else retrospectively by questionnaires of diet history over a period of time or of recall of the types and amounts of food and drink consumed over the previous 24 hours. Such surveys, especially using weighing methods, are obviously time consuming and costly and therefore cannot be carried out on a complete national sample at such regular intervals. In many countries they may be conducted each decade or less frequently.

In the United Kingdom a national survey of weighed food intake over several days of successive sections of the population, such as infants, schoolchildren, and the elderly, are conducted on a more or less continuous basis depending on the perceived vulnerability of the particular group and the need for information at a particular time. In the United States the National Health and Nutrition Examination Survey (NHANES) is conducted every few years and collects a single day's intake data for each participant. In the more recently initiated Continuing Survey of Food Intake in Individuals (CSFII) women and young children only provide a single day of data six times during a year. Other countries also collect data from individuals by varying methodology and frequency.

Food composition tables

To enable foods to be translated into nutrients a variety of national, regional, and international food composition tables exist. These are in the form of books or computer databases. Raw and cooked foods and dishes are listed along with values for their content of major constituents, vitamins and minerals, and in some cases composition of amino acids, fatty acids, and a few other components. The values are usually based on the laboratory analysis of a limited number of samples of each food and preparation. They are therefore only approximate as the nutrient content of foods, especially the micronutrients, can be quite variable, depending on cultivar, growing conditions, maturity, processing procedures, analytical procedures and, in the case of dishes, recipe. These factors account for discrepancies in nutrient values found in different tables.

Recommended daily amounts (RDAs)

Various expert committees in the United Kingdom and other countries, as well as international organizations, produce guidelines for the recommended intakes of energy and specific nutrients, and of major food components. These are given for different age and sex groups. There are differences between countries, not because of differences in nutritional science but due to different philosophies of presentation.

Up to the 1990s a single value for each nutrient was given as the recommended daily intake for each age group. For energy this was the average amount required for that particular group, but for the nutrients it was a more generous estimate, and set at the value adequate to cover the needs of people with the highest requirement. These were called RDAs (recommended daily amounts in the United Kingdom and recommended dietary allowances in the United States of America). The United States continued to use this approach.

A problem with this system is that individual variation in requirements is not explicit, so that the values are frequently misinterpreted. It is often incorrectly concluded that individuals with daily intakes of specific nutrients below the RDA value are deficient. The RDAs were established for use in planning diets for population groups to ensure adequacy even for members of the group with the highest requirements. They were not originally intended for assessing the intakes of individuals but are often used for this purpose.

In 1991 the United Kingdom introduced the concept of a range of values to reflect the distribution of requirements for each nutrient in a population of healthy individuals. They are therefore given as an average requirement value, and values at the lower and upper ends of the estimated normal range of requirements (two standard deviations below and above the average). These three values are collectively called dietary reference values (DRVs) and, individually, the estimated average requirement (EAR), the lower reference nutrient intake (LRNI), and the reference nutrient intake (RNI). The latter value corresponds to the RDA. Other advisory bodies, such as the European Union, World Health Organization, and the Food and Agricultural Organization, are adopting this concept. People with intakes below the LRNI would require further investigation.

Other guidelines for healthy diets have been developed that are concerned less with nutrient deficiencies than with dietary patterns likely to prevent the diseases now prevalent in affluent countries, including obesity, coronary heart disease, and hypertension. The guidelines in the United Kingdom, United States, and other countries are similar and specify goals for the intakes of total fat, saturated fat, polyunsaturated fat, cholesterol, carbohydrate, sugar, fibre, salt, and protein. Such goals are shown in Table 14.

To enable consumers to know what they are consuming, the United States, United Kingdom, and other countries have requirements for the nutrition labelling of packaged foods. The legislation is not the same but the principles are similar, and within Europe the previously national regulations are being harmonized to conform with European standards. It is mandatory to provide information on certain constituents and others can be given voluntarily. The form required may be quantity per 100 g, per serving, or as a percentage of RDA and nutrient goals.

Nutrition tables

Table 1 Major food groups as a percentage of total available kcalories (1986–88 averages)

Major food groups and items	United Kingdom	Australia	Bangladesh	Thailand	Japan
Cereals	21.1	23.5	82.1	61.2	41.8
Roots and tubers[a]	6.3	3.3	2.0	4.5	3.1
Sugar	13.5	14.8	4.0	9.7	8.7
Pulses and beans	1.0	0.5	2.3	2.5	4.4
Nuts and oilseeds	1.0	1.0	0.2	3.9	1.0
Vegetables + Fruits	3.5	4.2	0.6	3.9	4.3
Animal products[b]	28.4	30.3	2.6	7.8	18.1
Oils and fats[c]	18.1	15.1	5.6	2.9	12.1
Alcoholic beverages	6.4	5.3	0	3.0	5.4
Others	0.6	2.1	0.6	0.5	1.0
Plant foods	**63**	**62**	**96**	**92**	**78**
Animal foods	**37**	**38**	**4**	**8**	**22**
Total (kcal)	3317	3179	2019	2431	2902

[a] Includes bananas and plantains.
[b] Not including fats such as lard and butter.
[c] Added amounts only.
Source: Food and Agricultural Organisation (1991): Regional Office for Asia and the Pacific Report: 1991/16 Qureshi R. Report of regional expert consultation of Asia Pacific network for food and nutrition on progress in nutrition improvement during the decade of the 80s. Bangkok; Qureshi R. Personal communication.

Table 2 Composition of grains: cereals and pseudo-cereals per 100 g edible portion

					Carbohydrate					Minerals				Vitamins					
	Water	Energy	Protein	Fat	Total	Starch	Sugars	Fibre (Total)	Fibre (NSP)	Na	K	Ca	Fe	Carotene	E	B_1	B_2	C	Folate
Description	(g)	(kcal)	(g)	(g)	(g)	(g)	(g)	(g)	(g)	(mg)	(mg)	(mg)	(mg)	(μg)	(mg)	(mg)	(mg)	(mg)	(μg)
Barley, pearl	10.6	360	7.9	1.7	83.6	83.6	Tr	5.9	N	3.0	270	20	3.0	0.0	0.2	0.12	0.05	0.0	20
Barley, wholegrain	11.7	301	10.6	2.1	64.0	62.2	1.8	–	14.8	4.0	560	50	6.0	0.0	0.9	0.31	0.10	0.0	50
Buckwheat (groats)	13.2	364	8.1	1.5	84.9	84.5	0.4	–	2.1	1.0	220	12	4.9	–	Tr	0.28	0.07	0.0	–
Corn, sweet (kernels)	76.0	93	3.4	1.8	16.6	14.6	2.0	3.3	1.5	1.0	260	3	0.7	97.0	0.7	0.16	0.05	8.0	41
Maize (grits)	12.0	362	8.7	0.8	77.7	–	–	–	–	1.0	80	4	–	44.0	–	0.13	0.04	0.0	–
Millet (*Panicum miliaceum*)	8.7	378	11.0	4.2	72.9	–	–	1.0*	–	5.0	195	8	3.0	–	–	0.42	0.29	0.0	–
Oatmeal (raw)	8.9	401	12.4	8.7	72.8	72.8	Tr	6.3	6.8	33.0	370	55	4.1	0.0	1.7	0.50	0.10	0.0	{60}
Quinoa	9.3	374	13.1	5.8	68.9	–	–	–	–	–	740	60	9.3	–	–	0.20	0.40	0.0	–
Rice, brown (*Oryza*)	13.9	363	6.7	2.8	81.3	80.0	1.3	3.8	1.9	3.0	130	140	2.1	0.0	0.8	0.59	0.07	0.0	49
Rice, white (*Oryza*) (easy cook)	11.4	390	7.3	3.6	85.8	85.8	Tr	2.7	0.4	4.0	150	51	0.5	0.0	{0.1}	0.41	0.02	0.0	20
Rice, polished	11.7	361	6.5	1.0	86.8	86.8	Tr	2.2	0.5	6.0	110	4	0.5	0.0	0.1	0.08	{0.02}	0.0	{20}
Rice, wild (*Zizania*)	7.8	357	14.7	1.1	74.9	–	–	5.2	–	7.0	427	21	2.0	–	–	0.12	0.26	0.0	95
Rye	11.0	335	14.8	2.5	69.8	–	–	14.6	–	6.0	264	33	2.7	–	1.3	–	–	–	–
Sorghum	9.2	339	11.3	3.3	74.9	–	–	2.4*	–	–	350	28	4.4	–	–	0.24	0.14	0.0	–
Wheat (hard red spring) (*Triticum aestivum*)	12.8	329	15.4	1.9	68.0	–	–	2.3*	–	2.0	340	25	3.6	–	–	0.50	0.11	0.0	43
Wheat (soft white (*Triticum aestivum*)	10.4	340	10.7	2.0	75.4	–	–	–	–	–	435	34	5.4	–	–	0.41	0.11	0.0	–
Wheat durum (*Triticum durum*)	10.9	339	13.7	2.5	71.1	–	–	–	–	2.0	431	34	3.5	–	–	0.42	0.12	0.0	–
Wheat flour, white (bread-making)	14.0	347	11.5	1.4	75.3	73.9	1.4	3.7	{3.1}	3.0	130	140	2.1	0.0	{0.3}	0.32	0.03	0.0	31
Wheat flour, wholemeal	14.0	315	12.7	2.2	63.9	61.8	2.1	8.6	9	3.0	340	38	3.9	0.0	1.4	0.47	0.09	0.0	60

Sources:
(1) Royal Society of Chemistry/MAFF (1991) and supplements (1988–1994)
(2) United States Department of Agriculture. Handbook no. 8. Composition of Foods (1963) and supplements (1982–1989).
Fibre: NSP, non starch polysaccharides; *, crude fibre. Carotene, β-carotene equivalent; Tr, trace; { }, estimates; N, present in significant quantities but data not reliable; –, no data provided; Na, sodium; K, potassium; Ca, calcium; Fe, iron.

Table 3 Composition of nut and oilseeds per 100 g edible portion

Description	Water	Energy	Protein	Fat	Carbohydrate Total	Starch	Sugars	Fibre (Total)	Fibre (NSP)	Minerals Na	K	Ca	Fe	Vitamins Carotene	E	B_1	B_2	C	Folate
	(g)	(kcal)	(g)	(g)	(g)	(g)	(g)	(g)	(g)	(mg)	(mg)	(mg)	(mg)	(μg)	(mg)	(mg)	(mg)	(mg)	(μg)
Almonds	4.2	606	21.1	55.8	6.9	2.7	4.2	{12.9}	{7.4}	14	780	240	3.0	0	24.0	0.21	0.75	0	48
Brazil-nuts	2.8	673	14.1	68.2	3.1	0.7	2.4	8.1	4.3	3	660	170	2.5	0	7.2	0.67	0.03	0	21
Cashew-nuts	4.4	568	17.7	48.2	18.1	13.5	4.6	N	3.2	15	710	32	6.2	6	0.9	0.69	0.14	0	67
Chestnut (European) (*Castanea sativa*)	51.7	170	2.0	2.7	36.6	29.6	7.0	6.1	4.1	11	500	46	0.9	0	1.2	0.14	0.02	Tr	N
Chestnut, water (*Eleocharis dulcis*)	79.8	47	1.4	0.2	10.4	5.6	4.8	3.2	N	12	530	8	0.7	0	N	0.09	0.11	5	N
Coconut (flesh from kernel)	45.0	351	3.2	36.0	3.7	0.0	3.7	12.2	7.3	17	370	13	2.1	0	0.7	0.04	0.01	3	26
Coconut water	92.2	22	0.3	{0.2}	4.9	0.0	4.9	Tr	Tr	110	310	29	0.1	0	Tr	Tr	Tr	2	–
Hazel-nut	14.8	375	7.6	36.0	6.8	2.1	4.7	6.1	6.1	1	350	44	1.1	0	21.0	0.40	–	Tr	72
Olives, green (in brine)	76.5	103	0.9	11.0	Tr	0.0	Tr	4.0	2.9	2250	91	61	1.0	180	2.0	Tr	Tr	0	Tr
Peanuts (fresh)	6.3	560	25.6	46.1	12.5	6.3	6.2	7.3	6.2	2	670	60	2.5	0	10.1	1.14	0.10	0	110
Pecan nuts	3.7	680	9.2	70.1	5.8	1.5	4.3	N	4.7	1	520	61	2.2	50	4.4	0.71	0.15	0	39
Pine nuts	2.7	679	14.0	68.6	4.0	0.1	3.9	N	1.9	1	780	11	5.6	10	13.7	0.75	0.19	Tr	N
Pistachio nuts (roasted/salted)	2.1	594	17.9	55.4	8.2	2.5	5.7	N	6.1	530	1040	110	3.0	130	4.2	0.70	0.23	0	58
Queensland (macadamia) nut (salted)	1.3	748	7.9	77.6	4.8	0.8	4.0	N	5.3	280	300	47	1.6	0	1.5	0.28	0.06	0	N
Sesame seeds	4.6	112	18.2	58.0	0.9	0.5	0.4	N	7.9	20	570	670	10.4	6	2.5	0.93	0.17	0	97
Soya beans	8.5	371	35.9	18.6	10.3	4.8	5.5	N	15.7	5	1730	240	9.7	12	2.9	0.61	0.27	Tr	370
Sunflower seeds	4.4	576	19.8	47.5	18.0	16.3	1.7	N	6.0	3	710	110	6.4	15	37.8	1.60	0.19	0	N
Walnuts	2.8	678	14.7	68.5	3.3	0.7	2.6	5.9	3.5	7	450	94	2.9	0	3.8	0.40	0.14	0	66

Source: Royal Society of Chemistry/MAFF (1991) and supplements (1988–1994).
NSP, non-starch polysaccharides; carotene, β-carotene equivalent; Tr, trace; { }, estimates; N, present in significant quantities but data not reliable; –, no data provided; Na, sodium; K, potassium; Ca, calcium; Fe, iron.

Table 4 Composition of legumes per 100 g edible portion

Description	Water	Energy	Protein	Fat	Carbohydrate Total	Starch	Sugars	Fibre (Total)	Fibre (NSP)	Minerals Na	K	Ca	Fe	Vitamins Carotene	E	B_1	B_2	C	Folate
	(g)	(kcal)	(g)	(g)	(g)	(g)	(g)	(g)	(g)	(mg)	(mg)	(mg)	(mg)	(μg)	(mg)	(mg)	(mg)	(mg)	(μg)
Beans, black-eyed (cowpea) (seeds, dried)	10.7	32	23.5	1.6	50.4	47.5	2.9	N	8.2	16	1170	81	7.6	35	N	0.87	0.19	1	630
Beans, broad (seeds, fresh)	73.8	82	7.9	0.6	11.3	10.0	1.3	N	6.5	8	280	56	1.6	255	0.6	{0.03}	{0.06}	8	{32}
Beans, butter (seeds, dried)	11.6	278	19.1	1.1	49.8	46.2	3.6	21.6	21.6	62	1700	85	5.9	Tr	–	0.45	0.13	0	110
Beans, french (pods fresh) (*Phaseolus vulgaris*)	90.7	24	1.9	0.5	3.2	0.9	2.3	3.0	2.2	Tr	230	36	1.2	{330}	0.2	0.05	0.07	12	{80}
Beans, red kidney (seeds, dried) (*Phaseolus vulgaris*)	11.2	271	22.1	1.4	40.5	38.0	2.5	{23.4}	15.7	18	1370	100	6.4	11	0.5	0.65	0.19	4	130
Beans, haricot (seeds, dried)	11.3	275	21.4	1.6	45.5	42.7	2.8	25.4	25.4	43	1160	180	6.7	Tr	–	0.45	0.13	0	–
Beans, mung (seeds, dried)	11.0	284	23.9	1.1	42.4	40.9	1.5	13.9	10.0	12	1250	89	6.0	24	N	0.36	0.26	Tr	140
Beans, scarlet runner (pods, fresh)	91.2	22	1.6	0.4	3.2	0.4	2.8	2.6	2.0	Tr	220	33	1.2	145	0.2	0.06	0.03	18	60
Chick-peas (seeds, dried)	10.0	324	21.3	5.4	46.4	43.8	2.6	13.5	10.7	39	1000	160	5.5	60	2.9	0.39	0.24	Tr	180
Gram, black (seeds, dried)	11.5	280	24.9	1.4	38.9	37.6	1.3	17.9	N	40	800	150	6.3	38	N	0.42	0.37	Tr	132
Gram, sprouts (mung) (fresh)	90.4	31	2.9	0.5	4.0	1.8	2.2	{5.6}	1.5	5	74	20	1.7	40	N	0.11	0.04	7	61
Lentils, green (seeds, dried)	10.8	302	24.3	1.9	45.7	44.5	1.2	N	8.9	12	940	71	11.1	N	N	0.41	0.27	Tr	110
Lentils, red (seeds, dried)	11.1	324	23.8	1.3	53.2	50.8	2.4	10.5	4.9	36	710	51	7.6	{60}	N	0.50	0.20	Tr	35
Peas (seeds, fresh)	74.6	82	6.9	1.5	9.3	7.0	2.3	4.7	4.7	1	330	21	2.8	300	0.2	0.74	0.02	24	62
Peas, sugar snap (pods, fresh)	87.1	35	3.4	0.2	5.0	1.3	3.7	N	1.5	4	150	54	0.8	200	N	0.17	0.14	32	7
Pigeon-pea (seeds, dried)	10.0	306	20.0	2.0	{54}	{45}	{9}	{15}	{15}	29	1100	100	5.0	30	–	0.50	0.15	Tr	100

Source: Royal Society of Chemistry/MAFF (1991) and supplements (1988–1994).
NSP, non-starch polysaccharides; carotene, β-carotene equivalent; Tr, trace; { }, estimates; N, present in significant quantities but data not reliable; –, no data provided; Na, sodium; K, potassium; Ca, calcium; Fe, iron.

Table 5 Composition of temperate fruits per 100g edible portion

Description	Water (g)	Energy (kcal)	Protein (g)	Fat (g)	Carbohydrate: Total (g)	Starch (g)	Sugars (g)	Fibre (Total) (g)	Fibre (NSP) (g)	Minerals: Na (mg)	K (mg)	Ca (mg)	Fe (mg)	Vitamins: Carotene (μg)	E (mg)	B_1 (mg)	B_2 (mg)	C (mg)	Folate (μg)
Apple, 'Cox's Pippin'	83.3	47	0.5	0.1	11.4	Tr	11.4	{2.2}	2.0	3	130	4	0.2	{18}	0.6	0.03	0.03	9	4
Apple, 'Golden Delicious'	85.5	44	0.3	0.2	10.8	Tr	10.8	{1.9}	1.7	3	110	3	0.2	15	0.6	0.03	0.03	4	1
Apricot	87.2	32	0.9	0.1	7.2	0.0	7.2	1.9	1.7	2	270	15	0.5	405	N	0.04	0.05	6	5
Bilberry	85.9	32	0.6	0.2	6.9	0.0	6.9	{2.5}	1.8	3	88	12	0.5	30	N	0.03	0.03	17	6
Blackberry	85.0	25	0.9	0.2	5.1	0.0	5.1	2.0	0.7	2	160	41	0.7	80	2.4	0.02	0.05	15	34
Black currant	77.4	29	0.9	Tr	6.6	0.0	6.6	7.8	3.6	3	370	60	1.3	100	1.0	0.03	0.06	200	N
Cherry	82.8	49	0.9	0.1	11.5	0.0	11.5	1.5	0.9	1	210	13	0.2	25	0.1	0.03	0.03	11	5
Cranberries	87.0	16	0.4	0.1	3.4	0.0	3.4	3.8	3.0	2	95	12	0.7	22	N	0.03	0.02	13	2
Damson	77.5	39	0.5	Tr	9.6	0.0	9.6	3.7	{1.8}	2	290	24	0.4	{295}	0.7	0.10	0.03	{5}	{3}
Gooseberry	82.1	55	0.7	0.3	3.0	0.0	3.0	2.9	2.4	2	210	28	0.3	110	0.4	0.03	0.03	14	{8}
Grapes	81.8	61	0.4	0.1	15.4	0.0	15.4	0.8	0.7	2	210	13	0.3	17	Tr	0.05	0.01	3	2
Greengage	82.0	41	0.8	0.1	9.7	0.0	9.7	2.3	2.1	1	310	17	0.4	95	{0.7}	0.05	0.04	5	{3}
Lemon	86.3	19	1.0	0.3	3.2	0.0	3.2	4.7	N	5	150	85	0.5	18	N	0.05	0.04	58	N
Lime	89.7	9	0.7	0.3	0.8	0.0	0.8	N	N	2	130	23	0.4	12	N	0.03	0.02	46	8
Loganberry	85.0	17	1.1	Tr	3.4	0.0	3.4	5.6	{2.5}	3	260	35	1.4	{6}	{0.5}	{0.03}	{0.05}	35	{33}
Medlars	74.5	43	0.5	Tr	10.6	0.0	10.6	9.2	N	6	250	30	0.5	N	N	N	N	2	N
Mulberry	85.0	36	1.3	Tr	8.1	0.0	8.1	1.5	N	2	260	36	1.6	14	N	0.03	0.05	19	{33}
Orange, assorted	86.1	38	1.1	0.1	8.5	0.0	8.5	1.8	1.7	3	110	33	0.1	28	0.2	0.11	0.04	54	31
Peach	88.9	34	1.0	0.1	7.6	0.0	7.6	2.3	1.5	1	160	7	0.4	58	N	0.02	0.04	31	3
Pear	83.8	40	0.3	0.1	10.0	0.0	10.0	N	2.2	3	150	11	0.2	18	0.5	0.02	0.03	6	2
Plum	83.9	37	0.6	0.1	8.8	0.0	8.8	2.3	1.6	2	240	13	0.4	295	0.6	0.05	0.03	4	3
Prune	22.1	160	2.8	0.5	38.4	0.0	38.4	14.5	6.5	12	860	38	2.9	155	N	0.10	0.20	Tr	4
Quince	84.2	3	0.3	0.1	6.3	Tr	6.3	5.8	N	3	200	14	0.3	Tr	N	0.02	0.02	15	N
Raisin	13.2	277	2.1	0.4	69.3	0.0	69.3	6.1	2.0	60	1.2	46	3.8	12	N	0.12	0.05	1	10
Raspberry	87.0	26	1.4	0.3	4.6	0.0	4.6	6.7	2.5	3	170	25	0.7	6	0.5	0.03	0.05	32	33
Red currant	82.8	21	1.1	Tr	4.4	0.0	4.4	{4.5}	1.5	2	280	36	1.2	25	0.1	0.04	{0.06}	40	N
Strawberry	89.5	27	0.8	0.1	6.0	0.0	6.0	2.0	1.1	6	160	16	0.4	8	0.2	0.03	0.03	77	6
Tangerine	86.7	35	0.9	0.1	8.0	0.0	8.0	1.7	1.3	2	160	42	0.3	97	N	0.07	0.02	30	21

Source: Royal Society of Chemistry/MAFF (1991) and supplements (1988–1994).
NSP, non-starch polysaccharides; carotene, β-carotene equivalent; Tr, trace; { }, estimates; N, present in significant quantities but data not reliable; –, no data provided; Na, sodium; K, potassium; Ca, calcium; Fe, iron.

Table 6 Composition of tropical fruits per 100g edible portion

					Carbohydrate					Minerals				Vitamins					
	Water	Energy	Protein	Fat	Total	Starch	Sugars	Fibre (Total)	Fibre (NSP)	Na	K	Ca	Fe	Carotene	E	B_1	B_2	C	Folate
Description	(g)	(kcal)	(g)	(g)	(g)	(g)	(g)	(g)	(g)	(mg)	(mg)	(mg)	(mg)	(μg)	(mg)	(mg)	(mg)	(mg)	(μg)
Banana	75.1	96	1.3	0.3	23.2	2.3	20.9	3.1	1.1	1	400	6	0.3	21	0.3	0.04	0.06	11	14
Carambola	91.4	33	0.5	0.3	7.3	0.2	7.1	1.7	1.3	2	150	5	0.6	37	N	0.03	0.03	31	N
Dates	60.7	127	1.3	0.1	31.3	0.0	31.3	3.6	1.8	7	410	24	0.3	{18}	N	0.06	0.07	14	25
Durian	62.9	137	2.4	1.4	28.4	5.2	23.2	4.0	N	1	600	14	0.9	11	N	0.29	0.29	41	N
Figs (dried)	23.6	213	3.3	1.5	48.6	0.0	48.6	11.4	6.9	57	890	230	3.9	{59}	N	0.07	0.09	1	8
Grapefruit	89.0	30	0.8	0.1	6.8	0.0	6.8	{1.6}	1.3	3	200	23	0.1	17	{0.2}	0.05	0.02	36	26
Guava	84.7	27	0.8	0.5	5.0	0.1	4.9	4.7	3.7	5	230	13	0.4	435	N	0.04	0.04	230	N
Kiwifruit	84.0	49	1.1	0.5	10.6	0.3	10.3	N	1.9	4	290	25	0.4	37	N	0.01	0.03	59	N
Loquat	88.5	27	0.7	0.2	6.3	0.0	6.3	N	N	1	220	20	0.4	515	N	0.02	0.03	3	N
Lychee	81.1	59	0.9	0.1	14.3	0.0	14.3	1.5	0.7	1	160	6	0.5	0	N	0.04	0.06	45	N
Mango	82.4	59	0.7	0.2	14.1	0.3	13.8	{2.9}	2.6	2	180	12	0.7	1800	1.1	0.04	0.05	37	N
Mangosteen	81.0	73	0.6	0.5	16.4	0.3	16.1	1.3	N	1	130	14	0.3	0	N	0.05	0.01	3	N
Passion fruit	74.9	36	2.6	0.4	5.8	0.0	5.8	N	3.3	19	200	11	1.3	750	N	0.03	0.12	23	N
Pawpaw	88.5	36	0.5	0.1	8.8	0.0	8.8	2.3	2.2	5	200	23	0.5	810	N	0.03	0.04	60	1
Pineapple	86.5	42	0.4	0.2	10.1	0.0	10.1	1.3	1.2	2	160	18	0.2	18	0.1	0.08	0.03	12	5
Pomegranate	80.0	52	1.3	0.2	11.8	0.0	11.8	N	3.4	2	240	12	0.7	33	N	0.05	0.04	13	N
Rambutan	80.4	70	1.0	0.4	16.3	0.0	16.3	1.3	N	1	100	14	0.1	0	N	0.02	0.06	78	N
Sharon fruit	79.9	74	0.8	Tr	18.6	0.0	18.6	N	1.6	5	210	10	0.1	950	N	0.03	0.05	19	7

Source: Royal Society of Chemistry/MAFF (1991) and supplements (1988–1994).
NSP, non-starch polysaccharides; carotene, β-carotene equivalent; Tr, trace; { }, estimates; N, present in significant quantities but data not reliable; –, no data provided; Na, sodium; K, potassium; Ca, calcium; Fe, iron.

Table 7 Composition of vegetable fruits per 100g edible portion

Description	Water (g)	Energy (kcal)	Protein (g)	Fat (g)	Carbohydrate: Total (g)	Starch (g)	Sugars (g)	Fibre (Total) (g)	Fibre (NSP) (g)	Minerals: Na (mg)	K (mg)	Ca (mg)	Fe (mg)	Vitamins: Carotene (μg)	E (mg)	B_1 (mg)	B_2 (mg)	C (mg)	Folate (μg)
Aubergine	92.9	15	0.9	0.4	2.2	0.2	2.0	2.3	2.0	2	210	10	0.3	70	0.0	0.02	0.01	4	18
Avocado	72.5	187	1.9	19.5	0.5	Tr	0.5	N	3.4	6	450	11	0.4	16	3.2	0.10	0.18	6	11
Capsicum, green	93.3	16	0.8	0.3	2.5	0.1	2.4	{1.9}	1.6	4	120	8	0.4	265	0.8	0.01	0.01	120	36
Capsicum, red	90.4	32	1.0	0.4	6.2	0.1	6.1	{1.9}	1.6	4	160	8	0.3	3840	0.8	0.01	0.03	140	21
Courgette	93.7	18	1.8	0.4	1.8	0.1	1.7	–	0.9	1	360	25	0.8	610	N	0.12	0.02	21	52
Cucumber	96.4	10	0.7	0.1	1.5	0.1	1.4	{0.7}	0.6	3	140	18	0.3	60	0.1	0.03	0.01	2	9
Marrow	95.6	12	0.5	0.2	2.2	0.1	2.1	1.1	0.5	1	140	18	0.2	110	Tr	0.08	Tr	11	13
Melon, cantaloupe	92.1	19	0.6	0.1	4.2	0.0	4.2	0.9	1.0	8	210	20	0.3	1000	0.1	0.04	0.02	26	5
Melon, 'Galia'	91.7	24	0.5	0.1	5.6	0.0	5.6	{0.9}	1.0	8	210	20	0.3	1000	0.1	0.04	0.02	26	5
Melon, 'Honeydew'	92.2	28	0.6	0.1	6.6	0.0	6.6	0.8	0.6	32	210	9	0.1	4.8	0.1	0.03	0.01	9	2
Melon, water	92.3	32	0.5	0.3	7.1	0.0	7.1	0.3	0.1	2	100	7	0.3	230	{0.1}	0.05	0.01	8	2
Papaya (pawpaw) (*Carica papaya*)	88.5	36	0.5	0.1	8.8	0.0	8.8	2.3	2.2	5	200	23	0.5	810	N	0.03	0.04	60	1
Pumpkin	95.0	13	0.7	0.2	2.0	0.3	1.7	0.5	1.0	Tr	130	29	0.4	450	1.1	0.16	Tr	14	10
Squash, 'Acorn'	87.8	40	0.8	0.1	8.8	7.8	1.0	N	2.3	3	350	33	0.7	{204}	N	0.14	0.01	11	17
Squash, 'Butternut'	86.4	37	1.1	0.1	7.9	3.4	4.5	N	1.6	4	350	48	0.7	3650	1.8	0.10	0.02	21	27
Squash, 'Spaghetti'	91.6	27	0.6	0.6	4.2	0.6	3.6	N	2.3	17	110	23	0.3	{30}	N	0.04	0.02	2	12
Tomato	93.1	17	0.7	0.3	3.1	Tr	3.1	1.3	1.0	9	250	7	0.5	640	1.2	0.09	0.01	17	17

Source: Royal Society of Chemistry/MAFF (1991) and supplements (1988–1994).
NSP, non-starch polysaccharides; carotene, β-carotene equivalent; Tr, trace; { }, estimates; N, present in significant quantities but data not reliable; –, no data provided; Na, sodium; K, potassium; Ca, calcium; Fe, iron.

Table 8 Composition of salad and vegetable crops per 100 g edible portion

Description	Water (g)	Energy (kcal)	Protein (g)	Fat (g)	Carbohydrate Total (g)	Starch (g)	Sugars (g)	Fibre (Total) (g)	Fibre (NSP) (g)	Minerals Na (mg)	K (mg)	Ca (mg)	Fe (mg)	Vitamins Carotene (μg)	E (mg)	B_1 (mg)	B_2 (mg)	C (mg)	Folate (μg)
Broccoli	88.2	33	4.4	0.9	1.6	0.1	1.5	N	2.6	8	370	56	1.7	575	{1.3}	0.10	0.06	87	90
Brussels sprouts	84.3	42	3.5	1.4	3.9	0.8	3.1	3.8	4.1	6	450	26	0.7	215	1.0	0.15	0.11	115	135
Cabbage, average	90.1	26	1.7	0.4	4.1	0.1	4.0	{2.9}	2.4	5	270	52	0.7	385	0.2	0.15	0.02	49	75
Cabbage, white	90.7	27	1.4	0.2	5.0	0.1	4.9	2.4	2.1	7	240	49	0.5	40	0.2	0.12	0.01	35	34
Cabbage, red	90.7	21	1.1	0.3	3.4	0.1	3.3	3.1	2.5	8	250	60	0.4	15	0.2	0.02	0.01	55	39
Cabbage, savoy	88.1	27	2.1	0.5	3.9	0.1	3.8	{3.1}	3.1	5	320	53	1.1	995	0.2	0.15	0.03	62	150
Cauliflower	88.4	34	3.6	0.9	2.9	0.4	2.5	1.9	1.8	8	380	21	0.7	50	0.2	0.17	0.05	43	66
Celery	95.1	8	0.5	0.2	0.9	Tr	0.9	1.6	1.1	60	320	41	0.4	50	0.2	0.06	0.01	8	16
Chard	92.7	19	1.8	0.2	2.9	2.3	0.6	N	N	210	380	51	1.8	4625	N	0.04	0.09	30	165
Chicory	94.3	11	0.5	0.6	0.9	0.2	0.7	0.7	N	1	170	21	0.4	120	N	0.14	Tr	5	14
Curly kale	88.4	33	3.4	1.6	1.4	0.1	1.3	{3.3}	3.1	43	450	130	1.7	3145	{1.7}	0.08	0.09	110	120
Fennel	94.2	12	0.9	0.2	1.8	0.1	1.7	N	2.4	11	440	24	0.3	140	N	0.06	0.01	5	42
Globe artichoke	85.2	18	2.8	0.2	1.3	Tr	1.3	N	N	27	360	41	0.7	39	{0.1}	Tr	0.01	Tr	21
Leeks	90.8	22	1.6	0.5	2.5	0.3	2.2	2.8	2.2	2	260	24	1.1	735	0.9	0.29	0.05	17	56
Lettuce	95.1	14	0.8	0.5	1.7	Tr	1.7	1.3	0.9	3	220	28	0.7	355	0.6	0.12	0.02	5	55
Mustard & cress	95.3	13	1.6	0.6	0.4	Tr	0.4	3.3	1.1	19	110	50	1.0	{1280}	0.7	0.04	0.04	33	60
Okra	86.6	31	2.8	1.0	3.0	0.5	2.5	4.5	4.0	8	330	160	1.1	515	N	0.20	0.06	21	88
Onions	89.0	36	1.2	0.2	5.6	Tr	5.6	1.5	1.4	3	160	25	0.3	10	0.3	0.31	Tr	5	17
Pe-Tsai (Chinese cabbage) (*Brassica pekinensis*)	95.4	12	1.0	0.2	1.4	Tr	1.4	N	1.2	7	230	54	0.6	70	N	0.09	Tr	21	77
Spinach	89.7	25	2.8	0.8	1.6	0.1	1.5	{3.9}	2.1	140	500	170	2.1	3535	1.7	0.07	0.09	26	150
Watercress	92.5	22	3.0	1.0	0.4	Tr	0.4	3.0	1.5	49	230	170	2.2	2520	1.5	0.16	0.06	62	N

Source: Royal Society of Chemistry/MAFF (1991) and supplements (1988–1994).
NSP, non-starch polysaccharides; carotene, β-carotene equivalent; Tr, trace; { }, estimates; N, present in significant quantities but data not reliable; –, no data provided; Na, sodium; K, potassium; Ca, calcium; Fe, iron.

Table 9 Composition of root vegetables and sago per 100 g edible portion

Description	Water (g)	Energy (kcal)	Protein (g)	Fat (g)	Carbohydrate: Total (g)	Carbohydrate: Starch (g)	Carbohydrate: Sugars (g)	Carbohydrate: Fibre (Total) (g)	Carbohydrate: Fibre (NSP) (g)	Minerals: Na (mg)	Minerals: K (mg)	Minerals: Ca (mg)	Minerals: Fe (mg)	Vitamins: Carotene (μg)	Vitamins: E (mg)	Vitamins: B_1 (mg)	Vitamins: B_2 (mg)	Vitamins: C (mg)	Vitamins: Folate (μg)
Arrowroot (dried)	12.2	362	0.4	0.1	94.0	94.0	Tr	N	0.1	5	18	7	2.0	0	Tr	Tr	Tr	0	Tr
Beetroot	87.1	37	1.7	0.1	7.6	0.6	7.0	2.8	1.9	66	380	20	1.0	20	Tr	0.01	0.01	5	150
Carrots, new	88.8	30	0.7	0.5	5.8	0.2	5.6	{2.6}	2.4	40	240	34	0.4	5330	{0.6}	0.04	0.02	4	28
Carrots, old	89.8	35	0.6	0.3	7.7	0.3	7.4	2.6	2.4	25	70	25	0.3	8115	0.6	0.10	0.01	6	12
Cassava (raw)	64.5	142	0.6	0.2	36.8	35.3	1.5	1.7	1.6	5	330	18	0.5	Tr	N	0.06	0.02	31	19
Jerusalem artichoke (boiled)	80.2	41	1.6	0.1	1.6	Tr	1.6	N	3.5	3	420	30	0.4	20	0.2	0.10	Tr	2	N
Parsnip	79.3	65	1.8	1.1	11.9	6.2	5.7	4.3	4.6	10	450	41	0.6	30	1.0	0.23	0.01	17	87
Potatoes, new	81.7	71	1.7	0.3	16.1	14.8	1.3	1.3	1.0	11	320	6	0.3	Tr	{0.1}	0.15	0.02	16	25
Potatoes, old	79.0	76	2.1	0.2	17.2	16.6	0.6	1.6	1.3	7	360	5	0.4	Tr	0.1	0.21	0.02	11	35
Radish	95.4	12	0.7	0.2	1.9	Tr	1.9	0.9	0.9	11	240	19	0.6	Tr	0.0	0.03	Tr	17	38
Sago (dried)	12.6	355	0.2	0.2	94.0	94.0	Tr	N	0.5	3	5	10	1.2	0	Tr	Tr	Tr	0	Tr
Swede	91.2	24	0.7	0.3	5.0	0.1	4.9	2.4	1.9	15	170	53	0.1	350	Tr	0.15	Tr	31	31
Sweet potato (orange[a])	73.7	87	1.2	0.3	21.3	15.6	5.7	2.3	2.4	40	370	24	0.7	3930	4.6	0.17	Tr	23	17
Taro	68.3	106	1.4	0.2	26.2	25.1	1.1	3.5	2.4	4	360	25	0.8	37	N	0.08	0.03	13	N
Turnip	91.2	23	0.9	0.3	4.7	0.2	4.5	2.5	2.4	15	280	48	0.2	20	Tr	0.05	0.01	17	14
Yam	67.2	117	1.5	0.3	28.2	27.5	0.7	3.7	1.3	2	380	15	0.7	Tr	N	0.16	0.01	4	8

Source: Royal Society of Chemistry/MAFF (1991) and supplements (1988–1994).
NSP, non-starch polysaccharides; carotene, β-carotene equivalent; Tr, trace; { }, estimates; N, present in significant quantities but data not reliable; –, no data provided; Na, sodium; K, potassium; Ca, calcium; Fe, iron.
[a] Orange, carotene range 1820–16 000 μg/100 g; white, approx. 69 μg/100 g.

Table 10 Composition of mushrooms and algae per 100 g edible portion

Description	Water	Energy	Protein	Fat	Carbohydrate Total	Starch	Sugars	Fibre (Total)	Fibre (NSP)	Minerals Na	K	Ca	Fe	Vitamins Carotene	E	B_1	B_2	C	Folate
	(g)	(kcal)	(g)	(g)	(g)	(g)	(g)	(g)	(g)	(mg)	(mg)	(mg)	(mg)	(µg)	(mg)	(mg)	(mg)	(mg)	(µg)
Mushroom, common (raw) (*Agaricus campestris*)	92.6	13	1.8	0.5	0.4	0.2	0.2	2.3	1.1	5	320	6	0.6	0	0.1	0.09	0.31	1.0	44
Mushroom (*Agaricus bisporus*)	91.8	25	2.1	0.4	4.7	–	–	1.3	–	4	370	5	1.2	–	0.1	0.10	0.45	3.5	21
Mushroom, enoki (*Pholita nameko*)	89.4	35	1.5	0.4	7.9	–	–	–	–	3	381	1	–	–	–	0.09	0.11	11.9	30
Mushroom, Jew's ear (tender dried, soaked, raw)	92.0	29	0.7	0.7	5.7	N	N	N	N	1	60	23	5.1	10	N	0.01	0.07	0.0	N
Mushroom, oyster (raw)	90.1	8	1.6	0.2	Tr	Tr	Tr	N	N	77	230	29	1.9	0	N	N	0.43	Tr	N
Mushroom, shitake (*Lenitius edodus*) (cooked)	83.5	55	1.6	0.2	12.3	N	N	N	N	13	1530	11	1.7	0	N	0.04	0.17	Tr	N
Mushroom, straw (canned)	89.6	15	2.1	0.2	1.2	0.8	0.4	N	N	260	71	10	1.1	0	N	0.01	0.12	0.0	Tr
Seaweed, agar (*Euchema* spp.)	91.3	26	0.5	0.0	6.8	–	–	0.5*	–	9	226	54	1.9	–	–	0.01	0.00	0.0	–
Seaweed, Irish moss (*Chondrus crispus*) (raw)	81.3	8	1.5	0.2	Tr	0.0	Tr	N	12.3	67	63	72	8.9	N	N	0.01	0.47	N	N
Seaweed, kelp (*Laminaria* spp.)	81.6	43	1.7	0.6	9.6	–	–	1.33*	–	233	89	168	2.9	–	–	0.15	0.47	0.1	0
Seaweed, laver (*Porphyra laciniata*)	85.0	35	5.8	0.3	5.1	–	–	0.27*	–	48	356	70	1.8	–	–	0.10	0.45	39.0	–
Seaweed, wakame (*Undaria* spp.)	80.0	45	3.0	0.6	9.1	–	–	0.54*	–	872	50	150	2.2	–	–	0.06	0.23	3.0	–
Spirulina (*Spirulina* spp.)	90.7	26	5.9	0.4	2.4	–	–	0.34*	–	98	127	–	–	–	–	0.22	0.34	0.9	–

Sources:
(1) Royal Society of Chemistry/MAFF (1991) and supplements (1988–1994)
(2) United States Department of Agriculture. Handbook no. 8. Composition of Foods (1963) and supplements (1982–1989).
Fibre: NSP, non starch polysaccharides; *, crude fibre. Carotene, β-carotene equivalent; Tr, trace; { }, estimates; N, present in significant quantities but data not reliable; –, no data provided; Na, sodium; K, potassium; Ca, calcium; Fe, iron.

Table 11 Fat composition of selected vegetable and animal sources (Royal Society of Chemistry/MAFF 1991)

Foodstuff	Fat (g/100 g)	Saturated fat (%)	Monounsaturated fat (%)	Polysaturated fat (%)
Meat				
Beef	16.2	45.2	50.0	4.8
Chicken	4.3	35.0	45.0	20.0
Lamb	34.6	53.2	41.8	5.1
Pheasant	9.3	35.2	52.3	12.5
Pork	29.5	39.7	44.4	15.9
Turkey	2.2	35.0	45.0	20.0
Fish				
Cod fillets	0.7	20.0	20.0	60.0
Herring	18.5	31.5	50.0	18.5
Mackerel	16.3	22.6	54.8	22.6
Milk				
Channel Island milk	5.1	70.2	27.7	2.1
Goats' milk	3.5	71.9	25.0	3.1
Whole milk	3.9	66.7	30.6	2.8
Cereals				
Brown rice	3.6	29.2	29.2	41.7
Brown wheat flour	2.2	16.7	16.7	66.7
Cornflour	0.7	20.0	20.0	60.0
Oatmeal	9.2	18.6	38.4	43.0
White wheat flour	1.4	22.2	11.1	66.7
White rice	2.8	29.0	29.0	41.9
Nuts, oilseeds, and oils				
Coconut oil	99.9	91.1	7.1	1.8
Olive oil	99.9	14.8	73.4	11.8
Palm oil	99.9	47.6	43.7	8.7
Peanuts	46.1	18.8	48.4	32.8
Sunflower seeds	47.5	9.9	21.6	68.4
Walnut	68.5	8.5	18.9	72.5
Vegetables				
Avocado	19.5	22.3	65.8	12.0

Table 12 Vitamin sources and deficiency effects

Vitamin	Chemical name	Good natural sources	Deficiency effects	Prevalence
Fat soluble				
A	Retinol, β-carotene	Animal foods, meat, coloured fruit and vegetables	Night blindness, xerophthalmia, skin keratinization	Widespread in poor countries
D	Calciferol	Animal foods	Rickets, osteomalacia	Occurs in populations protected from sun, diets very high in phytate
E	Tocopherols, Tocotrienols	Widespread	Neurological dysfunction	Very rare
K	Phylloquinone, menaquinone	Green leafy vegetables	Impaired blood clotting, haemorrhage	Rare, newborn, patients on anticlotting drugs
Water soluble				
B_1	Thiamin	Wholegrains	Berberi, peripheral and central nerve damage	Exists where restricted diet. refined cereal alcoholics
B_{12}	Cobalamin	Only animal foods, some bacteria, no plant sources	Pernicious anaemia	Not uncommon, failure to absorb B_{12}
B_2	Riboflavin	Widespread, milk	Lesions of mouth, seborrhoeic dermatitis	Widespread
B_6	Pyridoxine, pyridoxal, pyridoxamine	Meat, wholegrains, vegetables, nuts	Disorders of amino acid metabolism, convulsions	Extremely rare, faults in manufacture of infant feed
Biotin	Same	Widespread	Impaired fat and carbohydrate metabolism, dermatitis	Unknown except TPN, consumers of large amounts of raw eggs
C	Ascorbic acid	Fruit and vegetables	Scurvy, impaired wound healing, subcutaneous haemorrhage	Not uncommon
Folic acid	Same	Widespread	Megaloblastic anaemia	Not uncommon
Niacin	Nicotinic acid, nicotinamide	From typtophan in proteins	Pellagra, light sensitivity, depression	Parts of Africa, Asia
Pantothenic acid	Same	Widespread	Peripheral nerve damage	Very rare

TPN, total parenteral nutrition—drip feeding.

Table 13 Typical losses (%) of vitamins in cooking (Royal Society of Chemistry/MAFF 1991)

	Vegetables		Fruit	Cereals	
	Boil	Fry	Fruit Stew	Boil	Fry
Vitamin E	0	0	–	–	–
Thiamin	35	20	25	40	25[a]
Riboflavin	20	0	25	40	15
Niacin	30	0	25	40	5
Vitamin B_6	40	25	20	40	25
Folate	40	55	80	50	50
Pantothenate	–	–	25	40	25
Biotin	–	–	25	40	0
Vitamin C	45	30	25	–	–

[a] 15% in bread making.

Table 14 Population nutrient goals

	Limits of population average intakes	
	Lower	Upper
Total energy[1]		
Total fat (% total energy)	15	30[2]
Saturated fatty acids (% total energy)	0	10
Polyunsaturated fatty acids (% total energy)	3	7
Dietary cholesterol (mg/day)	0	300
Total carbohydrate (% total energy)	55	75
Complex carbohydrate[3] (% total energy)	50	70
Dietary fibre (g/day):		
As non-starch polysaccharides	16	24
As total dietary fibre	27	40
Free sugars (% total energy)	0	10
Protein (% total energy)	10	15
Salt (g/day)	–	6

[1] Sufficient for normal childhood growth, the needs of pregnancy and lactation, for work and desirable physical activities, and to maintain appropriate body reserves.
[2] Interim goal for countries with high fat intakes, further benefits expected by reducing intake towards 15% of total energy.
[3] Including a daily minimum intake of 400 g of vegetables and fruits, of which at least 30 g should be pulses, nuts and seeds.
Source: World Health Organization (1990). *Diet, nutrition and the prevention of chronic diseases*. WHO technical report no 797. WHO, Geneva.

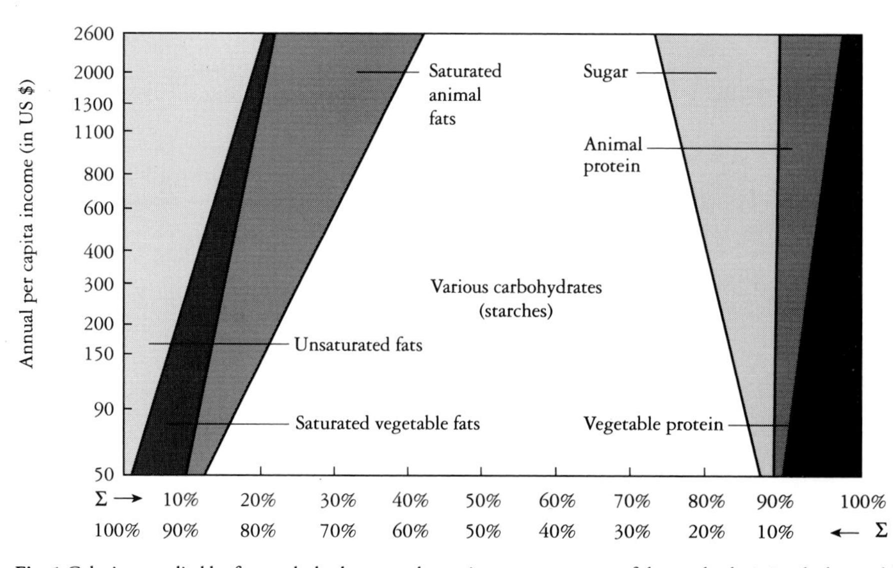

Fig. 1 Calories supplied by fats, carbohydrates, and proteins as a percentage of the total caloric intake by wealth of country. (After Food and Agriculture Organization (1969). La consommation, les perspectives nutritionelles et les politiques alimentaires. In *Plan Indicatif Mondiale Provisoire pour le Developpement de l'Agriculture*, Vol. 2, Chapter 13. FAO, Rome.)

Recommended reading

Anderson, E. N. (1988). *The food of China.* Yale University Press, New Haven.

Bender, A. E. (1978). *Food processing and nutrition.* Academic Press. London.

Bender, A. E. and Bender, D. A. (1995). *A dictionary of food and nutrition.* Oxford University Press, Oxford.

Bender, D. A. (1993). *Introduction to nutrition and metabolism.* University College London Press, London.

Blackburn, F. (1984). *Sugar-cane.* Longman, London.

British Nutrition Foundation (1988). *Vegetarianism.* Briefing Paper no. 13. BNF, London.

Brothwell, D. and Brothwell, P. (1969). *Food in antiquity. A survey of the diet of early peoples.* Thames and Hudson, London.

Carpenter, K. J. (1981). *Pellagra.* Cambridge University Press, Cambridge.

Carpenter, K. J. (1986). *The history of scurvy and vitamin C.* Cambridge University Press, Cambridge.

Davies, J. and Dickerson, J. W. T. (1991). *Nutrient content of food portions.* Royal Society of Chemistry, Cambridge.

De Rougemont, G. M. (1989). *A field guide to the crops of Britain and Europe.* Collins, London.

Drummond, J. C. and Wilbraham, A. (1969). *The Englishman's food. Five centuries of English diet.* Jonathan Cape, London.

Duckworth, R. B. (1966). *Fruits and vegetables.* Pergamon, Oxford.

Fieldhouse, P. (1986). *Food and nutrition: customs and culture.* Croom Helm, London.

Garrow, J. S. and James, W. P. T. (ed.) (1993). *Human nutrition and dietetics,* (9th edn). Churchill Livingstone, Edinburgh.

Geissler, C. and Oddy, D. J. (ed.) (1993). *Food, diet and economic change past and present.* Leicester University Press, Leicester.

Guiry, M. D. and Blunden, G. (ed.) (1991). *Seaweed resources in Europe: uses and potential.* Wiley, Chichester.

Hawkes, J. G. (1990). *The potato—evolution, biodiversity and genetic resources.* Belhaven Press, London.

Hawthorn, J. (1981). *Foundations of food science.* W. H. Freeman, Oxford.

Herklots, G. A. C. (1972). *Vegetables in South-East Asia.* George Allen & Unwin, London.

Jansen, P. C. M. and Westphal, E. (general ed.) *Plant resources of South-East Asia (PROSEA).* Pudoc, Wageningen/Prosea, Bagor/Backhuys Publishers, Leiden.

Vol. 1. Van der Maesen, L. J. G. and Somaatmadja, S. (ed.) (1990). *Pulses.*

Vol. 2. Verheif, E. W. M. and Coronel, R. E. (ed.) (1991). *Edible fruits and nuts.*

Vol. 8. Siemonsma, J. S. and Piluek, K. (ed.) (1993). *Vegetables.*

Vol. 9. Flach, M. and Rumawas, F. (ed.) (1996). *Plants yielding non-seed carbohydates.*

[The Prosea Foundation (Plant Resources of South-East Asia) is producing a series of some 20 volumes about the plants of economic importance in South-East Asia. Some of these volumes concern food plants. On a worldwide basis this project is of great importance as regards information concerning plant utilization. The volumes are produced in paperback and hardbound—further information can be obtained from Dr P. C. M. Jansen, Department of Plant Taxonomy, The Agricultural University, Wageningen, The Netherlands.]

Kent, N. L. and Evers, A. D. (1994). *Technology of cereals,* (4th edn). Pergamon, Oxford.

Kuhnlein, H. V. and Turner, N. J. (1991). *Traditional plant foods of Canadian indigenous peoples—nutrition, botany and use.* Gordon and Breach, Philadelphia.

Liener, I. E. (1980). *Toxic constituents of plant foodstuffs,* (2nd edn). Academic Press, London.

Mabey, R. (1972). *Food for free.* Collins, London.

Macrae, R., Robinson, R. K., and Sadler, M. J. (ed.) (1993). *Encyclopaedia of food science, food technology and nutrition,* Vols 1–8. Academic Press, London.

Morgan, J. and Richard, A. (1993). *The book of apples.* Ebury Press, London.

National Research Council (1989). *Lost crops of the Incas: little-known plants of the Andes for worldwide cultivation.* National Academy Press, Washington, DC.

Phillips, R. and Rix, M. (1993). *Vegetables.* Pan Books, London.

Purseglove, J. W. (1968). *Tropical crops—dicotyledons.* Longman, London.

Purseglove, J. W. (1972). *Tropical crops—monocotyledons.* Longman, London.

Purseglove, J. W., Brown, E. G., Green, C. L., and Robbins, S. R. J. (1981). *Spices.* Longman, London.

Rosengarten, F. (1984). *The book of edible nuts.* Walker, New York.

Royal Society of Chemistry/MAFF (1991). *McCance and Widdowson's 'The composition of foods',* (5th edn). Royal Society of Chemistry/Ministry of Agriculture, Fisheries and Food, Cambridge. Supplements: Cereals and cereal products (3rd suppl. to 4th edition, 1988); Vegetables, herbs and spices (5th suppl. to 4th edition, 1991); Fruits and nuts (1st suppl. to 5th edition, 1992); Miscellaneous food (4th suppl. to 5th edition, 1994).

Salaman, R. (1986). *The history and social influence of the potato.* Cambridge University Press, Cambridge.

Simmons, N. W. and Smartt, J. (ed.) (1995). *Evolution of crop plants,* (2nd edn). Longman, London.

Singer, R. and Harris, B. (1987). *Mushrooms and truffles. Botany, cultivation and utilization,* (2nd edn). Koeltz Scientific Books, Koenigstein.

Summerfield, R. J. and Roberts, E. H. (ed.) (1985). *Grain legume crops.* Collins, London.

Tannahill, R. (1975). *Food in history.* Paladin. Frogmore, St Albans, UK.

United States Department of Agriculture (1992). *Composition of foods* (Agriculture Handbook No. 8). USDA, Washington, DC. Including revisions: Fruits and fruit juices, 1982; Vegetables and vegetable products, 1984; Nuts and seed products, 1984; Legumes and legume products, 1986; Cereal grams and pasta, 1989.

Webb, G. P. (1995). *Nutrition: a health promotion approach.* Edward Arnold, London.

Yamaguchi, M. (1983). *World vegetables.* Ellis Horwood, Chichester.

Zohary, D. and Hopf, M. (1993). *Domestication of plants in the Old World,* (2nd edn). Clarendon Press, Oxford.

Index of plant names

Acer nigrum 18
Acer saccharum 18, 19
Achras zapota, see Manilkara zapota
Actinidiaceae 116
Actinidia chinensis, see A. deliciosa
Actinidia deliciosa syn. *A. chinensis* 116, 117
Adansonia 108
adlay (*Coix lachryma-jobi*) 10
Aframomum melegueta 138, 140
agar (*Euchema* spp.) 194, 223
Agaricus bisporus 196
Agaricus campestris 196, 197
Agave spp. 146
akee (*Blighia sapida*) 108, 109
alecost (*Chrysanthemum balsamita*) 154, 155
algae 194, 195, 201, 223
Allium ampeloprasum var. *porrum* syn. *A. porrum* 178, 179
Allium cepa 176, 177
Allium chinense 178
Allium fistulosum 176, 177
Allium porrum, see Allium ampeloprasum var. *porrum*
Allium sativum 178, 179
Allium schoenoprasum 176, 177
Allium scorodoprasum 178
Allium tuberosum 178
allspice (*Pimento dioica*) 142, 143
almond (*Prunus dulcis* syns. *P.amygdalus, Amygdalus communis*) xv, xviii, 30, 31
almond
  bitter (*Prunus dulcis* var. *amara*) 30
  sweet (*Prunus dulcis* var. *dulcis*) 30, 216
Amanita muscaria 209
Amaranthaceae xvii
Amaranthus caudatus 14, 170
Amaranthus spp. 170, 171
amaranth xvii
Amomum spp. 140
Amygdalus communis, see Prunus dulcis
Anacardium occidentale 34, 35
Ananas comosus 102, 103
Anethum graveolens 148, 149, 156, 157
angelica (*Angelica archangelica*) 158, 159
angelica, wild (*Angelica sylvestris*) 158
Angelica archangelica 158, 159
anise, star (*Illicium verum*) 146
anise/aniseed (*Pimpinella anisum*) 146, 147
Annona cherimola 102, 103
Annona diversifolia 102
Annona muricata 102, 103
Annona purpurea 102
Annona reticulata 102
Annona squamosa 102
Anthemis nobilis syn. *Chamaemelum nobile* 154
Anthriscus cerefolium 156, 157
Apiaceae, see Umbelliferae
Apium graveolens 158, 159
  var. *rapaceum* 184, 185
apple xviii
  'Braddick's Nonpareil' 56, 57
  'Calville Blanche d'Hiver' 58, 59
  'Christmas Pearmain' 54, 55
  'Coe's Golden Drop' 54, 55
  'Court Pendu Plat' 54, 55
  'Cox's Orange Pippin' 58, 59, 218
  crab 52
  'Discovery' 60
  'Edward VII' 56, 57
  'Egremont Russet' 58, 59
  'Ellison's Orange' 56, 57
  'Emneth Early' 56, 57
  'Golden Delicious' 58, 59, 218
  'Golden Hornet' 52, 53
  'Golden Spire' 56, 57
  'Ida Red' 60
  'John Downie' 52, 53
  'Kingston Black' 66, 67
  'Laxton's Fortune' 60, 61
  'May Queen' 54, 55
  'Merton Charm' 60, 61
  'Monarch' 56, 57
  'Orleans Reinette' 58, 59
  'Pitmaston Pine Apple' 54, 55
  'Spartan' 60, 61
  'Sweet Coppin' 66, 67
  'Transcendent' 52
  'Tremlitt's Bitter' 66, 67
  'Tydeman's Early Worcester' 60, 61
  see also Malus spp.
apricot (*Prunus armeniaca* syn. *Armeniaca vulgaris*) xix, 78, 79, 218
Araceae 190
Arachis hypogaea xviii, 26, 27
Araucaria angustifolia 34
Araucaria araucana 34
Araucaria bidwilli 34
Arbutus unedo 88
Arenga pinnata syn. *A. saccharifera* 18, 19, 21
Arenga saccharifera, see A. pinnata
Armeniaca vulgaris, see Prunus armeniaca
Armoracia rusticana 144, 145
arrowroot (*Maranta arundinacea*) 190, 191, 222
arrowroot
  East Indian (*Tacca leontopetaloides*) 190
  Indian (*Curcuma angustifolia*) 190
  Queensland (*Canna edulis*) 190
Artemisia abrotanum 154, 155
Artemisia absinthium 146, 147
Artemisia dracunculoides 154
Artemisia dracunculus 154, 155
artichoke
  globe (*Cynara scolymus*) xv, 174, 175, 221
  Jerusalem (*Helianthus tuberosus*) 188, 189, 222
Artocarpus altilis syn. *A. communis* 122, 123
Artocarpus communis, see A. altilis
Artocarpus heterophyllus syn. *A. integrifolia* 122
Artocarpus integrifolia, see A. heterophyllus
Ascophyllum nodosum 194, 195
ash-gourd, *see* wax-gourd
asparagus (*Asparagus officinalis*) 172, 173
Asparagus officinalis 172, 173
asparagus-pea (*Lotus tetragonolobus*) 38, 46, 47
asparagus-pea, *see also* bean, winged
Aspergillus flavus 26, 210
Aspergillus parasiticus 26
atemoya (*Annona cherimola* × *A. squamosa*) 102
Atriplex hortensis 170, 171, 198
Atropa belladonna 134
aubergine (*Solanum melongena*) xv, 134, 136, 137, 220
Aucuba japonica 142
Avena sativa 4, 5
Averrhoa carambola 108, 109
avocado pear (*Persea americana*) xix, 122, 123, 220, 224
azarole (*Crataegus azarolus*) 68, 69

Bactris gasipaes 20
balm, lemon (*Melissa officinalis*) 152, 153
balsam herb, *see* alecost
bamboo shoots 172, 173
Bambusa spp. 172
banana (*Musa* spp.) xviii, 114, 115, 219
  'Dwarf Cavendish' 114
  'Gros Michel' 114
  'Lady's Fingers' 114, 115
  red 114, 115
baobob (*Adansonia*) 108
Barbarea verna 162
barberry (*Berberis vulgaris*) 198, 199
barley (*Hordeum vulgare*) xv, xvi, xviii, 4, 5, 215
Basellaceae 188
basil
  bush (*Ocimum minimum*) 152
  sweet (*Ocimum basilicum*) 152, 153
bean
  adzuki (*Phaseolus angularis*) 38
  broad (*Vicia faba*) xviii, 208, 217
    see also bean, faba
  cluster (*Cyamopsis tetragonolobus*) 50, 51
  common (*Phaseolus vulgaris*) xviii, 40, 41, 217
  faba (*Vicia faba*) xv, xviii, 44, 45
  French, *see* bean, common
  Goa, *see* bean, winged
  haricot, *see* bean, common
  jack (*Canavalia ensiformis*) 44, 45
  kidney 208, 217
    see also bean, common
  lima xviii, 208
    see also butter-bean
  locust xvii
  Madagascar, *see* butter-bean
  mung, *see* gram, green
  scarlet runner (*Phaseolus coccineus* syn. *P. multiflorus*) 40, 41, 217
  snap, *see* bean, common
  soya (*Glycine max*) xvi, xvii, xviii, 28, 29, 203, 207, 216

bean (*cont.*)
tepary (*Phaseolus acutifolius*) 38
Windsor, *see* bean, faba
winged (*Psophocarpus tetragonolobus*) 38, 46
beetroot (*Beta vulgaris*) 180, 181, 222
belladonna (*Atropa belladonna*) 134
Benincasa hispida 128
beniseed, *see* sesame
Berberidaceae 198
Berberis vulgaris 198, 199
Bertholletia excelsa 34, 35
Beta vulgaris 170, 171, 180
var. *esculenta* xvii, 16, 17
var. *maritima* 16
bible leaf, *see* alecost
bilberry (*Vaccinium myrtillus*) 88, 89, 218
bindi, *see* okra
blackberry (*Rubus ulmifolius*) xv, 84, 85, 218
blackberry, American (*Rubus alleghanensis*) 84
blackthorn (*Prunus spinosa*) 72, 73
Blastophaga psenes 100
blewits (*Lepista* spp.) 196, 197
Blighia sapida 108, 109
blueberry
highbush (*Vaccinium corymbosum*) 88, 89
lowbush (*Vaccinium angustifolium*) 88, 89
Boletus edulis 196, 197
Bombacaceae 108
Borassus flabellifer 18, 19, 112, 113
borecole, *see* kale
boysenberry (*Rubus* spp.) 84
Brachiaria ramosa 10
bracken fern (*Pteridium aquilinum*) xx, 209
brambles (*Rubus* spp.) 84, 85
Brassica campestris 28
syn. *Brassica rapa* 182, 183
Brassicaceae, *see* Cruciferae
Brassica chinensis 164, 165
Brassica hirta, see Sinapis alba
Brassica juncea xv, 142, 164
Brassica napus 28, 29, 162
Brassica nigra 142, 143
Brassica oleracea 166, 167, 194
Brassica pekinensis 164, 165, 221
Brassica rapa 28, 29
see also Brassica campestris
Brassica spp. xix
Brazil-nut (*Bertholletia excelsa*) xiv, 34, 35, 216
breadfruit (*Artocarpus altilis* syn. *A. communis*) 122, 123
brinjal, *see* aubergine
broccoli 221
green sprouting (*Brassica oleracea*) 168, 169
sprouting (*Brassica oleracea*) 168, 169
Brussels sprouts (*Brassica oleracea*) 166, 167, 221
buck-wheat
common (*Fagopyrum esculentum*) xvi, xvii, 14, 15
perennial (*Fagopyrum cymosum*) 14
tartary (*Fagopyrum tartaricum*) 14
bullace (*Prunus insititia*) 72, 73
bullock's heart (*Annona reticulata*) 102
butter-bean (*Phaseolus lunatus*) 42, 43, 217
see also bean, lima
butternut, *see* walnut, white
Butyrospermum xiv

cabbage xviii, 221
flower 166, 167
head (*Brassica oleracea*) 166, 167, 221
red 166, 221
savoy 166, 221
spring 166
wild (*Brassica oleracea*) 166, 167
see also palm
Cajanus cajan 38, 39
calabrese, *see* broccoli, green sprouting
'Calmondin' (× *Citrofortunella microcarpa*) 92
Calvatia gigantea 196, 197
Camellia sinensis 120, 121
Canavalia ensiformis 44, 45
Canavalia gladiata 44
Cannabaceae 146
Canna edulis 190
Cantharellus cibarius 196, 197
Cape gooseberry (*Physalis peruviana*) 136, 137
Cape gooseberry, dwarf, *see* cherry, ground
capers (*Capparis spinosa*) 142, 143
Capparaceae 142
Capparis spinosa 142, 143
Caprifoliaceae 198
Capsicum annuum 138
Capsicum frutescens 138
Capsicum spp. xvi, xix, 134, 220
carambola (*Averrhoa carambola*) 108, 109, 219
caraway (*Carum carvi*) 148, 149
cardamom (*Elettaria cardamomum*) 140, 141
cardoon (*Cynara cardunculus*) 174, 175
Carica candamarcensis 122
Carica papaya 122, 123
carob (*Ceratonia siliqua*) xvii, 120
carrageen (*Chondrus crispus*) 194, 195
carrot (*Daucus carota*) xv, xviii, xix, 184, 185, 222
Carthamus tinctorius 28, 142
Carum carvi 148, 149
Carya illinoensis, see C. pecan
Carya pecan syn. *C. illinoensis* 32, 33
Caryota urens 18, 20
cashew-nuts (*Anacardium occidentale*) 34, 35, 216
cassava (*Manihot esculenta* syn. *M. utilissima*) xv, xviii, xix, 190, 191, 208, 222
cassia (*Cinnamomum* spp.) 140
Castanea dentata 30
Castanea mollissima 30
Castanea sativa xvii, 30, 31
Castanospermum australe 34, 35
castor (*Ricinus communis*) xvii
cauliflower (*Brassica oleracea*) 168, 169, 221
Ceiba sp. 108
Celenicereus sp. 116
celeriac (*Apium graveolens* var. *rapaceum*) 158, 184, 185
celery (*Apium graveolens*) 158, 159, 221
celery
leaf (*Apium graveolens* var. *secalinum*) 158
leaf-stalk (*Apium graveolens* var. *dulce*) 158, 159
cep (*Boletus edulis*) 196, 197
Ceratonia siliqua 120
ceriman (*Monstera deliciosa*) 102, 103
Chaerophyllum bulbosum 184
Chamaemelum nobile, see Anthemis nobilis
chamomile (*Anthemis nobilis* syn. *Chamaemelum nobile*) 154
chamomile
German 154
wild (*Matricaria recutica*) 154
champignon, fairy-ring (*Marasmius oreades*) 196, 197
chanterelle (*Cantharellus cibarius*) 196, 197
chard, *see* seakale-beet
chayote (*Sechium edule*) 130, 131
Chenopodiaceae xvii, 170
Chenopodium album 198
Chenopodium bonus-henricus 198, 199
Chenopodium pallidicaule 14
Chenopodium quinoa 14, 15
cherimoya (*Annona cherimola*) 102, 103
cherry
amarelle 70
bird (*Prunus padus*) 70
bladder (*Physalis alkekengi*) 136
'Early Rivers' 70, 71
ground (*Physalis pruinosa*) 136, 137
ground (*Prunus fruticosa*) 70
hybrid 70
morello 70, 71
'Napoleon' 70, 71
sour (*Prunus cerasus*) 70, 71
sweet (*Prunus avium*) 70, 71, 218
West Indian (*Malpighia glabra* syn. *M. punicifolia*) xix
chervil (*Anthriscus cerefolium*) 156, 157, 184
chervil, turnip-rooted (*Chaerophyllum bulbosum*) 184
chestnut
Australian, *see* chestnut, Moreton Bay
Moreton Bay (*Castanospermum australe*) 34, 35
sweet/Spanish (*Castanea sativa*) xv, xvii, 30, 31, 216
chickasaw (*Prunus angustifolia*) 72
chick-pea (*Cicer arietinum*) xv, xviii, 42, 43, 217
chicory (*Cichorium intybus*) 118, 119, 160, 161, 221
'Whitloof' 160
chilli (*Capsicum annuum*; *C. frutescens*) 138
chilli, bird (*Capsicum frutescens*) 138
Chinese lantern, *see* cherry, bladder
chives (*Allium schoenoprasum*) 176, 177
chives, Chinese (*Allium tuberosum*) 178
Chlorophyceae xix
Chondrus crispus 194, 195
christophine, *see* chayote
chrysanthemum, garland (*Chrysanthemum coronarium*) 164, 165
Chrysanthemum balsamita 154, 155
Chrysanthemum coronarium 164, 165
chufa, *see* tiger nut
ciboule, *see* onion, Welsh
Cicer arietinum 42, 43
Cichorium endivia 160, 161
Cichorium intybus 118, 119, 160, 161
Cinnamomum spp. 140
Cinnamomum verum syn. *C. zeylandicum* 140, 141
Cinnamomum zeylandicum, see C. verum
cinnamon (*Cinnamomum verum* syn. *C. zeylandicum*) 140, 141
citron (*Citrus medica*) 94, 95
Citrullus lanatus syn. *C. vulgaris* 128, 129
Citrullus vulgaris, see C. lanatus
Citrus aurantifolia 90, 92, 93
Citrus bergamia 90
Citrus hystrix 94

Citrus limon 90, 91
Citrus maxima 92
Citrus medica 94, 95
Citrus paradisi 90, 92, 93
Citrus reticulata 90, 92, 93
Citrus sinensis 90, 91
Citrus spp. hybrids 94
clementine (*Citrus* spp.) 94, 95
cloudberry (*Rubus chamaemorus*) 84, 85
cloves (*Syzygium aromaticum* syn. *Eugenia caryophyllus*) 142, 143
cob, *see* hazel
Coccoloba uvifera 98
cocoa (*Theobroma cacao*) 120, 121
coconut (*Cocos nucifera*) xix, 18, 20, 22, 23, 216, 224
Cocos nucifera xix, 18, 20, 22, 23
cocoyam, new, *see* tannia
Coffea arabica 118, 119
Coffea canephora 118, 119
Coffea liberica 118
coffee
  arabica (*Coffea arabica*) 118, 119
  liberica (*Coffea liberica*) 118
  robusta (*Coffea canephora*) 118, 119
Coix lachryma-jobi 10
Colchicum autumnale 142
collard, *see* kale
Colocasia antiquorum, see C. esculenta
Colocasia esculenta syn. *C. antiquorum* 190, 191
Compositae xix, 118, 146
  aromatic 154
  Jerusalem artichoke 188
  oriental leaf vegetables 164, 165
  root crops 182
  salad plants 160, 161
Convolvulaceae 192
Coprinus spp. 210
coriander (*Coriandrum sativum*) 148, 149, 156
Coriandrum sativum 148, 149, 156
Corylus avellana xvii, 30, 31
Corylus maxima 30, 31
Corypha elata 18
costmary, *see* alecost
cotton (*Gossypium* spp.) xv, 28
courgette (*Cucurbita* spp.) 130, 131, 220
cowberry (*Vaccinium vitis-idaea*) 88
cowpea (*Vigna unguiculata*) xvi, 48, 49, 217
Crambe maritima 172, 173
cranberry (*Vaccinium oxycoccus*) 88, 89, 218
cranberry
  American, *see* cranberry, large
  large (*Vaccinium macrocarpon*) 88, 89
  mountain, *see* cowberry
Crataegus azarolus 68, 69
cress (*Lepidium sativum*) 162, 163, 221
cress
  land, *see* cress, winter
  winter (*Barbarea verna*) 162
  see also watercress
Crithmum maritimum 156, 157
crocus, autumn (*Colchicum autumnale*) 142
Crocus sativus 142, 143
Cruciferae xviii, 142
  European brassicas 166–9
  horseradish 144
  oriental leaf vegetables 164, 165
  radish 180, 181
  root crops 182, 183
  salad plants 162, 163
  seakale 172, 173
cucumber (*Cucumis sativus*) 124, 125, 220
Cucumeropsis edulis 128
Cucumeropsis manii 128
Cucumis anguria 124, 125
Cucumis hardwickii 124
Cucumis longipes 124
Cucumis melo 126, 127
Cucumis sativus 124, 125
Cucurbitaceae 124
Cucurbita ficifolia 132
Cucurbita maxima 130, 131
Cucurbita mixta 132
Cucurbita moschata 130, 132
Cucurbita spp. 130, 131
cumin (*Cuminum cyminum*) 148
cumin, black (*Nigella sativa*) 148
Cuminum cyminum 148
Cupressaceae 146
Curcuma angustifolia 190
Curcuma domestica syn. *C. longa* 144, 145
Curcuma longa, see C. domestica
currant
  black (*Ribes nigrum*) 86, 87, 218
  red (*Ribes sativum*; *R. petraeum*; *Ribes rubrum*) 86, 87, 218
  white 86, 87
custard apple (*Annona* spp.) 102
custard-marrow (*Cucurbita* spp.) 130, 131
Cyamopsis tetragonolobus 50, 51
cycads (*Cycas* spp.) xx, 20, 34
Cycas circinalis 20
Cycas revoluta 20
Cycas spp. 34
Cydonia vulgaris 68, 69
Cynara cardunculus 174, 175
Cynara scolymus 174, 175
Cyperus esculentus 188
Cyphomandra betacea 134, 135

Daemonorops schmidtiana 20
damson (*Prunus damascena*) 72, 73, 218
dandelion (*Taraxacum officinale* agg.) 118, 119
date, *see* palm, date
Datura spp. 209
Datura stramonium 134
Daucus carota 184, 185
Dendrocalamus spp. 172
dewberry (*Rubus caesius*) 84, 85
Digitaria exilis 10
Digitaria iburua 10
dill (*Anethum graveolens*) 148, 149, 156, 157
Diospyros kaki 110, 111
Dolichos biflorus 48
Dolichos lablab, see Lablab niger
dulse (*Palmaria palmata* syn. *Rhodymenia palmata*) 194, 195
durian (*Durio zibethinus*) 108, 109, 219
Durio zibethinus 108, 109

Echinochloa frumentacea 12, 13
egg plant, *see* aubergine
einkorn (*Triticum monococcum*) 2
Elaeis guineensis 24, 25
Elaeis spp. xvi
elder (*Sambucus nigra*) 198, 199
Eleocharis dulcis syn. *E. tuberosa* 36, 37
Eleocharis tuberosa, see E. dulcis
Elettaria cardamomum 140, 141
Eleusine coracana 10, 11
emmer (*Triticum dicoccum*) 2, 3
endive (*Cichorium endivia*) 160, 161
ensete (*Ensete ventricosa*) 114
Ensete ventricosa 114
Eragrostis tef 10
Ericaceae 88–9
Eriobotyra japonica 110, 111
Eruca sativa 162, 163
Euchlaena mexicana 6
Eugeissona utilis 20
Eugenia caryophyllus, see Syzygium aromaticum
Euphorbiaceae 190
Euterpa oleracea 20

Fabaceae, *see* Leguminosae
Fagopyrum esculentum 14, 15
fat hen (*Chenopodium album*) 198
fennel (*Feoniculum vulgare*) 148, 149, 221
fennel, Florence/Florentine (*Foeniculum vulgare* var. *dulce*) 148, 158, 159
fenugreek (*Trigonella foenum-graecum*) 142
Ficus benghalensis 100
Ficus carica 100, 101
Ficus elastica 100
field-bean, *see* bean, faba
fig (*Ficus carica*) 100, 101, 219
filbert (*Corylus maxima*) 30, 31
Flammulina velutipes 196
Foeniculum vulgare 148, 149
  var. *dulce* 148
Foeniculum vulgare var. *dulce* 158, 159
fonio, *see* rice, hungry
Fortunella spp. 92, 94, 95
Fragaria chiloensis 80
Fragaria moschata 80
Fragaria spp. 80, 81
Fragaria vesca 80
Fragaria virginiana 80
Fragaria viridis 80
Fragaria × *ananassa* 80, 81
frijoles, *see* bean, common
Fungi xix, 196, 209–10, 223
Furcellaria lumbricalis 194
Fusarium graminearum 196

gage (*Prunus italica*) 72, 73
Galipea officinalis 146
Garcinia mangostana 106, 107
garlic (*Allium sativum*) xv, 178, 179, 206
Gelidium sesquipedale 194
gherkin (*Cucumis anguria*) 124, 125
gingelly, *see* sesame
ginger (*Zingiber officinale*) 144, 145
Ginkgo biloba 34
gingko (*Gingko biloba*) 34
Glycyrrhiza glabra 144, 145
Gnetum gnemon 34
gnetum (*Gnetum gnemon*) 34
Good King Henry (*Chenopodium bonus-henricus*) 198, 199
gooseberry (*Ribes grossularia* syn. *R. uva-crispa*) 86, 218
gooseberry
  Chinese, *see* kiwifruit

gooseberry (*cont.*)
currant (*Ribes hirtellum*) 86
Worcesterberry (*Ribes divaricatum*) 86
Gossypium spp. 28
gourd
bitter, *see* pear, balsam
bottle (*Lagenaria siceraria*) 128
fig-leaf (*Cucurbita ficifolia*) 132
Malabar, *see* gourd, fig-leaf
gram
Bengal, *see* chick-pea
black (*Phaseolus mungo* syn. *Vigna mungo*) 42, 43, 217
golden, *see* gram, green
green (*Phaseolus aureus* syn. *Vigna radiata*) 42, 43, 217
red, *see* pigeon-pea
Gramineae xvi, xvii
bamboo 172
granadilla
giant (*Passiflora quadrangularis*) 104, 105
purple, *see* passion fruit
sweet (*Passiflora ligularis*) 104
grape (*Vitis vinifera*) xv, xvii, 96–9, 218
grape
'Cabernet' 98, 99
'Chardonnay' 98, 99
cultivars 96, 98, 99
muscadine (*Vitis rotundifolia*) 96
'Pinot Noir' 98, 99
'Riesling' 98, 99
sea (*Coccoloba uvifera*) 98
slip skin (*Vitis labrusca*) 96
wild American (*Vitis spp.*) 98
grapefruit (*Citrus paradisi*) xviii, xix, 90, 92, 93, 219
grass-pea (*Lathyrus sativus*) xviii, 50, 51, 208
greengage (*Prunus italica*) 72, 73, 218
Grossulariaceae 86, 87
groundnut (*Arachis hypogaea*) xv–xvi, xvii, xviii, 26, 27, 38, 209, 216, 224
groundnut
bambara (*Voandzeia subterranea* syn. *Vigna subterranea*) 38
Hausa (*Kerstingiella geocarpa*) 38
Kersting's, *see* groundnut, Hausa
guar, *see* bean, cluster
guava (*Psidium guajava*) 104, 105, 219
guava, strawberry (*Psidium littorale*) 104
gumbo, *see* okra
Gyromitra spp. 210

hazel (*Corylus avellana*) xv, xvii, 30, 31, 216
Helianthus annuus 28, 29
Helianthus tuberosus 188, 189
henbane (*Hyoscyamus niger*) 134
herb patience (*Rumex patientia*) 198
Hibiscus esculentus 174, 175
Hibiscus sabdariffa 174
hoogly, *see* ugli
hop (*Humulus lupulus*) 146, 147
Hordeum sativum, see H. vulgare
Hordeum vulgare syn. *H. sativum* 4, 5
horse-bean, *see* bean, jack
horse-gram (*Dolichos biflorus*) 48
horse-mint (*Mentha longifolia*) 150
horse-radish (*Armoracia rusticana*) 144, 145
huckleberry, garden (*Solanum intrusum* syn. *Solanum nigrum* var. *guineense*) 136, 137
Humulus lupulus 146, 147
Hyoscyamus niger 134

ilama (*Annona diversifolia*) 102
Ilex paraguariensis 118
Illicium verum 146
Ipomoea batatas xv, 192, 193
Irish moss (*Chondrus crispus*) 194, 195, 223

jackfruit/jakfruit (*Artocarpus heterophyllus* syn. *A. integrifolia*) 122
jamberry, *see* tomatillo
Job's tears, *see* adlay
jojoba (*Simmondsia californica*) xvii
Juglans cinerea 32, 33
Juglans nigra 32, 33
Juglans regia xvii, xviii, 32, 33
Juglans sieboldiana 32
juniper (*Juniperus communis*) 146, 147
Juniperus communis 146, 147

kale (*Brassica oleracea*) xviii, 166, 167, 221
kale, Chinese (*Brassica alboglabra*) 164
kaniwa (*Chenopodium pallidicaule*) 14
kapok (*Ceiba* spp.) 108
katemfe (*Thaumatococcus daniellii*) xvii
kava (*Piper methysticum*) 146
kelp (*Laminaria* spp.) 194, 223
Kerstingiella geocarpa 38
kiwicha (*Amaranthus caudatus*) 14
kiwifruit (*Actinidia deliciosa* syn. *A. chinensis*) xiv, xvi, 116, 117, 219
kohlrabi (*Brassica oleracea*) 168, 169
kumquat (*Fortunella*) 92, 94, 95
kurrat 178

Labiatae xix, 150, 151, 152, 153
lablab (*Lablab niger* syn. *Dolichos lablab*) 48, 49
Lablab niger syn. *Dolichos lablab* 48, 49
lad's love, *see* southernwood
lady's fingers, *see* okra
Lagenaria siceraria 128
Laminaria spp. 194
Lathyrus sativus xviii, 50, 51, 208
Lactuca sativa 160, 161
Lauraceae 142
laurel
bay (*Laurus nobilis*) 142, 143
cherry (*Prunus laurocerasus*) 142
Japanese (*Aucuba japonica*) 142
Laurus nobilis 142, 143
laver (*Porphyra umbilicalis*) 194, 195, 223
leek (*Allium ampeloprasum* var. *porrum* syn. *A. porrum*) 178, 179, 221
Leguminosae xviii, 50, 51
beans 40–45
exotic 38
fenugreek 142
lentils 46, 47
liquorice 144, 145
peas 46, 47, 48, 49
tropical pulses 42, 43
lemon (*Citrus limon*) xv, xviii, 90, 91, 218
lemon, water (*Passiflora laurifolia*) 104
Lens culinaris syn. *L. esculenta* 46, 47
Lens esculenta, see L. culinaris
lentil (*Lens culinaris* syn. *L. esculenta*) xv, xviii, 46, 47, 217
Lentinus edodus 196
Lepidium sativum 162, 163
Lepista spp. 196, 197
lettuce (*Lactuca sativa*) xv, 160, 161, 221
Levisticum officinale 156, 157
Ligusticum scoticum 156
lime (*Citrus aurantifolia*) 90, 92, 93, 218
lime
Rangpur (*Citrus reticulata*) 92
wild, *see* papeda
lingonberry, *see* cowberry
Lippia graveolens 150
liquorice (*Glycyrrhiza glabra*) 144, 145
Litchi chinensis 110, 111
loganberry (*Rubus loganobaccus*) 84, 85, 218
loofah (*Luffa cylindrica*) 128
loquat (*Eriobotyra japonica*) 110, 111, 219
lotus (*Nelumbo nucifera*) 36, 37
Lotus tetragonolobus 38, 46, 47
lovage (*Levisticum officinale*) 156, 157
lovage, Scotch (*Ligusticum scoticum*) 156
Luffa cylindrica 128
lupin 50, 51
narrow-leaved (*Lupinus angustifolius*) 50
white (*Lupinus albus* syn. *L. termis*) 50, 51
yellow (*Lupinus luteus*) 50
Lupinus albus syn. *L. termis* 50, 51
Lupinus angustifolius 50
Lupinus luteus 50
Lupinus mutabilis 50
Lupinus spp. xviii, 50, 51
lychee (*Litchi chinensis*) 110, 111, 219
Lycopersicon esculentum xv, 134, 135
var. *cerasiforme* 134

Macadamia integrifolia 34, 35
Macadamia nut (*Macadamia integrifolia*) xvi, 34, 35, 216
Macadamia tetraphylla 34, 35
mace (*Myristica fragrans*) 140, 141
mace, *see also* alecost
Macrolepiota procera 196
Macrolepiota rhacodes 196, 197
maize (*Zea mays*) xv, xvi, xvii, xviii, 6, 7, 205, 215
dent-type 6, 7
hybrid xv
Malpighia glabra syn. *M. punicifolia* xix
Malpighia punicifolia, see M. glabra
Malus baccata 52
Malus hupehensis 52
Malus manchuria 52
Malus orientalis 52
Malus sieversii 52
Malus sylvestris 52, 53
Malus × *domestica* 52
Malvaceae 174
mandarin (*Citrus reticulata*) 90, 92, 93
Mangifera indica 106, 107
mango (*Mangifera indica*) xix, 106, 107, 219
mangosteen (*Garcinia mangostana*) 106, 107, 219
Manihot esculenta syn. *M. utilissima* 190, 191
Manihot utilissima, see M. esculenta
Manilkara zapota syn. *Achras zapota* 104, 105
manioc, *see* cassava

maple
black (*Acer nigrum*) 18
sugar (*Acer saccharum*) xvii, 18, 19
Maranta arundinacea 190, 191
Marantaceae 190
Marasmius oreades 196, 197
marjoram
knotted, *see* marjoram, sweet
sweet (*Marjorana hortensis* syn. *Origanum marjorana*) 150
wild, *see* oregano
Marjorana hortensis syn. *Origanum marjorana* 150
marrow, vegetable (*Cucurbita* spp.) 130, 131, 220
mashua, *see* ysaño
mat-bean (*Phaseolus aconitifolius*) 38
maté (*Ilex paraguariensis*) 118
Matricaria recutica 154
medlar (*Mespilus germanica*) 68, 69, 218
medlar, Japanese, *see* loquat
Melissa officinalis 152, 153
melon (*Cucumis melo*) 126, 127, 220
'canteloupe' 126, 127, 220
'netted' 126, 127
'ogen' 126, 127
'winter' 126, 127
melon
egusi (*Cucumeropsis edulis*/*C. manii*) 128
see also water-melon
Mentha aquatica 150
Mentha longifolia 150
Mentha spicata 150
Mentha × *piperita* 150, 151
Mespilus germanica 68, 69
Metroxylon rumphii 20
Metroxylon sagu 20, 21
millet xvi, 208, 215
brown-top (*Brachiaria ramosa*) 10
bulrush (*Pennisetum typhoideum*) 10, 11
common (*Panicum miliaceum*) 12, 13
finger (*Eleusine coracana*) 10, 11
foxtail (*Setaria italica*) 12, 13
Japanese (*Echinochloa frumentacea*) 12, 13
kodo (*Paspalum scrobiculatum*) 10
little (*Panicum miliare*) 12, 13
pearl, *see* millet, bulrush
mint xix
round-leaved (*Mentha rotundifolia*) 150
Momordica charantia 128, 129
monkey nut, *see* groundnut
monkey-puzzle tree, *see* pine, Chile
Monstera deliciosa 102, 103
mooli (*Raphanus sativus*) 180
Morchella esculenta 196, 197
morel (*Morchella esculenta*) 196, 197
Morus alba 100
Morus nigra 100, 101
moth-bean, *see* mat-bean
mountain ash, *see* rowan
mulberry (*Morus nigra*) 100, 101, 218
mulberry, white (*Morus alba*) 100
Musa acuminata 114
Musa balbisiana 114
mushroom xix, 201, 223
Chinese (*Volvariella volvacea*) 196, 223
common (*Agaricus bisporus*) 196, 223
field (*Agaricus campestris*) 196, 197
Japanese black forest (*Lentinus edodus*) 196, 223
oyster (*Pleurotus ostreatus*) 196, 223
parasol (*Macrolepiota procera*) 196
straw, *see* mushroom, Chinese
winter (*Flammulina velutipes*) 196
musk-melon 126, 127
mustard 221
black (*Brassica nigra*) 142, 143
brown (*Brassica juncea*) xv, 142
Indian (*Brassica juncea*) xv, 142
white (*Sinapis alba* syn. *Brassica hirta*) 142, 143, 162, 163
Myristica fragrans 140, 141
Myrrhis odorata 156, 157
Myrtaceae 142

naranjilla (*Solanum quitoense*) 136
Nasturtium microphyllum × *officinale* 162, 163
Nasturtium officinale 162, 163
nectarine (*Prunus persica* var. *nectarina*) 78, 79
Nelumbo nucifera 36, 37
Nephelium lapaceum 106, 107
Nephelium mutabile 106
nettle
small (*Urtica urens*) 198
stinging (*Urtica dioica*) 198, 199
Nicotiana tabacum 134
Nigella sativa 148
Nipa fruticans 18
nori, *see* laver
nutmeg-melon 126, 127
nutmeg (*Myristica fragrans*) 140, 141, 209

oats (*Avena sativa*) xvi, 4, 5, 215, 224
oca (*Oxalis tuberosa*) 188, 189
Ocimum basilicum 152, 153
Ocimum minimum 152
okra (*Hibiscus esculentus*) xvi, 174, 175, 221
olive (*Olea europa*) xvii, 26, 27, 216, 224
onion (*Allium cepa*) xv, 176, 177, 221
Catawissa 178
Egyptian 178
Japanese bunching, *see* onion, Welsh
tree 178
Welsh (*Allium fistulosum*) 176, 177
Opuntia ficus-indica 116, 117
orache (*Atriplex hortensis*) 170, 171, 198
orange xviii
blood 90, 91
Seville (*Citrus aurantium*) xv, 90, 91, 218
sweet (*Citrus sinensis*) 90, 91, 218
oregano (*Origanum vulgare*) 150, 151
oregano, Mexican (*Lippia graveolens*) 150
Oreodoxa oleracea 20
Origanum marjorana, *see Marjorana hortensis*
Origanum vulgare 150, 151
Oryza sativa 8, 9
Oxalidaceae 188
Oxalis tuberosa 188, 189
oyster plant, *see* salsify

Pachyrrhizus erosus 192, 193
Pachyrrhizus sp. 38
pak-choi (*Brassica chinensis*) 164, 165
palm xix
borassus, *see* palm, palmyra; palmyra
buri (*Corypha elata*) 18
coconut (*Cocos nucifera*) 22, 23
date (*Phoenix dactylifera*) 18, 112, 113, 219
fishtail (*Caryota urens*) 18
gomuti, *see* palm, sugar
nipa (*Nipa fruticans*) 18
oil (*Elaeis guineensis*) xvi, 24, 25, 224
palmyra (*Borassus flabellifer*) 18, 19
sago (*Metroxylon sagu*) 20, 21
sugar (*Arenga pinnata* syn. *A. saccharifera*) 18, 19, 21
toddy, *see* palm, fishtail
wild date (*Phoenix sylvestris*) 18, 19
Palmaceae xix
Palmaria palmata syn. *Rhodymenia palmata* 194, 195
palmyra (*Borassus flabellifer*) 112, 113
palusan (*Nephelium mutabile*) 106
Panicum miliaceum 12, 13
Panicum miliare 12, 13
papaya (*Carica papaya*) xix, 122, 123, 219, 220
papeda (*Citrus hystrix*) 94
paprika (*Capsicum annuum*) 138
parasol, shaggy (*Macrolepiota rhacodes*) 196, 197
parsley (*Petroselinum crispum*) xix, 156, 157
parsnip (*Pastinaca sativa*) xv, xviii, 184, 185, 222
Paspalum scrobiculatum 10
Passiflora edulis 104, 105
Passiflora flavicarpa 104
Passiflora foetida 104
Passiflora laurifolia 104
Passiflora ligularis 104
Passiflora quadrangularis 104, 105
passion fruit
purple (*Passiflora edulis*) 104, 105
yellow (*Passiflora flavicarpa*) 104
Pastinaca sativa 184, 185
patty pan, *see* custard-marrow
pawpaw 219, 220
mountain (*Carica candamarcensis*) 122
see also papaya
pea-bean, *see* bean, common
peach (*Prunus persica* syn. *Persica vulgaris*) 78, 79, 218
peanut, *see* groundnut
pea (*Pisum sativum*) xv, xviii, 46, 47, 217
pear xviii, 218
balsam (*Momordica charantia*) 128, 129
'Bergamotte d'Espéren' 64, 65
common (*Pyrus communis*) 62
'Conference' 62, 63
'Durondeau' 64, 65
'Fertility' 62, 63
'Hazel' 62, 63
nashi 62, 64
'Olivier de Serres' 64, 65
oriental (*Pyrus pyrifolia*) 62
'Packham's Triumph' 64, 65
'Passe Crasanne' 64, 65
prickly (*Opuntia ficus-indica*) 116, 117
'Red Glow Williams' 62, 63
'Red Pear' 66
'Taynton Squash' 66
'Thorn' 66
'Williams' Bon Chrétien' 62, 63
'Yellow Huffcap' 66
pecan (*Carya pecan* syn. *C. illinoensis*) 32, 33, 216
Pennisetum typhoideum 10, 11
pepino (*Solanum muricatum*) 136

pepper
  black (*Piper nigrum*) xvi, 138, 139
  Guinea (*Xylopia aethiopica*) 138
  hot (*Capsicum*) xvi
  melegueta (*Aframomum melegueta*) 138, 140
  sweet (*Capsicum* spp.) xix, 134, 138
  Szechuan (*Zanthoxylum* spp.) 138
  tree (*Schinus molle*) 138
  white (*Piper nigrum*) 138, 139
peppermint (*Mentha* × *piperita*) 150, 151
Persea americana 122, 123
Persica vulgaris, see Prunus persica
persimmon 219
  Japanese (*Diospyros kaki*) 110, 111
  oriental, *see* persimmon, Japanese
Petroselinum crispum 156, 157
pe-tsai (*Brassica pekinensis*) 164, 165, 221
Phaeophyceae xix
Phaseolus aconitifolius 38
Phaseolus acutifolius 38
Phaseolus angularis 38
Phaseolus aureus syn. *Vigna radiata* 42, 43
Phaseolus calcaratus 38
Phaseolus coccineus syn. *P. multiflorus* 40, 41
Phaseolus lunatus 42, 43
Phaseolus multiflorus, see P. coccineus
Phaseolus mungo syn. *Vigna mungo* 42, 43
Phaseolus trilobus 38
Phaseolus vulgaris 40, 41
Phoenix acaulis 20
Phoenix dactylifera 18, 112, 113
Phoenix sylvestris 18, 19
Phyllophora truncata 194
Phyllostachys spp. 172
Physalis alkekengi 136
Physalis ixocarpa 136, 137
Physalis peruviana 136, 137
Physalis pruinosa 136, 137
pigeon-pea (*Cajanus cajan*) 38, 39, 217
pillepesara (*Phaseolus trilobus*) 38
pimento, *see* allspice
Pimento dioica 142, 143
Pimpinella anisum 146, 147
pine
  bunya-bunya (*Araucaria bidwilli*) 34
  Chile (*Araucaria araucana*) 34
  kernel (*Pinus* spp.) xx, 34, 35, 216
  Parana (*Araucaria angustifolia*) 34
  Stone (*Pinus pinea*) 34, 35, 216
pineapple (*Ananas comosus*) 102, 103, 219
piñon (*Pinus edulis*) 34
piñon
  Mexican (*Pinus cembroides*) 34
  single-leaf (*Pinus monophylla*) 34
Pinus cembra 34
Pinus cembroides 34
Pinus edulis 34
Pinus gerardiana 34
Pinus koraiensis 34
Pinus monophylla 34
Pinus pinea 34, 35
Pinus pumila 34
Pinus sibirica 34
Pinus spp. 34, 35
Piperaceae 138
Piper methysticum 146
Piper nigrum xvi, 138, 139
pistachio (*Pistacia vera*) 32, 33, 216
Pistacia vera 32, 33
Pisum sativum 46, 47
pitahaya (*Celenicereus* spp.) 116
Pleurotus ostreatus 196
plum xviii
  American (*Prunus americana*) 72
  cherry (*Prunus cerasifera*) 72, 74, 75
  'Coe's Golden Drop' 76, 77
  'Czar' 74, 75
  date, *see* persimmon, Japanese
  European (*Prunus domestica*) 72, 74, 75, 218
  Japanese (*Prunus salicina* syn. *P. triflora*) 72
  'Jefferson' 76, 77
  'Kirke's Blue' 76, 77
  'Laxton's Delicious' 76, 77
  'Monarch' 74, 75
  Oregon (*Prunus subcordata*) 72
  'Pershore Egg' 74, 75
  'Pond's, Seedling' 74, 75
  'Prune d'Agen' 74, 75
  'River's Early Prolific' 74, 75
  Texan (*Prunus orthosepala*) 72
  'Victoria' 76, 77
Poaceae xvi
Polygonaceae xvii, 172, 198
pomegranate (*Punica granatum*) 100, 101, 219
pomelo, *see* pummelo
Porphyra umbilicalis 194, 195
potato (*Solanum tuberosum*) xv, xviii, xix, 134, 186, 187, 207–8, 222
potato, sweet (*Ipomoea batatas*) xv, xix, 192, 193
Prunus americana 72
Prunus amygdalus, see P. dulcis
Prunus angustifolia 72
Prunus armeniaca syn. *Armeniaca vulgaris* 78, 79
Prunus avium 70, 71
Prunus cerasifera 74, 75
Prunus cerasus 70, 71
Prunus damascena 72, 73
Prunus domestica 74, 75
Prunus dulcis syns. *P.amygdalus, Amygdalus communis* xviii, 30, 31
Prunus fruticosa 70
Prunus insititia 72, 73
Prunus italica 72, 73
Prunus laurocerasus 142
Prunus orthosepala 72
Prunus persica syn. *Persica vulgaris* 78, 79
Prunus persica var. *nectarina* 78, 79
Prunus salicina syn. *P. triflora* 72
Prunus serotina 70
Prunus spinosa 72, 73
Prunus subcordata 72
Psidium guajava 104, 105
Psidium guineense 104
Psidium littorale 104
Psophocarpus tetragonolobus 38, 46
Pteridium aquilinum xx, 209
puff-ball, giant (*Calvatia gigantea*) 196, 197
pummelo (*Citrus maxima* syn. *C. grandis*) 92, 94
pumpkin (*Cucurbita pepo*; *C. maxima*; *C. mixta*; *C. moschata*) xix, 130, 131, 220
Punica granatum 100, 101

Queensland nut, *see* Macadamia nut
quince (*Cydonia vulgaris*) 68, 69, 218
quinoa (*Chenopodium quinoa*) xvii, 14, 15, 215

radish (*Raphanus sativus*) 180, 181, 222
radish, rat-tailed (*Raphanus caudatus*) 180
rakkyo (*Allium chinense*) 178
rambutan (*Nephelium lapaceum*) 106, 107, 219
Ranunculaceae 148
rape
  oilseed (*Brassica napus*) xvii, 28, 29, 162
  swede, *see* rape, oilseed
  turnip (*Brassica rapa*) 28, 29
raspberry (*Rubus idaeus*) 82, 83, 218
raspberry
  black (*Rubus occidentalis*) 82
  Chinese (*Rubus kuntzeanus*) 82
  strawberry (*Rubus illecebrosus*) 82
  wild (*Rubus strigosus*) 82
rettich (*Raphanus sativus*) 180
Rheum raphonticum 172, 173
Rheum rhabarabarum, see R. raphonticum
Rhodophyceae xix
Rhodymenia palmata, see Palmaria palmata
rhubarb (*Rheum raphonticum* syn. *Rheum rhabarbarum*) xviii, 172, 173
Ribes divaricatum 86
Ribes grossularia syn. *R. uva-crispa* 86
Ribes hirtellum 86
Ribes petraeum 86, 87
Ribes rubrum 86, 87
Ribes sativum 86, 87
rice (*Oryza sativa*) xv, xvi, xviii, 8, 9, 215, 224
rice
  hungry (*Digitaria exilis*; *D. iburua*) 10
  wild (*Zizania aquatica*) xiv, 8, 215
rice-bean (*Phaseolus calcaratus*) 38
Ricinus communis xvii
rocambole (*Allium sativum*; *A. scorodoprasum*) 178
rocket (*Eruca sativa*) 162, 163
Rosa canina 68, 69
Rosaceae xviii
Rosa rugosa 68, 69
rose, dog (*Rosa* spp.) 68, 69
roselle (*Hibiscus sabdariffa*) 174
rosemary (*Rosmarinus officinalis*) 152, 153
Rosmarinus officinalis 152, 153
rowan (*Sorbus aucuparia*) 68
Rubus caesius 84, 85
Rubus chamaemorus 84, 85
Rubus idaeus 82, 83
Rubus illecebrosus 82
Rubus kuntzeanus 82
Rubus loganobaccus 84, 85
Rubus occidentalis 82
Rubus phoenicolasius 82, 83
Rubus spp. 84, 85
Rubus strigosus 82
Rubus ulmifolius 84, 85
Rumex acetosa 198, 199
Rumex patientia 198
Rumex scutatus 198
Rutaceae xviii, 90
rye (*Secale cereale*) xvi, 4, 5, 215

Saccharum barberi 16
Saccharum officinarum xvii, 16, 17
Saccharum robustum 16
Saccharum sinense 16
Saccharum spontaneum 16
safflower (*Carthamus tinctorius*) 28, 142

saffron (*Crocus sativus*) 142, 143
sage
  clary (*Salvia sclarea*) 150
  pineapple (*Salvia rutilans*) 150
sage (*Salvia officinalis*) 150
sago (*Metroxylon sagu*) 20, 21, 220
St John's bread, *see* carob
salsify (*Tragopogon porrifolius*) 182, 183
salsify, black, *see* scorzonera
Salvia officinalis 150
Sambucus nigra 198, 199
samphire (*Crithmum maritimum*) 156, 157
Sapindaceae 106
sapodilla (*Manilkara zapota* syn. *Achras zapota*) 104, 105
sarson, *see* rape, turnip
Satureja hortensis 152, 153
Satureja montana 152, 153
savory
  summer (*Satureja hortensis*) 152, 153
  winter (*Satureja montana*) 152, 153
Schinus molle 138
Scorzonera hispanica 182, 183
scorzonera (*Scorzonera hispanica*) 182, 183
sea-beet (*Beta vulgaris* var. *maritima*) 16
seakale-beet (*Beta vulgaris*) 170, 171, 221
seakale (*Crambe maritima*) 172, 173
sea-lettuce (*Ulva lactua*) 194
seaweed xix, 194, 195, 201, 223
Secale cereale 4, 5
Sechium edule 130, 131
service tree (*Sorbus domestica*) 68
sesame (*Sesamum indicum*) 26, 27, 216
Sesamum indicum 26, 27
Setaria italica 12, 13
Setaria viridis 12
shaddock, *see* pummelo
shallots 178, 179
Sharon fruit, *see* persimmon
shea-nut (*Butyrospermum*) xiv
shiitake, *see* mushroom, Japanese black forest
shungiku, *see* chrysanthemum, garland
Simmondsia californica xvii
simsim, *see* sesame
Sinapis alba syn. *Brassica hirta* 142, 143, 162, 163
sloe
  American (*Prunus alleghaniensis*) 72
  see also blackthorn
snake-gourd (*Trichosanthes cucumerina*) 128, 129
Solanaceae xviii, 134, 138, 186
Solanum aethiopicum 136
Solanum intrusum syn. *Solanum nigrum* var. *guineense* 136, 137
Solanum macrocarpon 136
Solanum melongena 136, 137
Solanum muricatum 136
Solanum nigrum var. *guineense* 136, 137
Solanum quitoense 136
Solanum tuberosum xv
soncoya (*Annona purpurea*) 102
Sorbus aucuparia 68
Sorbus domestica 68
Sorghum bicolor syn. *vulgare* 10, 11
sorghum (*Sorghum bicolor* syn. *vulgare*) xvi, xvii, 10, 11, 205, 215
sorrel (*Rumex acetosa*) 198, 199
sorrel
  red, *see* roselle
  round-leaved (*Rumex scutatus*) 198
sour sop (*Annona muricata*) 102, 103
southernwood (*Artemisia abrotanum*) 154, 155
soy/soya, *see* bean, soya
Spinacea oleracea 170, 171
spinach (*Spinacea oleracea*) xv, 170, 171, 221
spinach
  amaranthus (*Amaranthus* spp.) 170, 171
  New Zealand (*Tetragonia expansa*) 170, 171
spinach-beet (*Beta vulgaris*) 170, 171
spirulina (*Spirulina* spp.) 223
squash 220
  scalloped summer, *see* custard-marrow
  summer (*Cucurbita* spp.) 130, 131, 220
  winter (*Cucurbita maxima*; *C. mixta*; *C. moschata*) 130, 131, 132, 133
star fruit, *see* carambola
strawberry xviii, 80, 81, 218
  alpine (*Fragaria vesca* var. *semperflorens*) 80, 81
  cultivated (*Fragaria* × *ananassa*) 80, 81
  'Hautbois' (*Fragaria moschata*) 80
  West Coast Pine (*Fragaria chiloensis*) 80
  wild (*Fragaria vesca*) 80
  wild (*Fragaria viridis*) 80
strawberry tree (*Arbutus unedo*) 88
string-bean, *see* bean, common
sugar apple, *see* sweet sop
sugar-beet (*Beta vulgaris* var. *esculenta*) xvii, 16, 17, 201
sugar-cane (*Saccharum officinarum*) xv, xvii, 16, 17, 201
sunberry, *see* huckleberry, garden
sunflower (*Helianthus annuus*) xvii, 28, 29, 216, 224
swede xviii, 222
  see also turnip
sweet cicely (*Myrrhis odorata*) 156, 157
sweet sop (*Annona squamosa*) 102
sword-bean (*Canavalia gladiata*) 44
Syzygium aromaticum syn. *Eugenia caryophyllus* 142, 143

Tacca leontopetaloides 190
tamarillo, *see* tomato, tree
tamarind (*Tamarindus indica*) 50, 51
Tamarindus indica 50, 51
Tanacetum vulgare 154, 155
tangerine (*Citrus reticulata*) 90, 92, 93, 218
tangho, *see* chrysanthemum, garland
tannia (*Xanthosoma sagittifolium*) 190, 191
tansy (*Tanacetum vulgare*) 154, 155
tapioca, *see* cassava
Taraxacum officinale agg. 118, 119
taro (*Colocasia esculenta* syn. *C. antiquorum*) 190, 191, 222
tarragon (*Artemisia dracunculus*) xix, 154, 155
tarragon
  false, *see* tarragon, Russian
  Russian (*Artemisia dracunculoides*) 154
tarwi (*Lupinus mutabilis*) 50
tea (*Camellia sinensis*) 120, 121
teff (*Eragrostis tef*) 10
teosinte (*Euchlaena mexicana*) 6
Tetragonia expansa 170, 171
Tetrapanax papyrifer 8
Thaumatococcus daniellii xvii
Theobroma cacao 120, 121
thorn-apple (*Datura stramonium*) 134
thyme
  breckland, *see* thyme, wild
  caraway (*Thymus herba-barona*) 150
  common (*Thymus vulgaris*) 150, 151
  lemon (*Thymus citriodorus*) 150, 151
  wild (*Thymus serpyllum*) 150
Thymus citriodorus 150, 151
Thymus herba-barona 150
Thymus serpyllum 150
Thymus vulgaris 150, 151
tick-bean, *see* bean, faba
tiger nut (*Cyperus esculentus*) 188
til, *see* sesame
tobacco (*Nicotiana tabacum*) 134
tomatillo (*Physalis ixocarpa*) 136, 137
tomato (*Lycopersicon esculentum*) xv, xviii, xix, 134, 135, 220
tomato
  cherry (*Lycopersicon esculentum* var. *cerasiforme*) 134, 135
  husk, *see* cherry, ground
  strawberry, *see* cherry, ground
  tree (*Cyphomandra betacea*) 134, 135
toria, *see* rape, turnip
Tragopogon porrifolius 182, 183
Trapa spp. 36, 37
Trichosanthes cucumerina 128, 129
Trigonella foenum-graecum 142
triticale (× *Triticosecale*) 4
Triticum aestivum syn. *T. vulgare* xv, 2, 3
Triticum dicoccum 2, 3
Triticum durum 2, 3
Triticum monococcum 2
Triticum turgidum 2, 3
Triticum vulgare, *see T. aestivum*
Tropaeolaceae 188
Tropaeolum tuberosum 188, 189
truffle (*Tuber aestivum*; *T. melanosporum*) xix, 196, 197
Tuber aestivum 196, 197
Tuber melanosporum 196, 197
turmeric (*Curcuma domestica* syn. *C. longa*) 142, 144, 145
turnip (*Brassica campestris* syn. *Brassica rapa*) xviii, 182, 183, 222

ugli (*Citrus*) spp. 94, 95
ulluco (*Ullucus tuberosus*) 188, 189
Ullucus tuberosus 188, 189
Umbelliferae xviii, xix, 146, 148, 149
  leaf stalks 158, 159
  leaves 156, 157
  root crops 184, 185
Undaria spp. 194
urd, *see* gram, black
Urtica dioica 198, 199
Urtica urens 198

Vaccinium angustifolium 88, 89
Vaccinium corymbosum 88, 89
Vaccinium myrtillus 88, 89
Vaccinium oxycoccus 88, 89
Vaccinium vitis-idaea 88
Vanilla fragrans syn. *V. planifolia* 140, 141
Vanilla planifolia, *see V. fragrans*

vanilla (*Vanilla fragrans* syn. *V. planifolia*) 140, 141
Verbenaceae 150
vetch
  bitter xv
  chickling, *see* grass-pea
Vicia faba 44, 45, 208
Vigna mungo, see Phaseolus mungo
Vigna radiata, see Phaseolus aureus
Vigna subterranea, see Voandzeia subterranea
Vigna unguiculata 48, 49
Vigna vexillata 48
Vitis berlandieri 98
Vitis labrusca 96
Vitis riparia 98
Vitis rotundifolia 96
Vitis rupestris 98
Vitis vinifera xvii, 96–9
Voandzeia subterranea syn. *Vigna subterranea* 38
Volvariella volvacea 196

wakame (*Undaria* spp.) 194, 223
walnut
  black (*Juglans nigra*) 32, 33
  common (*Juglans regia*) xv, xvii, xviii, 32, 33, 216, 224
  Japanese (*Juglans sieboldiana*) 32
  Persian, *see* walnut common
  white (*Juglans cinerea*) 32, 33
water-chestnut (*Eleocharis dulcis* syn. *E. tuberosa*; *Trapa* spp.) 36, 37, 216
watercress (*Nasturtium officinale*; *N. microphyllum* × *officinale*) xv, 162, 163, 221
water-melon (*Citrullus lanatus* syn. *C. vulgaris*) 128, 129, 220
water-melon, wild (*Passiflora foetida*) 104
wax-gourd (*Benincasa hispida*) 128
wheat xv, xvi, xviii, 209, 215, 224
  bread (*Triticum aestivum* syn. *T. vulgare*) xv, 2, 3, 4
  club (*Triticum clavatum*) 2
  cone, *see* wheat, rivet
  durum (*Triticum durum*) 2, 3, 4
  English, *see* wheat, rivet
  rivet (*Triticum turgidum*) 2, 3
wineberry (*Rubus phoenicolasius*) 82, 83
winter greens, *see* kale
woolly pyrol, *see* gram, black
wormwood (*Artemisia absinthium*) 146, 147
wrack, knotted (*Ascophyllum nodosum*) 194, 195

Xanthosoma sagittifolium 190, 191
Xylopia aethiopica 138

yam 222
  cush-cush (*Dioscorea trifida*) 192
  greater (*Dioscorea alata*) 192
  lesser (*Dioscorea esculenta*) 192
  white Guinea (*Dioscorea rotundata*) 192
  yellow Guinea (*Dioscorea cayensis*) 192
yam-bean (*Pachyrrhizus erosus*) 38, 192, 193
yam (*Dioscorea* spp.) xvi, xix, 192, 193
yautia, *see* tannia
ysaño (*Tropaeolum tuberosum*) 188, 189

Zanthoxylum spp. 138
Zea mays 6, 7
Zingiberaceae 140
Zingiber officinale 144, 145
Zizania aquatica xiv, 8
zucchini, *see* courgette

Subject index

aflatoxin xvii, 26, 209, 210
agar 194, 201
AIDS virus 34
alcohol(s) xix, 203
aldehydes xix
algae xix
alginates 201
alkaloids xviii, 207, 209
allergens 209
allergies xviii, 203
  gluten 2
alliins 176, 206
amino acids 203
  neurotoxic 208
amygdalin 30, 208
animal feed xix
  barley 4
  citrus waste 90
  lentils 46
  maize 6
  oats 4
  soya 28
anthocyanin xix, 90, 96, 160, 166
anthraquinone 172
antinutrients 207–10
antioxidants 207
apiin 158
apples 52–61, 66–7
  cider 66–7
  classification 52
  cultivars
    of flavour/quality 58–9
    historical 54, 55
    modern 60–1
    through the season 56, 57
archaeological information xiv, xviii
  cereals xvi
  Cucurbita spp. 130
  fig 100
  fruits xviii
  grapes 96
  maize 6
  oats 4
  olives 26
  plum 72
  potato 186
  vegetables xviii
  wheat 2
  food plant dispersal xv
arrack 18, 22
atropine 209

bagasse 16
barley flour/water 4
bergamot oil 90
beriberi 204–5
β-glucans 201
biofuels xvii
blanching 210
Bligh, Captain William 108, 122
blight, potato 186
bran xvi
  oat 4
bread making 2, 4
bromelain 102
bulgar 2

calcium 205, 206, 215–23
  binding 207
calories, source 215–23, 226
cancer 206–7
canning 210–11
caprification 100
capsaicinoids 138
carbohydrate 200–2, 211, 215–23
  absorption 202
  digestion 202
  source 226
carcinogens 209
carotenes xix, 6, 134, 138, 160, 204, 207, 215–23
  blanching effects 210
  palm oil 24
carrageenan 194, 201
cashew-nut shell liquid (CNSL) 34
castanospermine 34
catechin 207
cereals xvi–xvii
  composition 215
  crops 2–13
  dietary fibre 201, 215
  dispersal xv
  fat content 200, 215
  flour enrichment 211
  grain structure xvi
  milling xvi–xvii, 6, 211
  nutritional value xvi, xvii, 215
  water content 200, 215
chicle 104
chloride 205, 206
chlorophyll xix
cholesterol 201–2, 206
cider 66–7
citric acid xix, 90
classification of plants xiv
coeliac disease 2
coir 22
colchicine xv
colorectal cancer 201
Columbus, Christopher xv, xvi, xix, 6, 96
  Capsicum spp. 138
  cocoa 120
  second voyage 102
  sweet potato 192
constipation 201
contaminants 210
Cook, Captain James 170
cooking 210
  vitamin loss 225
copper 205, 207
copra 22
corn oil xvi, 6
coumestans 207
couscous 2
crop selection xiv, xv
cucurbitacins 124
cultivar xiv
  high-yielding xv
  miracle xv
cultivation xiv
currants 98
cyanogens 208
cycasin 209

daidzein 207
deoxyribose 201
depressants 209–10
dhurrin 208
dietary fibre 201–2
dietary patterns, international 211–12
disaccharides 201
dispersal xv–xvi
domestication
  cereals xvi
  food plants xiv
  parallels between cereals and legumes xviii
  wheat 2
drying 210–11
dwarf forms xv

elemicin 140
energy 200, 215–23
  food groups 214
enrichment 211
enzyme inhibitors 207–8
ergot 4, 210
erucic acid 28
essential oils xix, 90
esters xix
ethyl alcohol xvii
eugenol 142
European settlers in N. America xvi

F_1 seed xv
fabism xviii
family xiv
famine foods 212
fat 202, 215–23
  cereals content 200
  consumption levels 211–12
  energy 200
  source 226
  vegetable/animal source composition 224
fatty acids 202, 224
  essential 202
  trans xvii, 202
favism 44, 208
ferns xx
Fertile Crescent of Near/Middle East xv, xvi, 2, 4
fiddleheads xx
fireblight 62
flatus-producing substances 210
flavonoids 207

flour, white xvii
folate 210, 215–23
folic acid 205, 210
food
  consumption 211
  groups 214
  labelling 213
  tables 212
Food Balance Sheets 212
freezing 211
fructans 201
fructose xvii, xix, 201
  cherry 70
  syrup 6
fruits xviii–xix
  annonaceous 102, 103
  citrus 90–5
  composition 218, 219
  nutritional value xix
  temperate 218
  tropical 219
fruit trees xvi
furfural 4

galactose 201
Gama, Vasco da xix
genetic engineering xv, 134
genetic variation xv
genistein 207
genus xiv
gliadin 203
glucose xvii, xix, 201
  cherry 70
  syrup xvii, 6
glucosinolates xvii, xix, 28, 168, 180, 207, 208
gluten 2, 203
  maize 6
glycosides 198, 208
  cyanogenic xviii
goitrogens xviii, 10, 206, 207, 208
gossypol 28, 209
grain crops, *see* cereals
grapes, dried 98
grass family xvi, xvii
Greek Empire xix
Green Revolution xv, 2
  rice production 8
groundnut xvii
guar gum 201
gum arabic 201
gums 201
gymnosperms xx

haemagglutinin xviii, 208
hallucinogens 209–10
harvesting 211
herbs xix
hexoses 200, 201
honey xvii, 201
household surveys 212
hybridization xv
  cell level xv
  triticale 4
  wheat 2
hyoscyamine 209

Incas xvi
inositol 201
insect resistance xv
inulin 201
iodine 205, 206, 208
iron 205–6, 215–23
isoflavonoids 203
isothiocyanate 142, 207

kaempferol 207
kaki-tannin 110

lathyrism xviii, 208
lecithin 28
lectins xviii, 40, 208
legumes
  antinutritional substances xviii, 208
  composition 217
  dispersal xv
  exotic 38, 39
  nutritional content xviii, 217
  toxic substances xviii, 208
  tropical 42–3
legumes xviii
lignans 206–7
linamarin 208
linoleic acid 28
lotaustralin 208
lycopene 134

magnesium 205
Maillard reaction 210
malic acid xix, 96
malting 4
maltose 201
mannitol 201
metal binders 207
micro-fungi xx
milling xvi–xvii, 211
  maize 6
minerals 200, 205–6, 215–23
molasses 16
monosaccharides 200–1
mutation, wheat 2
mycotoxins 209
myristicin 140, 209

naringin xix
neohesperidin 90
Neolithic people xv, xvi, xviii
niacin 6, 204, 205, 210
nitrogen fixation xviii
non-flowering plants xix–xx
non-starch polysaccharide (NSP) 200, 201, 215–23
nopitos 116
nutrient goals 226
nutrition 200–6, 214–26
  surveys 212
nuts xvii–xviii
  composition 216
  nutritional value xvii–xviii, 216
nut trees 30–5

oatmeal 4
oestrogens 209
oil 202
  bitter almond 30
  clove 142
  coconut 22, 224
  grapeseed 98
  groundnut 26, 224
  maize 6
  olive 26, 224
  palm/palm kernel 24, 224
  rape 28
  saturated xvii, 202
  sesame 26
  soya 28
  sunflower 28, 224
  sweet almond 30
  unsaturated xvi, xvii, 202
  walnut 224
oilseed xvii–xviii, 22, 26, 28
  composition 216, 224
oleic acid 32, 34
oleoresin xix, 138, 144
oleuropein 26
oligosaccharides 201
oxalic acid 172, 188, 207

palm
  heart xix, 20, 22, 24
  oil 24, 224
  wine xix, 18, 112
pantothenic acid 204
papain 122
parboiling 211
patulin 210
pears
  cultivars 62–5
  perry 66–7
  stone cells 62, 64
pectin 90, 201
pellagra 205
pentoses 200, 201
perry 66–7
phenolic compounds 207
phosphorus 205
phycocolloids 194
phytates 207, 210
phytic acid 201, 207
phytochemicals 206
phytoestrogens 206–7
plant breeding xv
plums 72–7
  cooking cultivars 74–5
  dessert cultivars 76, 77
pollutants 210
polysaccharides 201
pomace 66
Portugese empire xv–xvi
potassium 205, 206, 215–23
pressor amines 209
processing 210–11
protease 203
  inhibitors 208
protein 200, 203, 215–23, 226
prussic acid 30
pseudo-cereals xvi, xvii, 14–15, 215
  Amaranthus spp. 170
pulses, *see* legumes

quercitin 207
quinoa xvi
Quorn 196

radiocarbon dating xiv–xv

raffinose xviii, 201, 210
raisins 98, 218
recommended daily amounts (RDAs) 212–13
retinol 204
riboflavin 210, 215–23
ribose 201
rice, parboiling 211
rice-paper 8
Roman Empire xv, xix
root nodules xviii
rust 198
rutin 14

safrole 209
sago xx, 20
  composition 222
salad
  crop composition 221
  plants xviii
  roots 180
saponins xviii, 14, 209
scopolamine 209
seaweed xix, 194, 195, 201
selenium 205, 207
semi-dwarf forms xv
semolina 2
Senecio alkaloids 209
sex hormone binding globulin (SHBG) 207
sodium 205, 206, 215–23
solanine 186, 207–8
sorbitol 201
sour-dough 4
Spanish empire xv–xvi
species xiv
  ancestral xv
spices xix
stachyose xviii, 201, 210
starch 201
stimulants 209–10
storage 210, 211
sucrose xvii, xix, 201
  palm sap 18
sudden death fungal disease 142
sugar xvii, 201, 215–23
  alcohols 201
  coconut sap 22
  crops xvii, 16–19
  invert xvii
  mushroom 201
  palm xix
  production methods 16
sultanas 98
sweet corn 6, 7, 215
synonyms xiv

tannin 110, 207
tartaric acid 96
thaumatin xvii
thiamin 210, 211, 215–23
thiaminase 209
thioallyl compounds *see* alliins
thioglycosides 208
tissue culture methods xv
toddy 22
tomatine 134
toxicants, natural 207–10
trans fatty acids xvii, 202
trehalose 201
triglycerides 202, 206

ubiquinone 207

variety, *see* cultivar
vegetables xvi, xviii–xix
  composition 220, 221, 222
  fruits 220
  nutritional value xix
  roots 222
verbascose 201
vine leaves 98
vitamin A 204
vitamin C 200, 204, 205, 210, 211, 215–23
vitamin D 204, 206, 207
vitamin B_{12} deficiency 205
vitamin B 204–5, 215–23
vitamin E 204, 207, 215–23
vitamin K 204
vitamins 200, 203–5, 215–23
  cooking 210
  cooking loss 225
  deficiencies 225
  fat-soluble 204
  sources 225
  storage effects 211
  water-soluble 204–5

water 200, 215–23
water plants, exotic 36–7
wild plants xiv, 198, 199
  comparison of cultivated plants xv

zein 203
zinc 205, 206, 207